CHEMICAL ENGINEERING METHODS AND TECHNOLOGY

MICROWAVE ASSISTED CYCLOADDITION REACTIONS

Chemical Engineering Methods and Technology

Additional books in this series can be found on Nova's website
under the Series tab.

Additional E-books in this series can be found on Nova's website
under the E-books tab.

CHEMICAL ENGINEERING METHODS AND TECHNOLOGY

MICROWAVE ASSISTED CYCLOADDITION REACTIONS

DAVOR MARGETIĆ

Nova Science Publishers, Inc.
New York

For permission to use material from this book please contact us:
Telephone 631-231-7269; Fax 631-231-8175
Web Site: http://www.novapublishers.com

Library of Congress Cataloging-in-Publication Data

Margetic, Davor.
Microwave assisted cycloaddition reactions / Author, Davor Margetic.
p. cm.
Includes index.
ISBN 978-1-61324-109-7 (hardcover)
1. Ring formation (Chemistry) 2. Microwave heating. I. Title.
QD281.R5M37 2011
547'.2--dc22

2011008470

Published by Nova Science Publishers, Inc. † New York

Dedicated to the memory of Professor Douglas N. Butler,
an unforgettable teacher and friend

CONTENTS

PREFACE

The modern World imposes serious environmental challenges to chemists, in reducing the human impact on the nature. In this respect, organic reactions conducted in microwave conditions offer an important 'green chemistry' method, which significantly enhances reaction efficiency, reduces the energy consumption and the use of the toxic organic solvents. This book demonstrates the importance of the use of microwave technique in organic synthesis, in particular its application to cycloaddition reactions. Several excellent books and reviews which only partially cover microwave cycloadditions are dealing with microwave reactions, cycloadditions or green chemistry methods, on the other hand this is the first comprehensive cover of literature from the early accounts in the field of microwave assisted cycloadditions until now. The number of research publications on microwave assisted organic reactions has greatly increased over the past years, giving enough book material for this specialized subject. The book is intended for advanced undergraduate and graduate chemistry students and organic research chemists involved in synthesis in academic and industrial research. The book begins with a chapter on Diels-Alder reactions, one of the most important reactions in organic synthesis. It is divided in three sections: all-carbon, hetero and intramolecular Diels-Alder reactions. Chapter 2 is devoted to a discussion of advances in the field of microwave assisted 1,3-dipolar cycloadditions. In separate sections, reactions of azides, nitrones and other 1,3-dipoles are covered. Other cycloaddition reactions are covered in Chapter 3. These chapters are complemented by chapter 4 that describes retrocyclization reactions. Chapter 5 gives a comprehensive review of applications of microwave assisted cycloadditions in material sciences, in particular for functionalization of fullerenes and nanotubes. Throughout the book, the special emphasize was given to reactions carried out in unconventional media (solid supported, solvent free, in aqueous media or ionic liquids), reaction mechanisms and reactive intermediates, domino and tandem reactions, metal catalyzed cycloadditions as well as multicomponent reactions involving cycloadditions. The final chapter gives practical advices for choice of experimental conditions for microwave assisted cycloadditions, making this part a small laboratory manual. Data presented here will allow synthetic chemist to conduct their own work by consulting this book only, without reading a great number of primary publications.

Funding for our own research described partially in this book was provided by Croatian Ministry of Science, Education, and Sport (grants no. 098-0982933-3218 and 098-0982933-2920). Leave of absence from Ruđer Bošković Institute allowed me to be in Japan during the early stages of manuscript preparation. I am very grateful to Professor Tsutomu Ishikawa and the Faculty of Pharmaceutical sciences at Chiba University for inviting me as visiting academic and providing office space and library facilities. Professor Takuya Kumamoto is also thanked for his help and hospitality. Many thanks to Mario Vazdar, PhD for his help in providing difficult to obtain literature, as well as Pavle Trošelj, BSc and Anamarija Briš, BSc, who carried out many of the experiments mentioned in this book. I would also like to thank to my family for being a source of strength, support and inspiration throughout the writing process.

ABBREVIATIONS

Δ	conventional (classical) heating
Ac	acetyl
Ad	adamantyl
AcOH	acetic acid
Bn	benzyl
Boc	*t*-butoxycarbonyl
b.p.	boiling point
i-Bu	*iso*-butyl
Bu	*n*-butyl
t-Bu	*tert*-butyl
Bz	benzoyl
Cbz or	Z benzyloxycarbonyl
Chex	cyclohexyl
CNT	carbon nanotube
COD	1,5-cyclooctadiene
Cy	cyclohexyl
DA	Diels-Alder
DABCO	1,4-diazabicyclo[2.2.2]octane
DBU	1,8-diazabicyclo[5.4.0]undec-7-ene
DCB	dichlorobenzene
DCE	dichloroethane
DCM	dichloromethane
de	diastereomeric excess
DEAD	diethylazodicarboxylate
DIBAL-H	diisobutylaluminum hydride
DIPEA	diisopropylethyl amine, Hünig's base
DMA	*N,N*-dimethylacetamide
DMAD	dimethylacetylene dicarboxylate
DMAP	4-(dimethylamino)pyridine
DMAS	*N,N*-dimethylsulfamoyl
DMB	2,5-dimethoxybenzyl
DMF	dimethylformamide
DMSO	dimethyl sulfoxide

DMTMM	4-(4,6-dimethoxy[1,3,5]triazin-2-yl)-4-methylmorpholinium chloride
DPPA	diphenyl phosphoryl azide
dr	diastereomeric ratio
EDDA	ethylenediaminediacetate
equiv.	equivalent
Et	ethyl
ETPPB	ethyltriphenylphosphonium bromide
Fmoc	9-fluorenylmethoxycarbonyl
Fur	furyl
g	gram
h	hour
hex	hexyl
HMPA	hexamethylphosphoramide
HP	high pressure
hν	photochemical irradiation
IBX	*o*-iodoxybenzoic acid
IMDA	intramolecular Diels-Alder
i-Pr	isopropyl
M	mol/L
Me	methyl
min	minute
MOM	methoxymethyl
MW	microwave
MWNT	multi-walled carbon nanotube
NBS	*N*-bromosuccinimide
NCS	*N*-chlorosuccinimide
NMM	*N*-methylmorpholine
NMP	*N*-methylpyrrolidone
Ph	phenyl
PMB or MPM	*p*-methoxybenzyl
POM	pivaloyloxymethyl
PTSA	*p*-toluenesulfonic acid
Pr	propyl
Py	pyridine
r.t.	room temperature
SWNT	single-walled carbon nanotube
TBAB	tetrabutylammonium bromide
TBDMS	*t*-butyldimethylsilyl
TBDPS	*t*-butyldiphenylsilyl
TBTA	tris(benzyltriazolylmethyl)amine
TCB	1,2,4-trichlorobenzene
TEA	triethylamine
TEAA	triethylammonium acetate
TEBAC	triethylbenzylammonium chloride
Tf	trifluoroacetyl
TFA	trifluoroacetic acid

TFE	trifluoroethanol
THF	tetrahydrofuran
TIPS	triisopropylsilyl
TMS	trimethylsilyl
Tol	tolyl
Tos	tosyl
Tr	triphenylmethyl, trityl
Ts	*p*-toluenesulfonyl
Tse	2-(*p*-toluenesulfonyl)
Tsv	*p*-toluenesulfonylvinyl

INTRODUCTION

Cycloaddition reactions are among the most important tools for organic synthesis. [1, 2] Two of the most widely used methods are Diels-Alder and 1,3-dipolar cycloadditions. Few chemical reactions rival these processes in the number of bonds that undergo transformation during the reaction, affording products of great complexity. These reactions are characterized by having cyclic transition structures, and also have highly predictable stereochemical features. Reaction conditions used in microwave reactions obey Woodward-Hoffmann selection rules for pericyclic reactions. The synthetic utility of Diels-Alder reactions for construction of six membered rings is well documented and numerous molecules possessing cyclic structures prepared by its application. The 1,3-dipolar cycloaddition reaction is the single most important synthetic method for the construction of heterocyclic five-membered rings in organic chemistry. Therefore, it is not surprising that cycloadditions were amongst the first organic transformations which were carried out in microwave conditions. [3] Since these pioneering days when the domestic microwave oven was used, microwave assisted organic reactions (often called MAOS) became an invaluable tool in synthetic organic laboratory. [4] Today, in the time of modern sophisticated microwave reactors, the whole reaction is much easier to control, with high reproducibility of results.

As the number of experimental results increased, it was realized that advantageous use of MAOS leads to a dramatic reduction of the reaction time, as compared to conventional reflux conditions. The yields obtained after microwave irradiation are comparable with those reached after thermal heating, or the yield is markedly improved. Usually, reactions are much cleaner, affording much less of side products, therefore making the purification process more efficient. The other important advantages of microwave technology include: more consistent results, safer application of heat, decreased sample preparation, solvent reduction, cost savings and reduced impact on environment. The majority of cycloaddition reactions follow the same regioselectivity and stereoselectivity as conventional reactions, which could be changed in some instances. Ocassionally the application of microwave irradiation led to formation of unprecedented products of great interest, which could not be obtained by classical methods. Especially attractive are multicomponent, tandem and domino cycloadditions, where in the single reaction pot several reactions, including cycloadditions take place, as rapid and efficient route to structures of considerable complexity.

A diverse reaction conditions were used for cycloaddition reactions: in organic solvent, water [5], ionic liquids [6], without the use of solvent, on solid support, catalyzed by bases, acids or metal complexes. Temperature covers the wide range from 50 to 400°C. Additives were also employed as an energy transfer medium to promote cycloadditions, such as active carbon or graphite, and less polar solvent were doped with ionic liquids or other polar solvents and salts in order to increase the dielectric constant of solvent medium. Depending on the nature of reagents, they are either conveniently mixed together and reacted, or in the case of unstable and reactive species (especially 1,3-dipoles), generated *in situ* and subsequently trapped without isolation. The presence of both reactants in the single molecule often benefits being more reactive and selective.

Microwave assistance has been used extensively in a promotion of the cycloaddition reactions. The synthetic utility of the cycloaddition reactions is evident from the number and scope of synthetic targets that can be prepared by this chemistry. Cycloadditions described here include small ring heterocycles often found in natural products. [7] Examples are triazoles, oxazoles, triazines, and pyridines, which are important classes of heterocycles, and many of such compounds have considerable utility as chemical intermediates, pharmaceuticals and biologically active compounds. Remarkable physical and chemical properties possess fullerene-based functionalized organic materials prepared by MAOS. [8]

References

[1] Carruthers, W. *Cycloadditions Reactions in Organic Synthesis*, Pergamon Press, Oxford, *1990*.

[2] Kobayashi, S.; Jorgensen, K. A. (eds.), *Cycloaddition Reactions in Organic Synthesis*, Wiley-VCH, Weinheim *2001*.

[3] Giguere, R. J.; Bray, T. L.; Duncan, S. M.; Majetich, G. *Tetrahedron Lett. 1986*, *27*, 4945-4948.

[4] For a recent review on microwave assisted cycloaddition reactions see: Appukkuttan, P.; Mehta, V. P.; Van der Eycken, E. V. *Chem. Soc. Rev. 2010*, *39*, 1467-1477.

[5] Dallinger, D.; Kappe, C. O. *Chem. Rev. 2007*, *107*, 2563-2591.

[6] Wasserscheid, P.; Welton, T. (eds.), *Ionic Liquids in Synthesis* Wiley-VCH, Weinheim *2008*.

[7] Quin, L. D.; Tyrell, J. *Fundamentals of Heterocyclic Chemistry: Importance in Nature and in the Synthesis of Pharmaceuticals*, Wiley, Chichester, *2010*.

[8] Müller, T. J. J.; Bunz, U. H. F. (eds.), *Functional organic materials: synthesis, strategies and applications*, Wiley-VCH, Weinheim, *2007*.

Chapter 1

DIELS-ALDER REACTIONS

INTRODUCTION

Diels-Alder is the organic reaction that has a great synthetic importance. With its broad scope and simplicity of operation, the Diels-Alder is the most powerful synthetic method for unsaturated six-membered rings. [1] Diels-Alder reaction is a conjugate reaction of a conjugated diene to an alkene (the dienophile) to produce a cyclohexene ring. The [4+2]-cycloaddition of a conjugated diene and a dienophile (an alkene or alkyne), an electrocyclic reaction that involves the 4 π-electrons of the diene and 2 π-electrons of the dienophile. The driving force of the reaction is the formation of new σ-bonds, which are energetically more stable than the π-bonds. There are Diels Alder reactions with inverse electronic demand that involve the overlap of the HOMO of the dienophile with the unoccupied MO of the diene. [2] Most well-studied inverse electron demand Diels-Alder reactions fall into the category of hetero Diels-Alder reaction, [3] where either 1,3-diene or dienophile possess an heteroatom, in which heterocyclic six-membered ring is produced. [4] Even in the first account on the use of microwave irradiation for organic synthesis by Giguere and Majetich [5] one of the key organic transformations was Diels-Alder reaction. Since then, a number of researchers have studied the efficiency of microwave irradiation on Diels-Alder reactions, and made applications in different fields, such as drugs and natural product synthesis, biomedicine and material science. Amongst the variety of reaction conditions applied, microwave irradiation in solvent-free conditions has well demonstrated its utility as energy source to dramatically improve yields and/or reactions conditions in Diels-Alder cycloadditions. [6]

Although Diels-Alder reactions are generally facilitated by microwave irradiation, there are several examples when Diels-Alder reaction was not successful in microwave conditions (reaction of maleic anhydride with substituted imidazolidin-4-ones), [7] but reaction works in classical conditions. In another example, high pressure and heating give product, but microwave conditions failed.

1.1. Intermolecular All-carbon Diels-Alder Reactions

1.1.1. Aromatic Dienes

Giguere and Majetich [5] published the first account on the application of commercial microwave ovens to organic synthesis. It drastically reduced the reaction time (at comparable yield) of Diels-Alder, Claisen and ene-reactions. Safety precautions as sealed tube or screw-cap pressure tubes placed either in a Malgene desiccator or a Corian box, each packed with vermiculine, which serves as adsorbent of reaction content in the event of an explosion. Examples of anthracene/maleic anhydride *1*/*2* and cyclohexenone/2,3-dimethylbutadiene *4*/*5* cycloadditions are illustrative (Scheme 1). The reaction of *4* with *5* also represents the fist example of solvent-free microwave conducted cycloadditions.

Commercial domestic microwave oven was also used without alteration in open system by Bose [8] for anthracene/maleic anhydride cycloaddition employing chlorobenzene, *o*-dichlorobenzene and diglyme. After 2-3 min of irradiation, cycloadduct *2* was obtained in 40, 85 and 90% yield, respectively. Anthracene/dimethyl fumarate reaction in the same solvents (5-10 min) gave corresponding cycloadduct in 5, 20 and 79% yield, respectively.

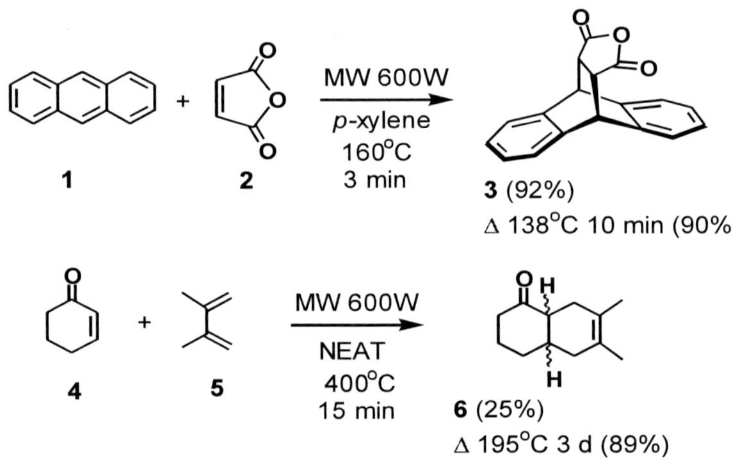

Scheme 1.

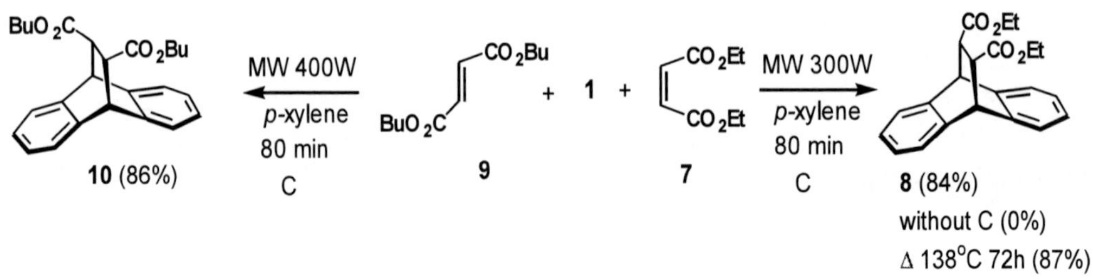

Scheme 2.

Phutdhawong [9] used different reaction conditions for anthracene/maleic anhydride cycloaddition (xylene 800W, 129-138°C, 8 min) to obtain cycloadduct in 97% yield, and a classical sealed tube (100-110°C, 24 h) gave 85% yield. In addition, Organ [10] carried out Diels-Alder cycloadditions in continuous flow microwave reactor, with thin Pd films deposited on the inside of capillary-sized tubes through which compounds have been flowed while being heated with microwave. Dramatic rate accelerations are observed in the presence of the film, playing both a heating and catalytic role. For instance, anthracene *1* and maleic anhydride *2* in DMSO were flowed together through a clear glass capillary while being irradiated at 205°C (flow rate 10 μL/min) to obtain 93% conversion.

Li [11] showed that active carbon enhanced the microwave-assisted organic reactions in low polar or nonpolar medium. Active carbon has strong absorption of microwave and can be heated up very rapidly under microwave irradiation, therefore it can be used as energy transfer medium to heat the reaction mixtures. Other advantages is that active carbon is inexpensive, commercially available. After reaction, it can be easily separated from product by filtration, column chromatography or distillation and can be also very easy and environmentally friendly disposed after use. Diels-Alder reaction of anthracene with diethyl maleate. The Diels-Alder reaction of anthracene with diethyl maleate took 72 h in xylene or toluene by conventional method, where both the reactant and solvent (xylene) have very low polarity. At 300W irradiation in the presence of active carbon, after 80 min cycloadduct *8* was obtained in 84% yield (Scheme 2). The reaction proceeded cleanly without byproduct formation, while without active carbon added reaction does not take place. With stronger irradiation power (500W), the yield was increased to 90% after 76 min. Analogously, dibutyl fumarate reacted with anthracene to obtain corresponding cycloadduct *10* in high yield.

Kazemi-Ani [12] modified the anthracene/dimethyl maleate reaction conditions and further accellerated the reaction by the addition of $AlCl_3$ catalyst (CH_2Cl_2, 600W, 12x10 s, 90%). Furthermore, Raner [13] found in kinetic study of reaction anthracene with diethyl maleate in o-xylene that rate constants were the same at the identical temperatures. No kinetic differences between microwave irradiated and conventionally heated reactions was found, and more rapid reaction proceedings in microwave is the consequence of more efficient heat energy transfer to reactants.

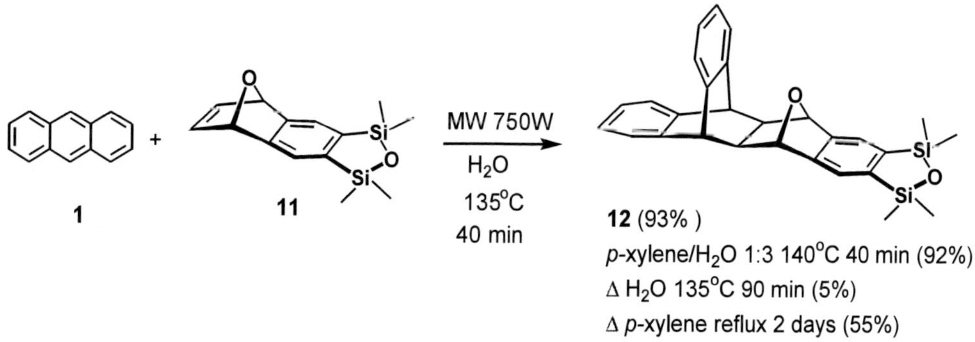

Scheme 3.

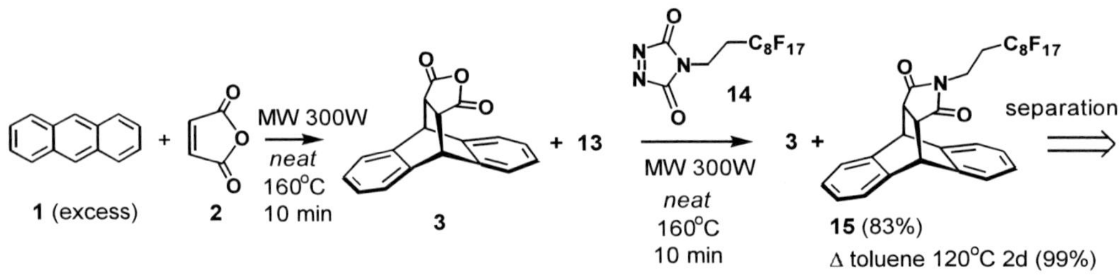

Scheme 4.

Synthesis of extended substituted triptycenes of type *12* was achieved by Lee [14] via Diels-Alder cycloaddition of anthracene and oxadisilole-benzofused endoxide *11* in water under microwave irradiation [15] (Scheme 3). Optimal reaction conditions were 135°C, 40 min, while equally suitable conditions were the mixture of water and *p*-xylene 1:3. In control experiment by classical heating of reactants in water (90 min), only small amounts of products were obtained (5%). By classical heating, the reaction was not completed and the cycloadduct *12* could be isolated in 55% yield after refluxing in *p*-xylene for two days. The reaction could be also carried out in an open beaker using a domestic microwave oven with isolated yield of 56% in 30 min.

Curran [16] proved usefulness of the fluorous dienophiles as diene scavengers. The fluorous [1,2,4]triazoline-3,5-dione *14* shows superior reactivity comparing to maleic anhydride and standard protocol was developed for the use of fluorous dienophiles such as *14* in typical scavenging reactions (Scheme 4). An excess of a diene *13* was reacted with maleic anhydride in solvent free conditions. When dienophile was consumed, a fluorous scavenger *14* was added, followed by efficient fluorous solid-phase extraction of adducts. The Diels-Alder reactions of the diene *13* with maleic anhydride under microwave conditions were complete after 10 min of irradiation in a microwave reactor at 160°C under solvent free conditions. Under the same conditions, the scavenger *14* reacted with the excess diene, and the Diels-Alder adduct *15* were obtained after FSPE in 79-83% yield and with a purity of 90-93%.

Sessler and Cavaleiro [17] showed that sapphyrin could be used as dienophile in microwave enhanced [4+2] cycloadditions. The Diels-Alder reaction between sapphyrin *16* and pentacene under thermal conditions by heating a 1,2,4-trichlorobenzene solution containing a 1:5 molar ratio of these two reactants provided cycloadduct *17* (21% yield), bisadduct (traces), and compound *18* (5% yield), which resulted from the dehydrogenation of *17* (Scheme 5). Low yields of the sapphyrin cycloadducts were significantly improved by the use of microwave irradiation. When a solution of sapphyrin *16* and pentacene (1.4 equiv) was irradiated, compound *17* was obtained in 51% yield, while compound *18* and the bisadduct were not formed. Thus, the use of a microwave irradiation-based procedure allows for a shorter reaction time and obviates the need to use a large excess of pentacene, a cleaner reaction mixture is obtained facilitating product purification.

A series of substituted 6,11-ethanobenzo[b]quinolizinium derivatives *22* was synthesized by Sasaki and Maeda [18] through the Diels-Alder reaction between azoniaanthracene *21* and the corresponding 1,1-disubstituted olefin (Scheme 6). The bistiophene olefin *20* was generated by elimination from precursor *19 in situ*, because it was unstable and easily polymerized. Then, *in-situ* trapping by the Diels-Alder reaction with *21* afforded cyclodduct

22. The most efficient reaction took place in 10% AcOH- trifluoroethanol in microwave irradiation enhanced conditions. These conditions were about 200 times faster than compared to classical reflux CH_3CN. Under the condition of microwave irradiation, almost 100 % conversion from *19* to *20* was observed in 10% AcOH-TFE after 1.5 min, but no olefin *20* was present in TFE after 10 min. These results suggest that microwave irradiation mainly effected acceleration of the Diels-Alder reaction but also the formation of olefin *20*. In classical heating, *19* requires heating in CH_3CN under reflux for 24 h to give *22*, the same reaction proceeded much faster by reflux in H_2O, but the addition of water in microwave conditions was less successful, indicating that the presence of acid is essential for the formation of *20*.

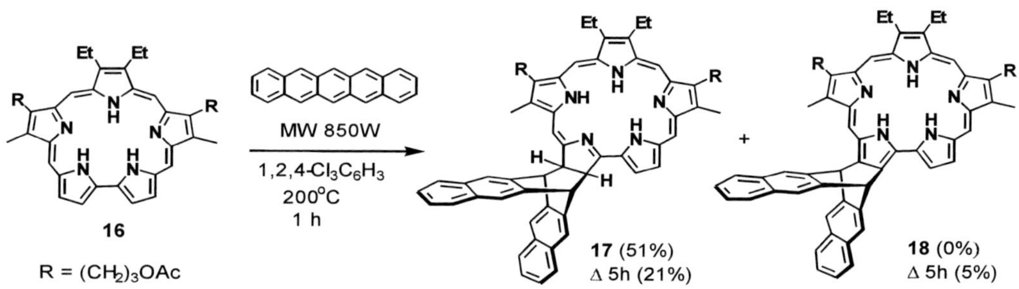

Scheme 5.

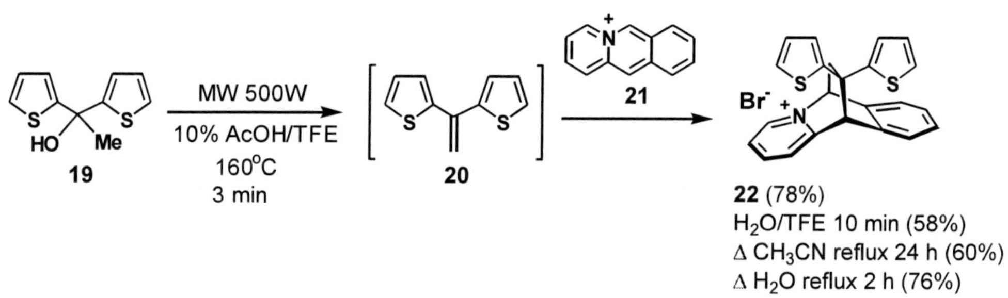

Scheme 6.

1.1.2. Furan

In one of the pioneering papers on the microwave assisted cycloadditions, Berlan and Giboreau [19] found that microwave irradiation greatly enhance the speed of Diels-Alder reactions of furan with DMAD; 2,3-dimethylbutadiene with methyl vinyl ketone; and anthracene with diethyl maleate, in xylene or dibuthyl ether as solvent (Scheme 7). Addition of the $AlCl_3$ catalyst further improves yield and shortens reaction time (CH_2Cl_2, 600W, 10x10 s, 95%).

The Diels-Alder cycloaddition reaction of 1,2-difluoro-1-chlorovinylphenylsulfone *26* with cyclopentadiene and diphenylisobenzofuran proceeds in microwave assisted reaction, as well as under conventional heating in toluene, although in lower yields. On the other hand, in conventional reaction furan failed even when refluxed in chlorobenzene for three days. However, Rao [20] has shown that under microwave irradiation, in solvent-free conditions sulfone *26* is remarkable reactive forming cycloadducts *27* and *28* in very short reaction times (Scheme 8). Using the same reaction conditions, Rao [21] also increased reactivity of 2,2-dichloro-1-fluorovinyl phenylsulfone in reaction with furan (25 min, 71%, 0.9:1.1 ratio) and 1,2-difluoro-2-chlorovinylphenylsulfone (25 min, 80%, 1:2 ratio).

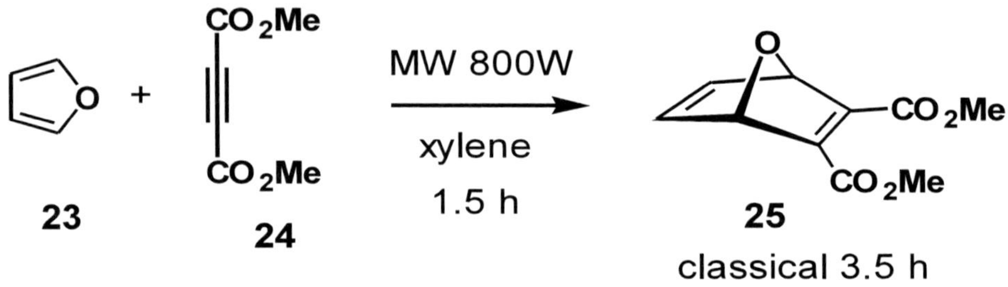

Scheme 7.

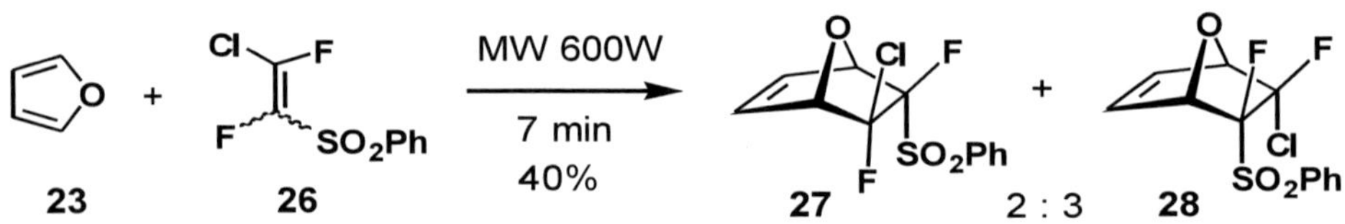

Scheme 8.

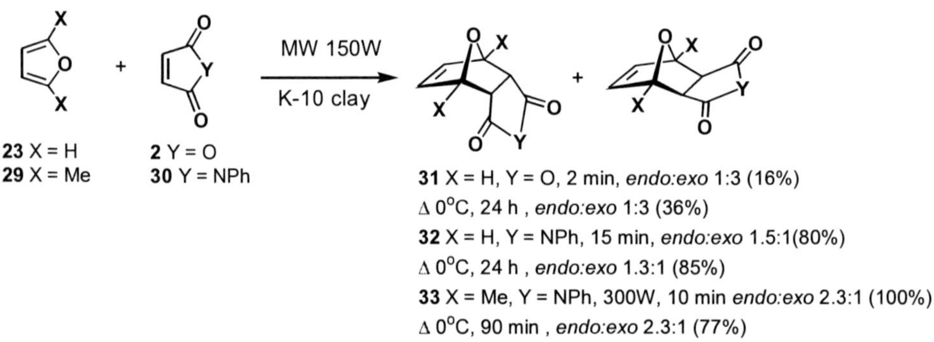

Scheme 9.

Clay-catalyzed solventless cycloaddition reaction of furans *23* and *29* with α,β-unsaturated carbonyl compounds *2* and *30* were carried out by Cintas. [22] Comparison of reactions carried out with K10 Montmorillonite at 0°C or in microwave conditions (Scheme 9). Comparable yields were obtained in both reaction conditions, with the preference of *endo*-products in the case of *N*-phenylmaleimide *30*, while maleic anhydride preferably forms *exo*-adducts. The change of reaction conditions (0°C to microwave) does not affect *endo/exo* ratio, while the reaction time is significantly reduced, from 24 h to less than 15 min.

De la Hoz [23] combined heterogeneous catalysis and microwave irradiation and showed a synergistic effect in one-pot synthesis of benzene derivatives such as *36* via ring-opening of Diels-Alder cycloadducts of substituted furans. The catalysts used were obtained by treatment of silica with $ZnCl_2$, Et_2AlCl or $TiCl_4$. Silica supported Lewis acids are the good catalysts for the Diels-Alder reactions of furan derivatives with methyl acrylate, acrylonitrile or maleimides to form 7-oxabicyclo[2.2.1]heptenes such as *endo-35/exo-35*. The resultant oxa bridge was removed in the single reaction step in the course of this solvent-free reaction (Scheme 10). [24] The use of Si(Ti) gives the best results to obtain the arene derivatives in good yields, and the order af catalyst activity is Si(Ti) > Si(Al) > Si(Zn). The use of classical reaction conditions (temperature and reaction time) leads to dramatic decrease in the yield of the product. In addition, the use of silica-supported Lewis acids as catalysts under microwave irradiation promotes regiospecific opening of the 7-oxabicyclo[2.2.1]heptene bridge of Diels-Alder cycloadducts of furans such as *38* and produces polysubstituted phenols *39* in a single synthetic step. [25]

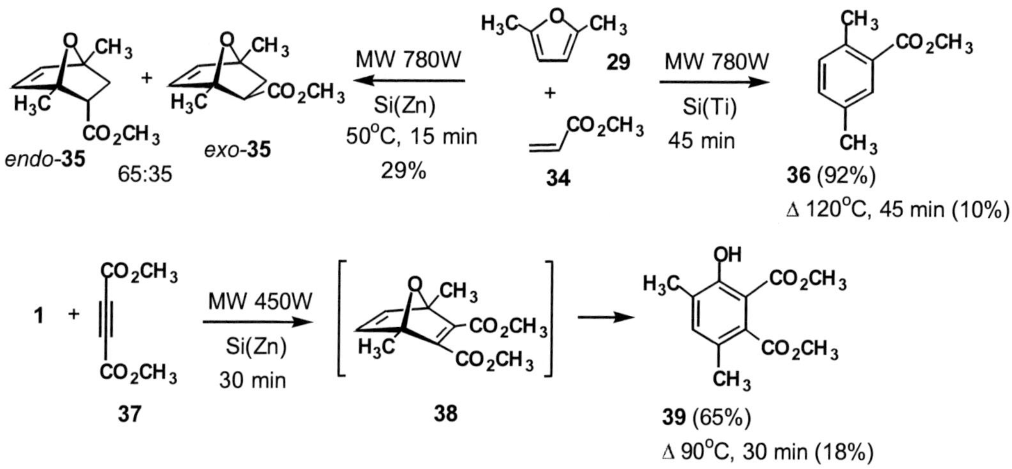

Scheme 10.

1.1.3. Cyclopentadiene

Cyclopentadiene and cyclohexadiene cycloadditions could also be facilitated by combined effects of microwave irradiation and $AlCl_3$ catalyst (0.25 equiv) in a heterogenous reaction, as found by Sridar. [26] The poor reactivity of 2-cycloalkenones as dienophiles was greatly improved by reactions using microwave irradiation in order to make the Diels-Alder intermediates. Thus, the Diels-Alder adducts of cycloalkenones *42a-d* and cyclic dienes *40* or *41* were synthesized in open Erlenmeyer flasks using a microwave with good yields (Scheme 11). It makes these reactions far more simple to carry out compared to the previously known hazardous procedures. Microwave irradiation also increased the selectivity of *endo* products. In the case of cyclopentadiene adducts *43a-d*. The synthesis was carried out with toluene and chlorobenzene, whereas chlorobenzene provided better yields than toluene. This may be attributed to the higher dielectric constant of chlorobenzene. Particularly significant acceleration was observed in the case of cyclohexadiene additions.

Reactions of benzyl 2-fluoroacrylate *45* and 2-fluorooct-1-en-3-one *46* as moderate dienophiles in [4+2]-cycloadditions with cyclopentadiene were investigated by Haufe [27] in thermally, microwave- assisted, and Lewis acid-mediated ($TiCl_4$) conditions (Scheme 12). By conventional heating, the esters *45* and *47*, either fluorinated or nonfluorinated, reacted significantly slower than the ketones *46* and *48*. Nonfluorinated Diels-Alder products are generally formed *endo-* selectively, while the cycloadditions of fluorinated dienophiles *45* and *47* proceeded with *exo-* selectivity. The microwave assisted reactions in toluene and without any solvent did not show a big difference. Compared to the conventional thermal Diels-Alder-reactions, slightly diminished yields were found in the case of the fluorinated dienophiles *45* and *46*, while *50b* was furnished with increased yields. The reaction times were significantly shortened with respect to the conventional thermal reaction, especially in the reactions without solvent. Reactions carried out in the presence of $TiCl_4$ (stoichiometric or slightly substoichiometric amounts) gave cycloadducts *49a* and *49b* and also some chlorine-containing side products formed from the primary cycloadducts. In $TiCl_4$ catalyzed reactions the *exo-* products *49a,b* were solely obtained in the case of fluorinated dienophiles, while the non fluorinated compounds, reacted with enhanced *endo-* diastereoselectivity (Table 1).

Table 1. Cyclopentadiene reactions with dienophiles 45-49

X	R	Method	t	cyclo-adduct	*endo/ exo*	yield (%)
F	OBn	Toluene 110°C	16 h	49a	31:69	73
F	C_5H_{11}		0.9 h	49b	27:73	59
H	OBn		1.5 h	50a	78:22	91
H	C_5H_{11}		0.5 h	50b	75:25	54
F	OBn	MW 500W toluene	66 min	49a	30:70	54
F	C_5H_{11}		46 min	49b	22:78	44
H	OBn		36 min	50a	76:24	88
H	C_5H_{11}		12 min	50b	78:22	61
F	OBn	MW 500W neat	72 min	49a	29:71	44
F	C_5H_{11}		16 min	49b	22:78	48
H	OBn		12 min	50a	74:26	91
H	C_5H_{11}		10 min	50b	79:21	64
F	OBn	$TiCl_4$ CH_2Cl_2 -55°C	18 h	49a	0:100	53
F	C_5H_{11}		14 h	49b	38:62	48
H	OBn		28 min	50a	96:4	86
H	C_5H_{11}		15 min	50b	96:4	73

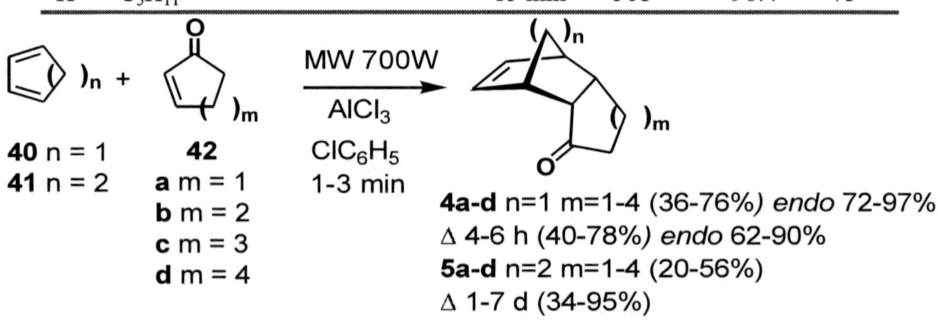

Scheme 11.

A key step in the synthesis of cyclic α-amino acids was the Diels-Alder reaction of methyl and (-)-menthyl-2-acetamidoacrylates *51a* and *51b* with cyclopentadiene. Mayoral [28] carried out these cycloadditions in the presence of heterogeneous catalysts. The best results were obtained when the reactions were carried out in the absence of a solvent using SiO_2 treated with $AlEt_2Cl$ as a catalyst. The selectivities obtained are similar to those for the reactions catalyzed by homogeneous Lewis acids, in particular a complete diastereofacial selectivity is obtained (Scheme 13). Microwave activation increases the reaction rate without decreasing the selectivity.

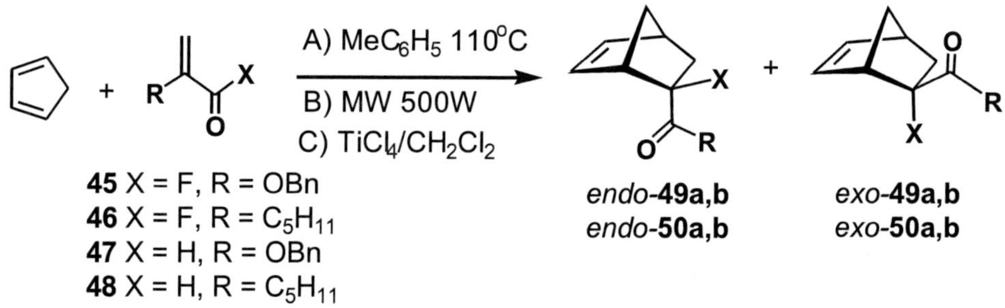

Scheme 12.

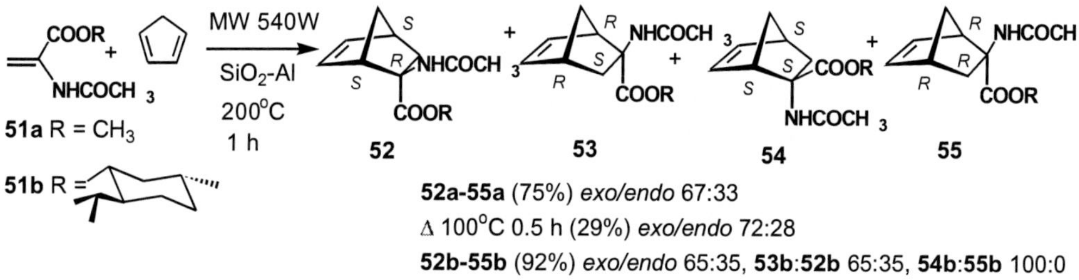

Scheme 13.

The water-soluble, organotungsten Lewis acid catalyst $[O]P(2\text{-}Py)_3W(CO)(NO)_2](BF_4)_2$ *57* was used by Yu [29] in a series of Diels-Alder reactions under conventional heating or microwave heating conditions. The cycloaddition reactions were efficiently conducted in either water or in an ionic liquid, 1-butyl-3-methylimidazolium hexafluorophosphate ($bmimPF_6$). Dramatic rate acceleration via microwave flash heating as compared to thermal heating was observed, as illustrated for addition of cyclopentadiene *40* to methylvinyl ketone *56* (Scheme 14).

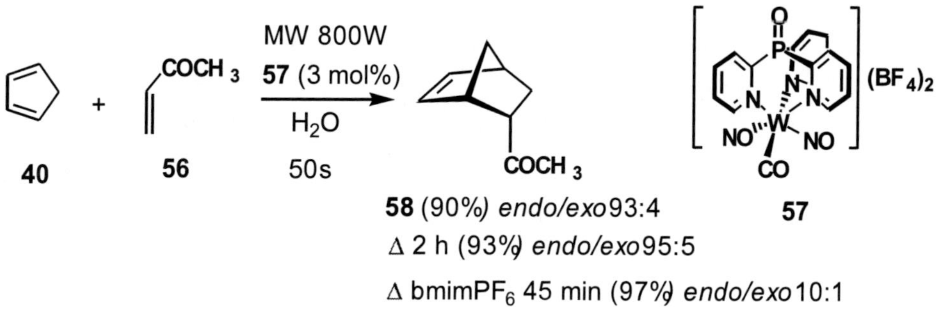

Scheme 14.

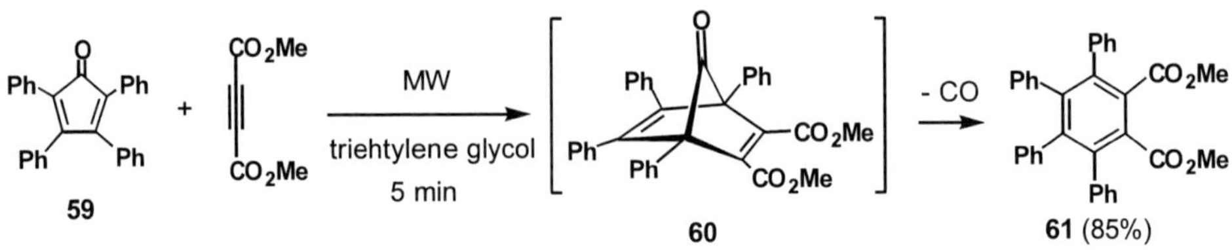

Scheme 15.

1.1.4. Cyclopentadienone

Microwave heating was found useful in the synthesis of Diels-Alder adduct of tetraphenylcyclopentadienone *59* and dimethylacetylene dicarboxylate. Elder [30] reported that reaction is faster and simpler to perform and product *61* of high purity (Scheme 15). Here, cheletropic elimination of CO from an intermediate cycloadduct *60* is an thermally facile process which does require catalyst, as in the previous example.

The trisubstituted (5-pyrimidyl)pentaphenylbenzene *64* was synthetitised Gourdon [31] in microwave ehnanced [2+4] cycloaddition of 5-((4-tert-butylphenyl)ethynyl)pyrimidine *62* and 3,4-bis(4-tert-butylphenyl)-2,5-diphenylcyclopenta-2,4-dienone *63* in 52 % yield (Scheme 16). The Diels-Alder reactions, and the subsequent *in situ* elimination of CO were carried out in diphenyl ether at 260°C giving the hexaaryl compounds in 40-50% yields.

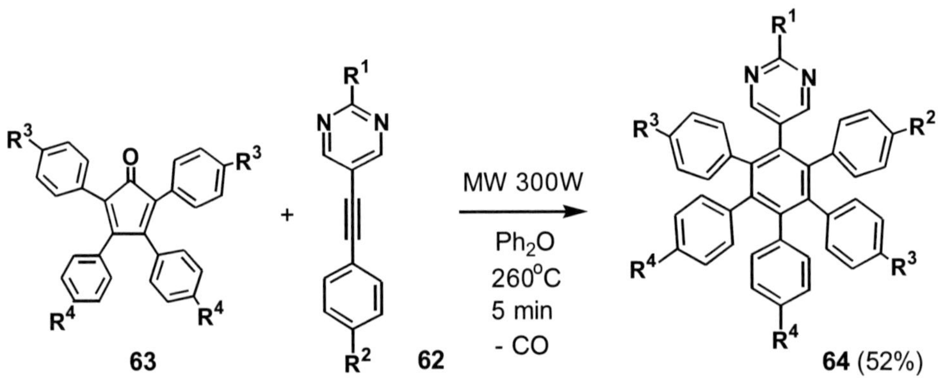

Scheme 16.

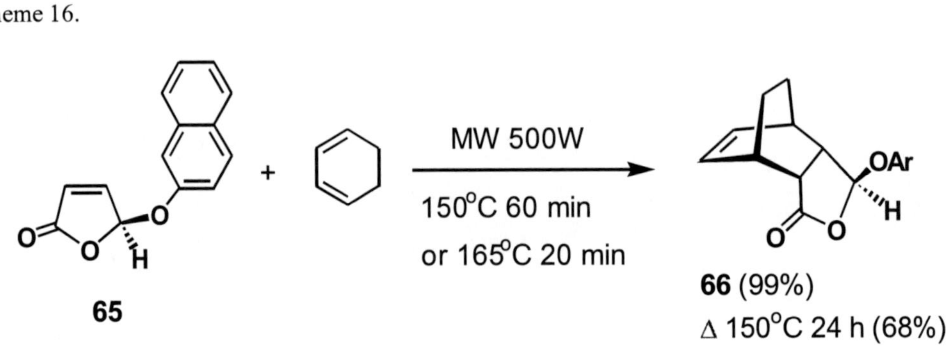

Scheme 17.

1.1.5. Cyclohexadiene

Trost [32] used 4-aryloxybutenolides as "chiral aldehyde" equivalents for an efficient enantioselective synthesis, where the naphthalene substituent has important role in control of the diastereoselectivity of 1,3-dipolar and Diels-Alder cycloadditions. The yields The yields of cycloaddition reactions with 4-naphthoxybutenolide *65* were excellent in the case of Diels-Alder reaction. While performed the Diels-Alder reaction in a normal thermal fashion 68% yield was obtained, but nearly quantitative yields were obtained using a microwave at 150°C (60 min) or 165°C (20 min) (Scheme 17).

1.1.6. Other Dienes

1.1.6.1. Cyclic Dienes

Formation of products after thermal elimination of small fragments from the initial cycloadducts (such as CO_2, ethene, or S) is often characteristic in [4+2] cycloadditions of cyclic dienes. Loupy [33] made an improvement in reaction of ethyl 1,3-cyclohexadiene carboxylate *67* with ethyl propiolate. In the absence of solvent or on silica gel as a support, substituted benzenes *71a* and *71b* were produced, with notable improvements (no solvent, atmospheric pressure, shorter reaction times and lower temperatures) and with enhanced yields (Scheme 18). Solvent-free reaction in classical or microwave conditions produced identical yields and product ratio. Furthermore, classical reaction of 3-carbomethoxy-2-pyrone *72* with methyl propiolate occur for 30 h in refluxing xylene, while microwave results obtained under solvent-free conditions are by far better within short reaction times. Selectivity of regiosimers is only slightly modified in the case of from *76a/76b* 42:58 (Δ) to 31:69 (MW). The progress of reaction of 2-methoxythiophene *77* and DMAD is deeply affected by the nature of the medium. When performed in xylene, it leads to Diels-Alder cycloaddition, whereas, when carried out in acetic acid, Michael addition takes place. Serious improvement in yields for *79a* and *79b* was obtained when compared to those of classical heating. Further improvement in yield was obtained when reaction was performed in solvent-free conditions under microwave activation.

Piettre compared classical heating, microwave heating and high pressure reaction conditions. [34] Electron-poor 6-oxo-1-sulfonyl-1,6-dihydropyridine-3-carboxylates such as *80* undergo cross-Diels-Alder reactions with electron-rich dienes such as *40* and *81* under hyperbaric conditions, reacting either as dienophiles to yield normal-electron-demand cycloadducts *82* or as dienes to give inverse-electron demand cycloadduct *83*. For instance, pyridone *80* reacts with 2,3-dimethylbuta-1,3-diene *81* in high pressure or classical thermal conditions to afford cycloadduct *82*. On the other hand, microwave heating failed to produce *82*, which is a rare example of failure of microwave irradiation to promote cycloaddition reaction. In the inverse electron-demand reaction of *80* with cyclopenta-1,3-diene *40*, cycloadduct *83* was produced, employing either high pressure, classical or microwave heating (Scheme 19). In this example, microwave and high pressure conditions gave comparable yields, while the classical heating proceeded in low yield.

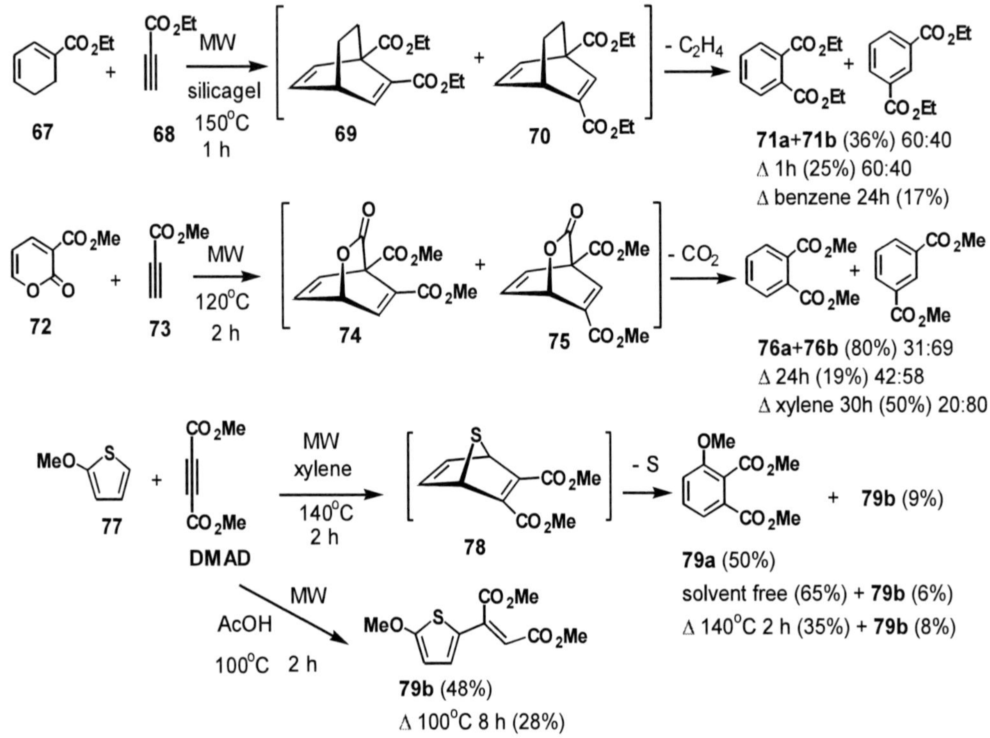

Scheme 18.

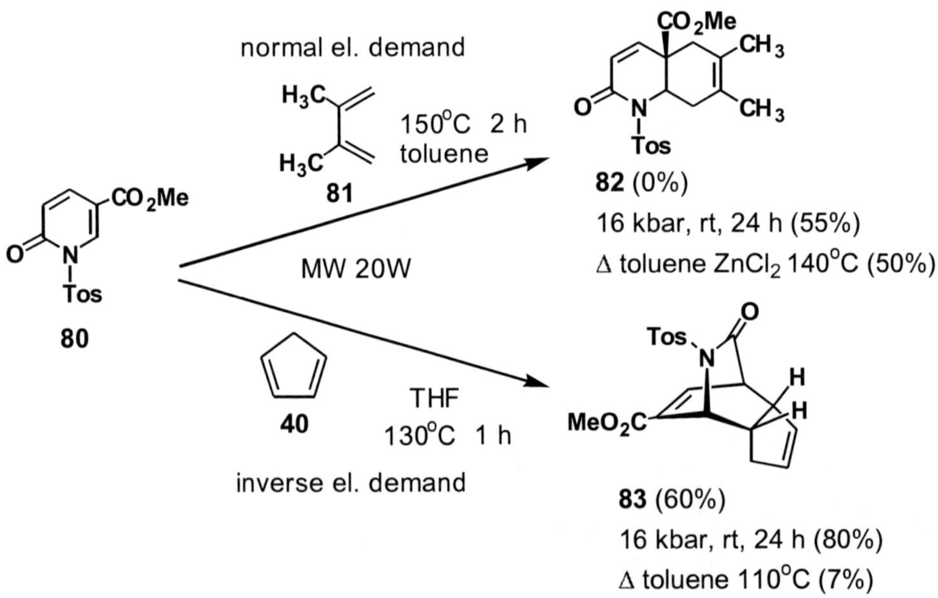

Scheme 19.

It was shown by Kočevar [35] that Diels-Alder reactions between differently substituted 3-acylamino-2*H*-pyran-2-ones such as *84* as diene and (*Z*)-1-methoxybut-1-en-3-yne *85* after a short microwave irradiation (45-180 min) at 150°C afford cycloadduct (*Z*)-*3* (Scheme 20).

Single cycloadduct was obtained, for the formation of which the triple bond had reacted in a completely regioselective way to give compounds containing the two hydrogen atoms on the aromatic ring in *p*-position to one another. For the cycloaddition between *84* and *85* under classical thermal conditions, the necessary reaction times were long, and reflux in toluene required 5.5 h. In the high-pressure conditions (13-15 kbar) the necessary reaction time in the dichloromethane solution was even longer (up to 36 days). In addition, after an extremely long reaction time (138 days) under high pressure a new reaction was observed giving product *87* by the addition of a molecule of MeOH to the previously formed 2-methoxyethenyl moiety of cycloadduct *86*.

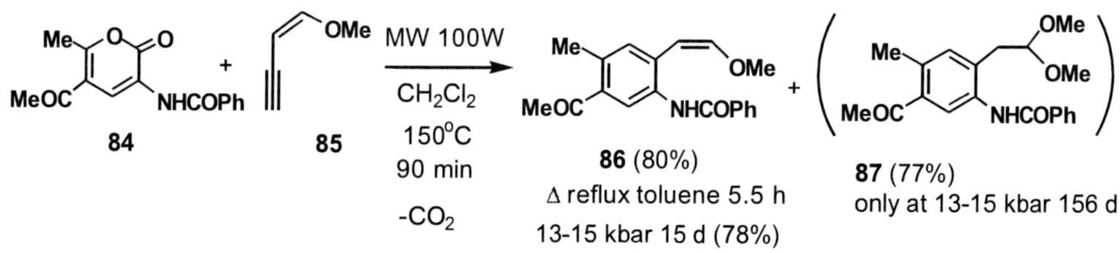

Scheme 20.

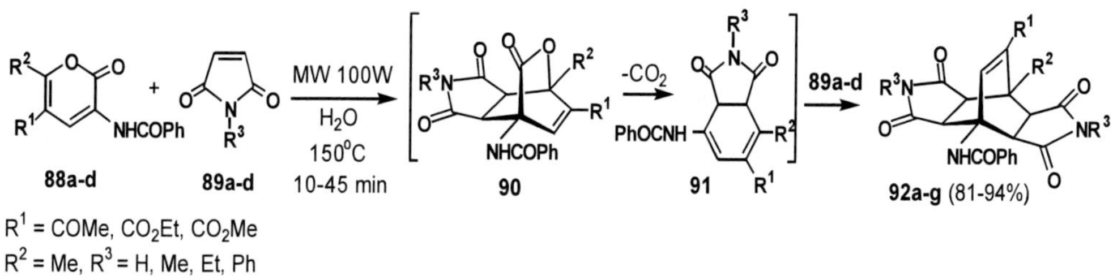

Scheme 21.

The efficient and green synthesis of sterically constrained prochiral didehydroamino acid derivatives of type *90*, bearing both bicyclo[2.2.2]oct-7-enes and an unsaturated amino ketone or ester functionality was developed by Kočevar and Michelet. [36] The synthesis of *90* is based on the Diels-Alder reaction of 2*H*-pyran-2-ones *88a-d* with maleimides *89a-d*. The cycloaddition of some fused pyran-2-ones with maleimides under drastic conditions (in boiling decalin or toluene) resulted in the synthesis of fused isoindoles accompanied in a few cases by double cycloadducts (fused bicyclo[2.2.2]octene derivatives); the latter were mainly detected as intermediates or side-products. Despite the low solubility of substrates *88* in water, cycloaddition was complete after microwave irradiation for 45 minutes affording the bicyclic derivatives *90* in excellent yields (Scheme 21). The conventional synthesis of the cycloadducts *90b* and *90e*, required higher temperatures and longer reaction times (boiling decalin, bp 189-191°C, 90-120 min) to furnish the products in 81% and 76% yield, respectively. When an aqueous mixture of *88a* and *89c* was conventionally heated at reflux, approximately only 60% of the starting 2*H*-pyran-2-one *88a* was transformed to *90c* after two hours. Later was found that solvent-free microwave conditions with a small amount of butan-1-ol additive increase yields up to 100%. [37] It was rationalized that addition of a minimal amount of a liquid with appropriate boiling point, which could vaporise prevents the deposition of volatile reactants in the upper colder part of the microwave ampoule. The liquid might also produce better contact between the reactants. [38]

The microwave absorbing ability of toluene is increased in the presence of quaternary ammonium or phosphonium salts. As found by Keglevich, [39] these additives somewhat accelerated the Diels-Alder reaction of 1-phenyl-1,2-dihydrophosphinine oxide *93* with *N*-phenylmaleimide *94* and maleic acid anhydride to afford phosphabicyclo[2.2.2]octene derivatives *95a,b* (Scheme 22). Toluene better absorbs microwave heat in the presence of onium salts, like triethylbenzylammonium chloride (TEBAC), tetrabutylammonium bromide (TBAB) or ethyltriphenylphosphonium bromide (ETPPB). The most efficient onium salt was TEBAC, with a small increase in yields when small amount of methanol was added to obtain homogeneous system.

1.1.6.2. Acyclic Dienes

Silica gel can be used as an excellent support for regioselective Diels-Alder reactions of myrcene under microwave irradiation. Oskooie [40] demonstrated that these conditions have the advantages being a practical and environmentally benign protocol that decreases the time, gives a cleaner reaction and leads to easier work up. Diels-Alder reactions of myrcene (7-methyl-3-methene-1,6-octadiene) *96* with carbonyl containing dienophiles such as acrolein and methyl acryate *97a,b* supported on silica gel under microwave irradiation give a high para/meta regioselectivity and the reaction rates are also enhanced dramatically. In the thermal reaction, the yield of the *para* adduct *98* is greater than that of the *meta* adduct *99*, while the selectivity of the cycloaddition is decreased with increase of the reaction temperature (Scheme 23).

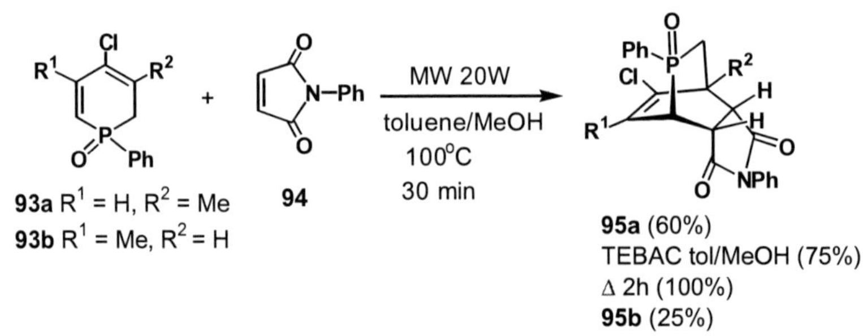

Scheme 22.

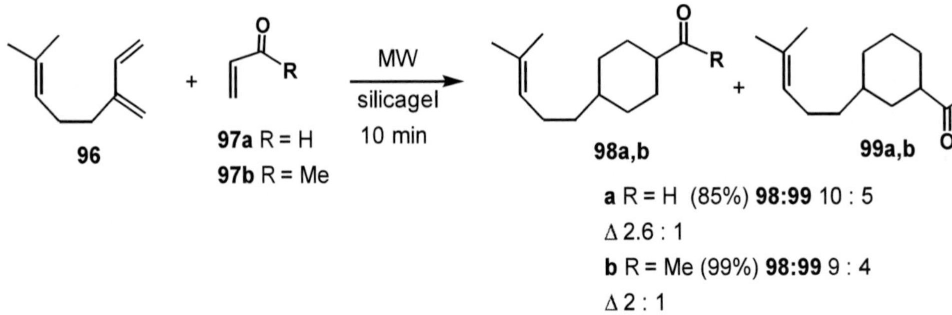

Scheme 23.

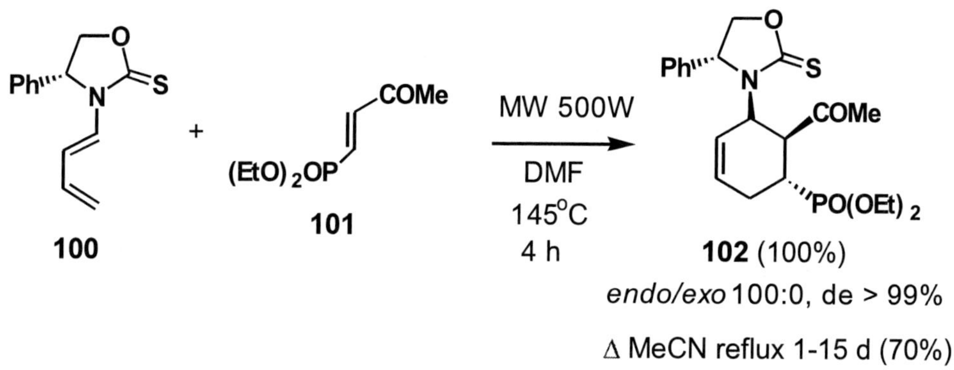

Scheme 24.

It was found by Marchand-Brynaert [41] that microwave activation method provides a useful way to overcome inherent low reactivity of the phosphonodienophile *101.* An increased efficiency to reach the cycloadduct *102* with excellent regio- and stereoselectivity was achieved by the reaction of chiral aminodiene *100* and dienophile *101.* Using toluene or dichloromethane as solvent (totally permeable to microwaves), very low conversion was observed. However, in acetonitrile, conversion occurred to the same extent as under classical heating conditions and in DMF at 145°C, complete conversion was observed after only 4 h. These observations highlight the lack of specific microwave activation effect as a consequence of an isopolar mechanism [33, 42] (which in good agreement with the predicted concerted synchronous character of these cycloadditions). There is no change of the observed selectivities under microwave activation compared to classical heating (Scheme 24).

Choshi and Hibino [43] have shown that regioselective Diels-Alder reaction of the isoquinoline-5,8-dione *104* with diene *105* in toluene at 110 ^{o}C gave 6-deoxybostrycoidine *107* with the elimination of a methoxy group from the initially formed cycloadduct *105* (Scheme 25). The microwave-irradiated conditions were more effective than the conventional conditions for the yield improvement (73/86%) and for the decrease of the reaction time (30/10 min).

Multicomponent reaction [44] of aldehydes, amides, and dienophiles was successful for the straightforward synthesis of a large variety of carbo- and heterocyclic amides. Beller [45] used a mixture of oxazolidin-2-one *106*, aldehyde *107* and maleimide *108* in the presence of Ac_2O, and *p*-TsOH under microwave irradiation to produce 4-(oxazolidin-2-on)-2-methyl-*cis*-3a,4,7,7a-tetrahydroisoindole-1,3-dione *110* in 73% yield (Scheme 26). The underlying mechanism involves an Oppolzer-Overman type 1-(oxazolidin-2-on)-1,3-butadiene *109*, which easily undergoes Diels-Alder addition to an electron-deficient dienophile. While generally microwave synthesis in solvent-free conditions is beneficial to yield increase, synthesis of *110* is an example in which solvent-free conditions led to decrease in yield to 52%. Variety of three-component adducts was synthetized by this method in 26-96% yield, based upon structural variations of carboxamide, aldehyde, and olefin components.

Microwave irradiation was also used by Díaz-Ortiz [46] as an efficient tool for the generation of *N*-heterocyclic *o*-quinodimethanes, which were then used in the synthesis of polyheterocyclic compounds by Diels–Alder reaction. An example is irradiation of tetrabromide-1,2,4-triazole *111* in the presence of NaI and a small amount of DMF (necessary to dissolve the sodium salt) which gave the corresponding *o*-quinodimethane *113*. Subsequent cycloaddition with electron-deficient dienophile and debromination afforded the corresponding cycloadduct *112* (Sheme 27). When acetylenic dienophiles such as DMAD or enamine *116* were used, subsequent elimination of two equivalents of HBr from initial cycloadducts afforded aromatic products such as *117*. [47]

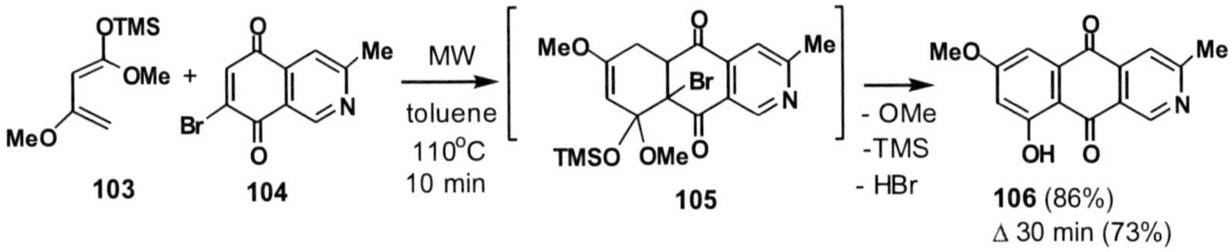

Scheme 25.

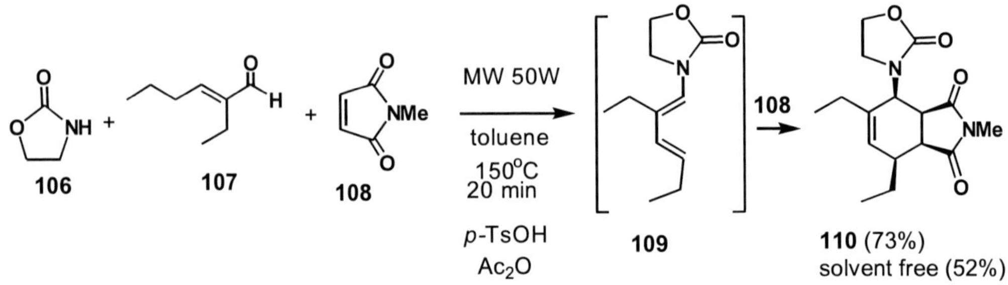

Scheme 26.

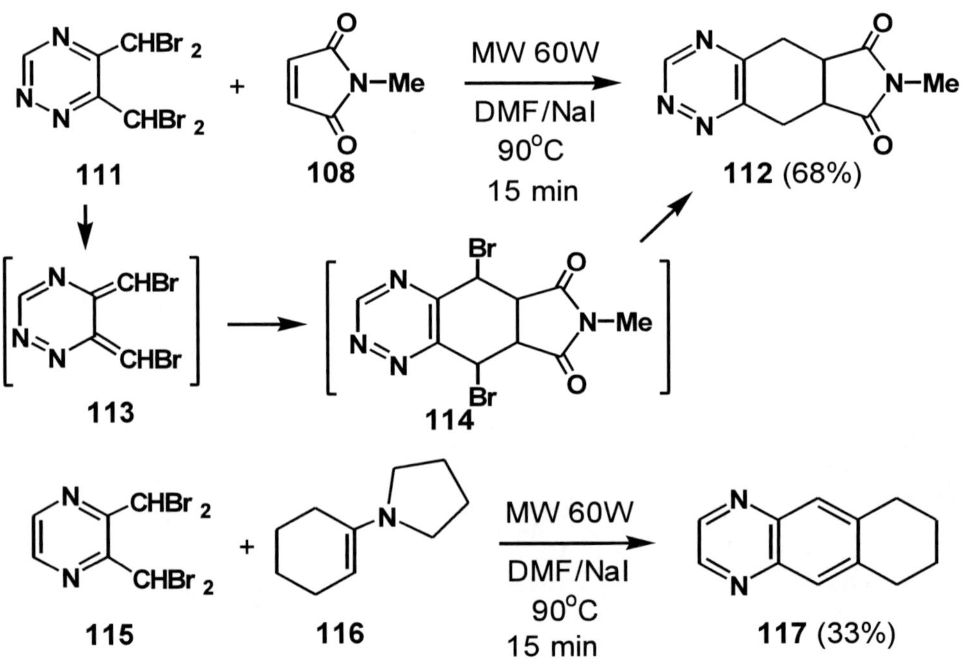

Scheme 27.

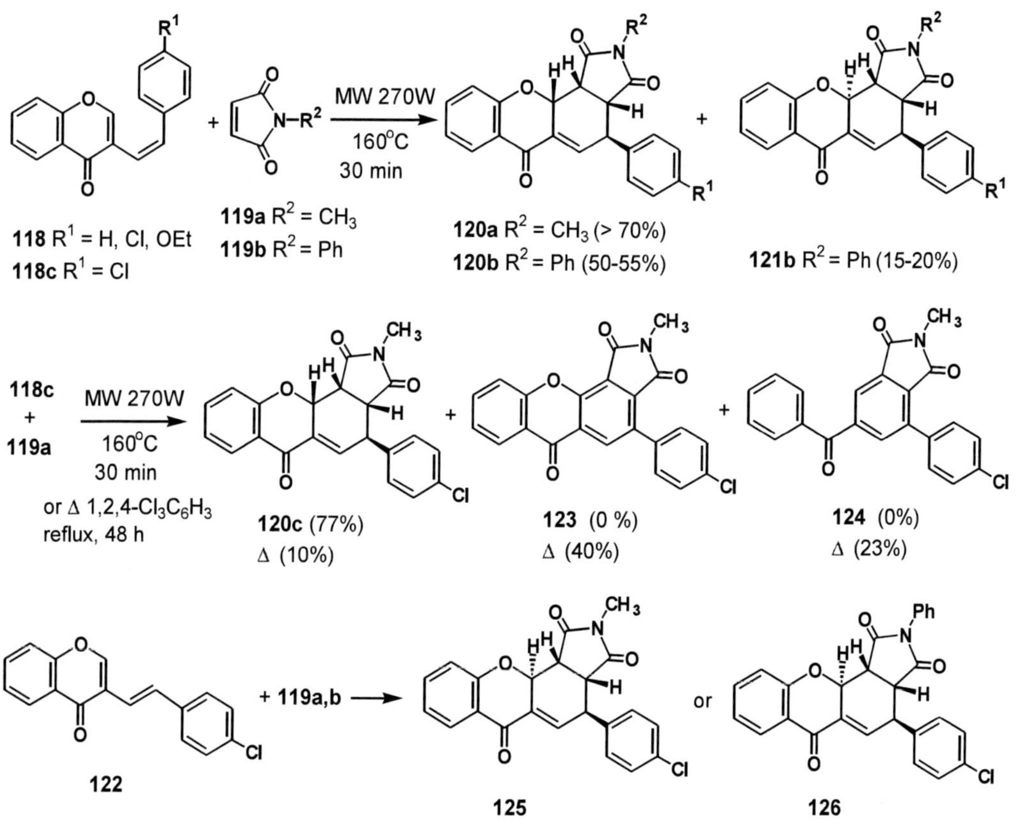

Scheme 28.

Solvent-free reactions carried out by Silva and de la Hoz [48] were highly efficient when microwave irradiation was applied in conjuction with a very reactive dienophiles. The cycloadduct 4-aryl-2-methyl-1,3-dioxopyrrolo[3,4-*c*]-3a,4,11a,11b-tetrahydroxanthone *120a* was obtained from *N*-methylmaleimide and (*Z*)-3-styrylchromones *118* in good yield under microwave irradiation (Scheme 28). Under classical heating conditions Diels-Alder reaction of *118c* with *119a* only occurs at high temperatures. Total consumption of the reactants occurred after 24 hours under reflux and the reaction produced a mixture of major cycloadduct *120c* and xanthone *123*. Another important aspect of this reaction under classical heating was that longer reaction times did not improve yields but caused degradation to give another byproduct, benzophenone *124*. [49] Microwave reactions carried out in the absence of solvent take the advantage of the synergy between microwave irradiation and solvent-free reactions, where the radiation would be directly absorbed by the reagents. By using this approach, cycloadduct *120c* was obtained as the only reaction product. In the case of the Diels-Alder reaction of (*Z*)-3-styrylchromones *118* with *N*-phenylmaleimide, the *endo* cycloadduct tetrahydroxanthone *120b* was obtained together with a small amount of isomer *121b*. Products *121b* and *124* are the *exo* cycloadducts of the reaction between *N*-phenylmaleimide and (*E*)-3-styryl-chromone *122* obtained by thermal isomerisation of (*Z*)-3-styrylchromones *118* and subsequent reaction with *N*-phenyl-maleimide giving the *exo* cycloadducts *121b*. This mechanistic rationale was proven by the Diels-Alder reactions of

(*E*)-3-styrylchromone *122* and *N*-methyl and *N*-phenylmaleimide, under microwave irradiation in solvent-free conditions gave *exo* adducts *125* and *126* in good yields.

Vinylpyrazoles take part as dienes in [4+2] cycloaddition reactions, whereas one of the π-bonds is embedded in pyrazole aromatic ring. When Díaz-Ortiz [6] reacted vinylpyrazoles *127* or *128* with DMAD under microwave irradiation in solvent-free conditions, the products were afforded in 72 and 45% yield, respectively (Scheme 29). Reaction of *127* was performed in a teflon vessel in a domestic oven, whereas the reaction of *128* was carried out at atmospheric pressure in a focused microwave reactor. This is one example in which the change from the classical heating to microwave also induces the formation of products, which were not obtained previously. Indazoles *130* and *133* resulted from a further ene reaction of the primary Diels-Alder adduct with a second molecule of DMAD, followed by two [1,3]-sigmatropic hydrogen shifts. At shorter times, the intermediate products *134* and *135* were isolated. These intermediates rearranged spontaneously in a few minutes to *130* and *133*. Contrary to the data reported by classical heating, ene reactions were also observed for compound *128*.

Similarly, Silva [50] found that reaction of (*E* and *Z*)-1-acetyl-3-(2-hydroxyphenyl)-4-styrylpyrazoles *136a-d* and *137a-c* with *N*-methylmaleimide enhanced by microwave irradiation in solvent free conditions gave cycloadducts 1-acetyl-3-(2-hydroxyphenyl)-7-methy-5-phenyl-6,8-dioxopyrrolo[3,4-*g*]-5,5a,8a,8b-tetrahydroindazoles *138a-d* or *139a-c* in moderate to very good yields. Products *138a,b,d* and *139c* were oxidised with DDQ with subsequent *N*-deacylation to afford 3-(2-hydroxyphenyl)-7-methyl-5-phenyl-6,8-dioxopyrrolo[3,4-*g*]indazoles *140a-d* (Scheme 30).

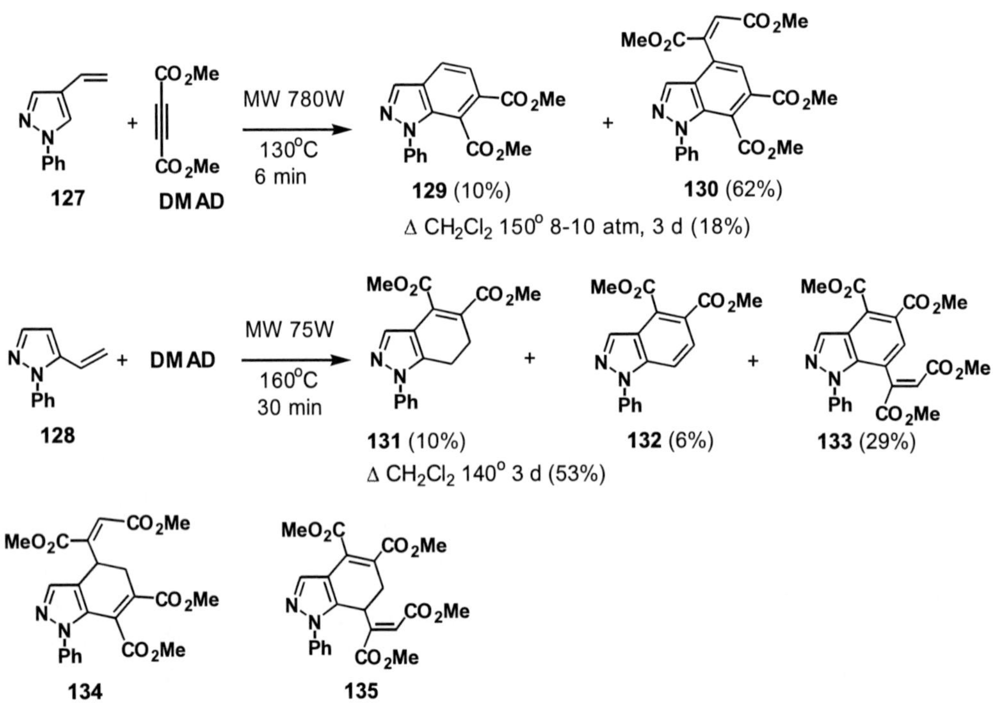

Scheme 29.

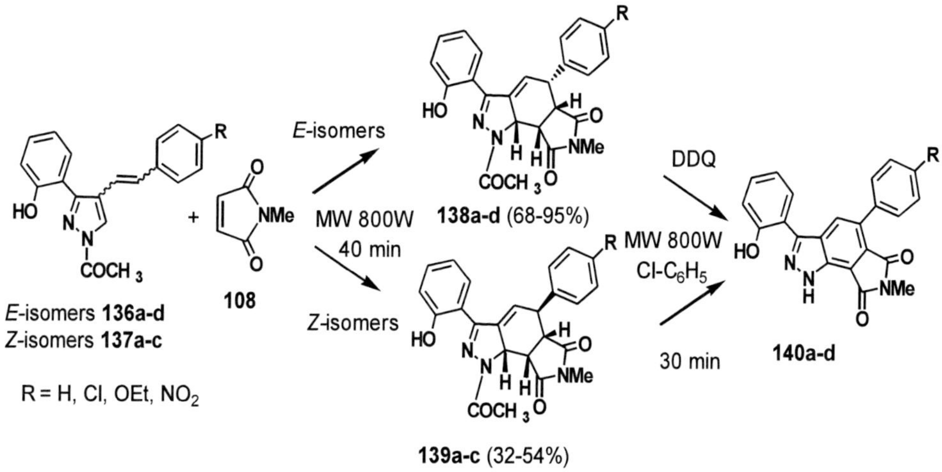

Scheme 30.

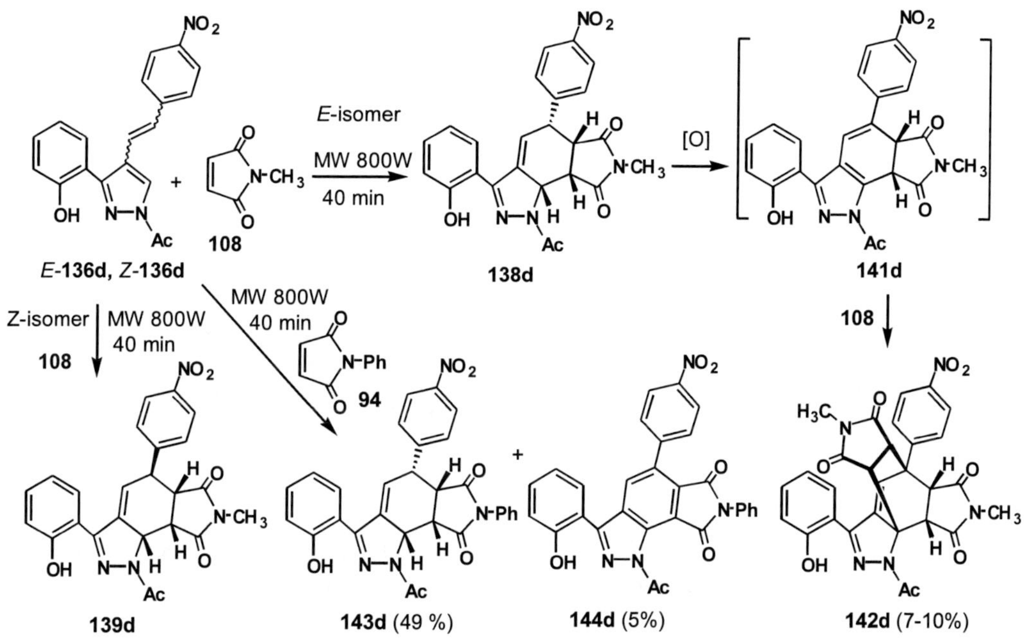

Scheme 31.

Further study by Silva [51] showed that the *Z* isomers of styrylpyrazoles are less reactive than the corresponding *E* isomers, since the adducts *139* were obtained in lower yields (54-60% for *139*, in comparison with 68-95% for *138*). In the reactions between *N*-methylmaleimide and *Z-136d*, another compound besides the expected adduct *138d* was isolated (7-10%). The formation of *142d* can be explained by partial oxidation of the adduct *138d* in the reaction medium, giving the intermediate *141d*. Cycloadditions between the formed diene moiety in this intermediate with another molecule of *N*-methylmaleimide, which was used in excess, afforded compound *142d* (Scheme 31). Interestingly, cycloadditions between the (*E*)-1-acetyl-4-nitrostyrylpyrazole *E-136d* and the other electron deficient dienophile *N*-phenylmaleimide (3 equiv) under microwave irradiation and solvent-free conditions led to the formation of the cycloadduct *143d* and to the oxidised indazole *144d*. On the other hand, cycloaddition with another electron deficient dienophile DMAD failed, producing only traces of Michael-type conjugate addition of the pyrazole nitrogen to triple bond.

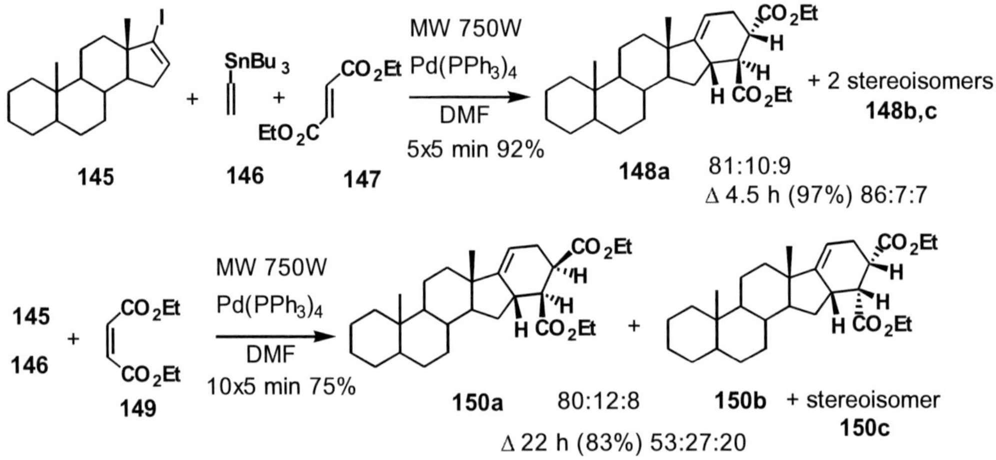

Scheme 32.

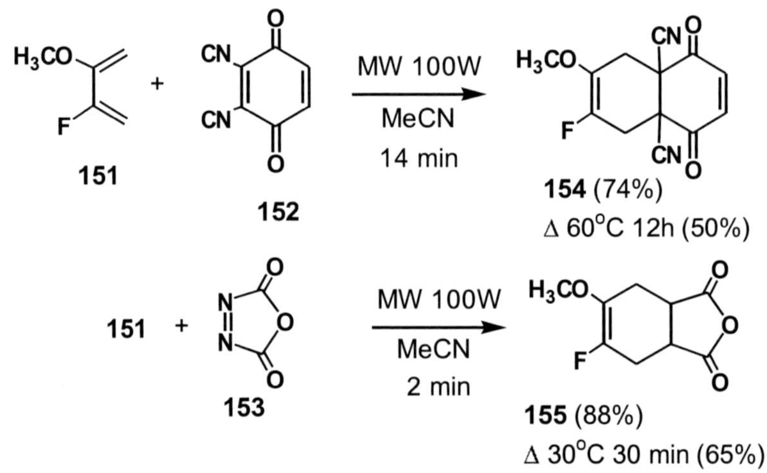

Scheme 33.

One-pot Stille-coupling/Diels-Alder reaction has been employed as a novel synthetic methodology by Skoda-Földes [52] to 17-iodo-androst-16-ene substrate *145*. Cycloaddition reaction of *145* with vinyltributyltin and diethyl fumarate or diethyl maleate in the presence of $Pd(PPh_3)_4$ catalyst produced mixtures of three stereoisomers *148a-c* and *150a-c* (Scheme 32). The same isomers were produced both with thermal and microwave heating. More rapid reactions could be observed with the use of microwave heating. With diethyl fumarate as the dienophile, distribution of the three stereoisomers was the same as with thermal heating. In the case of diethyl maleate selectivity of cycloaddition could be greatly improved in microwave conditions. The main product *150a* is obtained by the approach of the dienophile from the α side of the steroidal skeleton and through the less favorable *exo* transition state due to the steric hindrance. The *150a* is thermodynamically more stable than *150b*, which is produced through the sterically more favorable *endo* transition state. The improved selectivity in formation of the thermodynamically more stable product with microwave irradiation was explained in terms of direct rapid heating and superheating and possibility of retro-Diels-Alder/Diels-Alder reaction sequence. With microwave irradiation usually no change of selectivity is observed except for a few instances. [53]

Patrick [54] has found that 2-fluoro-3-methoxy-1,3-butadiene *151* reacts with various dienophiles in moderate yields (20-65%) under thermal activation (Scheme 33). When the cycloadditions are performed in closed vacuum hydrolysis tubes under 100W microwave irradiation, the yields are greatly improved (70-90%) and the reaction times (5-25 min) are much shorter.

Jankowski [55] studied the cycloaddition reactions on arecoline *157* and its isomer *158* with Danishefsky's diene. Under thermal conditions (160°C, toluene, 5 days) reaction gave the methyl *cis-N*-methylisoquinolidin-8-en-7-one-10-carboxylate *159* and *160* for the ethyl isomer *158*. Reactions are accompanied with a significant amount of diene decomposition and polymerization as well as dienophile dimers. When the same reaction was carried out using the microwave ehnacement, products *159* and *160* were obtained within 5 min, in 73 and 63% yield, respectively (Scheme 34). Another advantage of microwave conditions is that the amount of dimerization product was lowered, and the remaining diene could be recovered. The reaction of *156* with *158* under microwave also afforded an unexpected product (~50%), identified as *161*, which is formed by the isomerization of the double bond of *158* from the conjugate position, and its subsequent reaction with the diene.

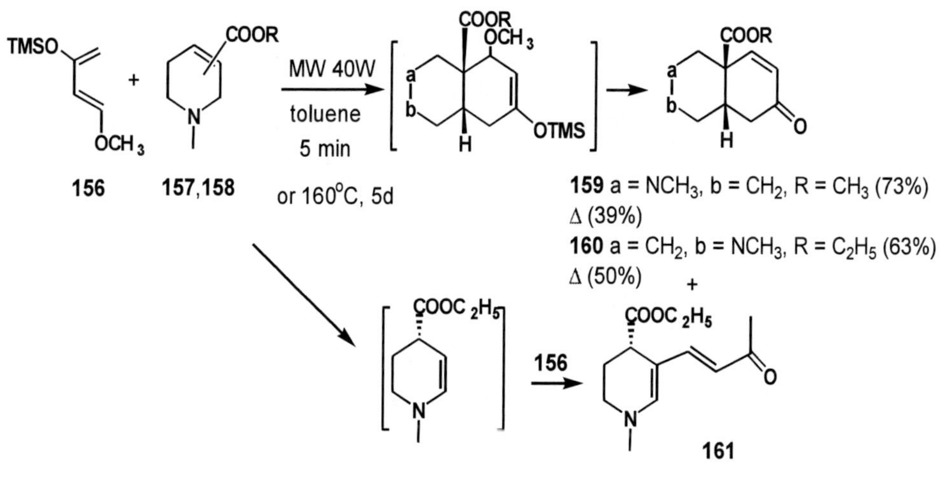

Scheme 34.

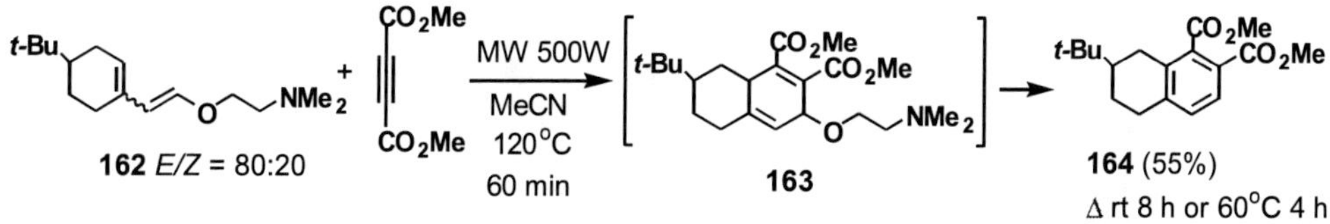

Scheme 35.

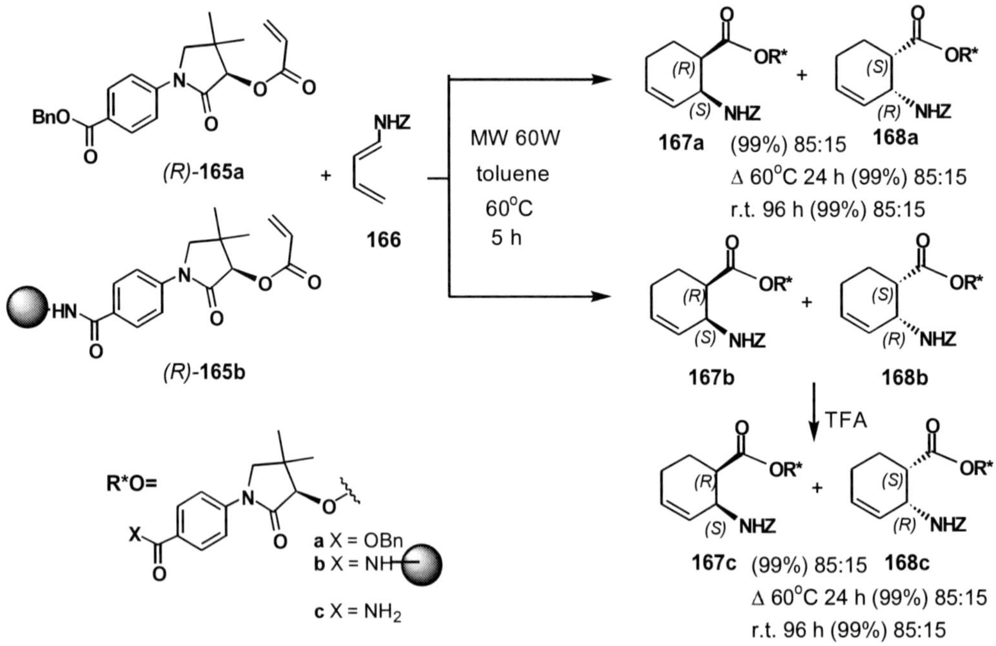

Scheme 36.

Microwave-mediated synthesis of tetrahydronaphthalene and dihydrophenanthrene derivatives via Diels-Alder intermediates was conducted by Larhed. [56] The *E/Z*-mixtures of *162* were treated with dimethyl acetylenedicarboxylate in order to obtain the corresponding Diels-Alder adducts. The thermal reactions (8 h at room temperature or 4 h at 60°C) were sluggish, furnishing significant amounts of side products. However, the microwave protocol proceeded well with complete conversion of the *E/Z*-diene mixture, but yielding aromatized tetrahydronaphthalene derivative *164* in moderate to good yield (Scheme 35). The formation of the resulting product *164* could be explained by immediate elimination of 2-dimethylamino ethanol from the primarily formed, intermediate Diels-Alder adduct *163*. Performing the microwave-mediated reaction with *162* under reductive conditions (by addition of hydroquinone) did not lead to the aromatic compound *164* but rather to intermediate *163* (in *163 : 162* ratio = 82:18).

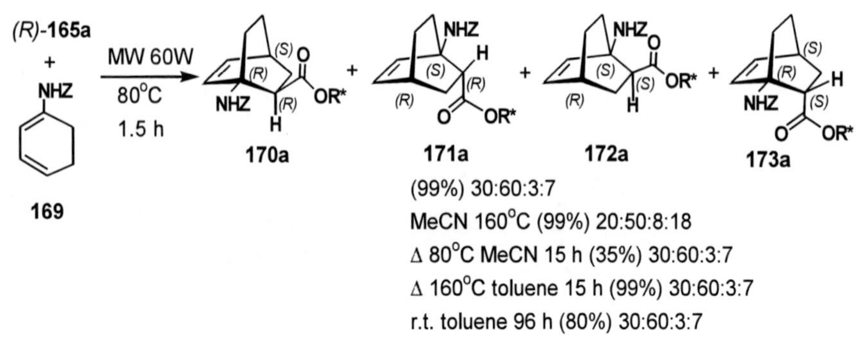

Scheme 37.

Calmès [57] applied (3*R*)-4-(3-hydroxy-4,4-dimethyl-2-oxopyrrolidin-1-yl)benzoic acid as a chiral auxiliary in asymmetric Diels-Alder reaction of aminodienes with a nonracemic acrylate. In addition, this auxiliary has a carboxylic acid function that was used for attachment to an amine-functionalized insoluble polymer (Rink resin) to afford (*R*)-*165b*. The asymmetric Diels-Alder reactions between (3*R*)-4,4-dimethyl-2-oxopyrrolidin-3-yl acrylate derivative *165a* and *N*-Z-protected 1-aminodiene *166* were carried out both in solution and on a solid support. Comparable results for each reaction conditions were observed with the formation of the corresponding constrained cyclic β-amino acids *167a-c* and *168a-c* in high yield and moderate to good selectivity (Scheme 36). A comparison of solution and solid-support synthesis by conventional heating and microwave activation revealed that the microwave technique is often advantageous. Reaction on a solid support offered good resolution of the problem of low reactivity and instability of the reagents.

With cyclic diene *169*, mixture of four diastereomeric cycloadducts *170a-173a* was obtained in cycloaddition reaction with (*R*)-*165a* (Scheme 37). [58] Comparable yields were obtained in solvent free conditions and in acetonitrile, but in less vigorous conditions. More interestingly, microwave heating at 160°C in acetonitrile caused the significnat change in distribution of diastereoisomers, which was not observed by the classical heating in toluene, where the change or ratio of products occurs only by the addition of catalysts such as $Eu(fod)_3$ or $LiNTf_2$.

1.1.7. Natural Product Synthesis

The total synthesis of complex diterpenes is an example of using vinylallene as diene in an intermolecular Diels-Alder reaction enhanced by microwave irradiation. Barriault [59] found out that tandem oxy-Cope/Claisen-ene/Diels-Alder reaction of vinylallene *174* with dienophiles such as methyl vinyl ketone afforded tetracycle *181* in 40% yield (Scheme 38). The reaction can be performed in one pot in toluene using a vinylmagnesium bromide. Mechanistically, heating *174* gives the macrocyclic allene ether *175* which is poised to undergo a Claisen rearrangement to generate enone *176*. The latter is converted via a transannular carbonyl ene reaction to vinylallene *177* which, in the presence of an activated dienophile, undergoes an intermolecular Diels-Alder reaction to form tricycle *181* as a single diastereoisomer. Of the two possible transition states (*178* and *179*), reaction takes place through transition state *178*, due to minimized steric interactions, were magnesium serves as a temporary tether.

Nice application of microwave heating in natural product chemistry of opium alkaloids was published by Maat. [60] Heating of 6-demethoxy-β-dihydroxythebaine *182* in an excess of methyl vinyl ketone under reflux for 60 h yielded two adducts *184* and *185* in 3:2 ratio. Main product 7β-acetyl-6β,14β-ethenomorphinan *184* was formed by the Diels-Alder addition at the α-side of the diene system. In conventional conditions extensive polymerization of the dienophile occurs, and dramatic improvement was achieved when reaction was carried out using a modified microwave oven at atmospheric conditions (Scheme 39). Reaction was complete within 24 h with substantially less formation of polymeric material.

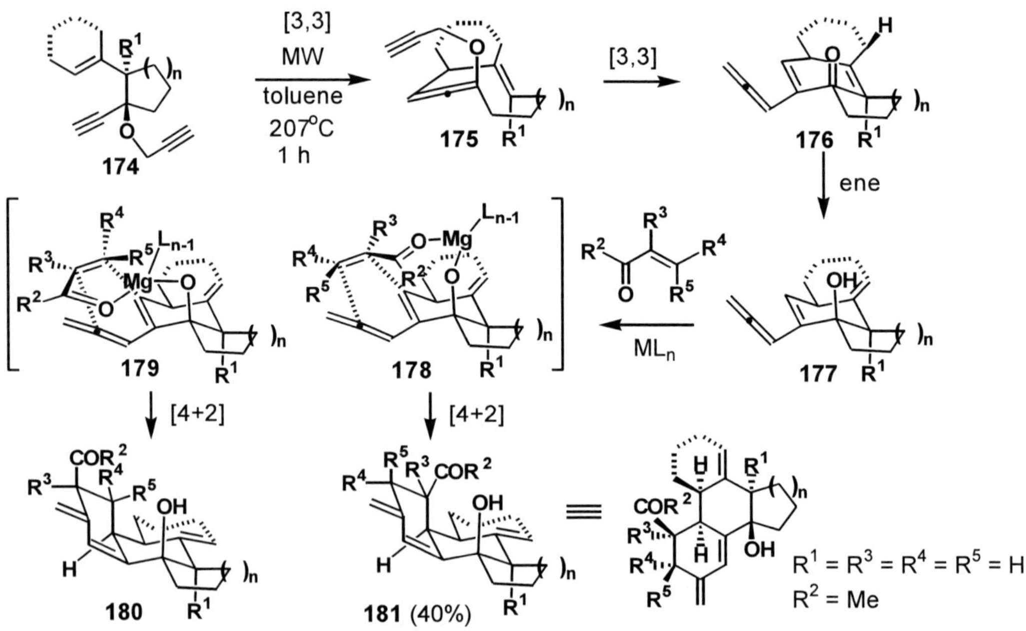

Scheme 38.

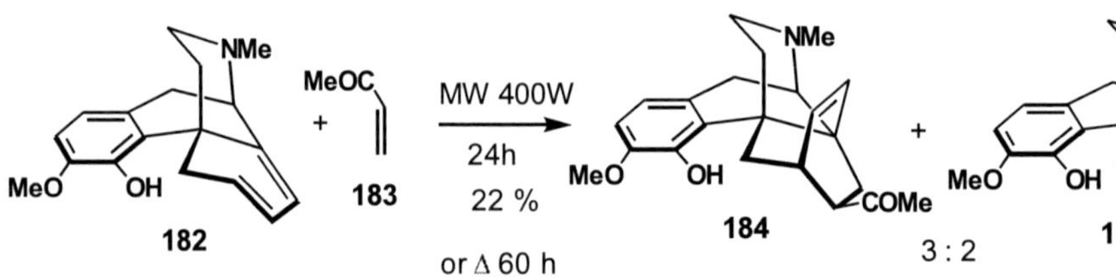

Scheme 39.

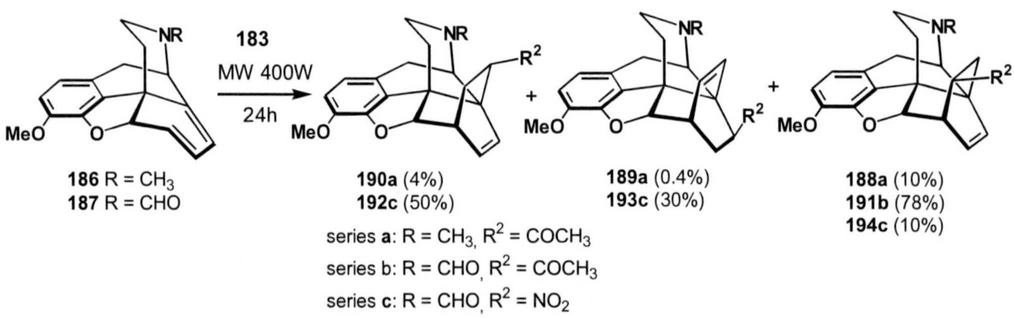

Scheme 40.

Diels-Alder reactions of 6-demethoxythebaine *186* with methyl vinyl ketone gave more complex mixture of products: the main isomer is 7α-acetyl-6α,14α-ethenoisomorphinan *188a*, followed by 8α-isomer *189a* (10%), the third product 8β-acetyl-6β,14β-ethenomorphinan *190a* (0.4%), and the fourth, minor product (<0.5%) isomeric 7β-substituted-6α,14α-ethenoisomorphinan (Scheme 40). [61] Cycloaddition of 6-demethoxy-*N*-formyl-*N*-northebaine *187* with methyl vinyl ketone gave similar results: two products were formed in 9:1 ratio, major product 7α-acetyl-6α,14α-ethenoisomorphinan *191b*. On the other hand, nitroethene favored attack of the dienophile from the more hindered α-face of the diene system. Cycloaddition of *187* to nitroethene gave *192c* as major product, followed by the second *193c* (30%), and the third product *194c* (10%).

Banwell [62] employed microwave activated Diels-Alder cycloaddition in the total synthesis of phellodonic acid, a natural product, functionalized hirsutane derivative. Acetonide diene derivative *195* fails to engage in a thermally-induced Diels-Alder cycloaddition with the dienophile *196*. Such *cis*-1,2-dihydrocatechol derivatives are especially prone to acid-catalyzed aromatization, which means that Lewis acids cannot be used to promote the desired cycloaddition process. This conversion was effected leading to adduct *197* at extremely high pressure in an apparatus that has a relatively small working volume (13 mL) and that is difficult to set up and maintain. When a mixture of diene *195* and dienophile *196* was heated (neat) using a microwave reactor then compound *197* could be obtained in high yield (Scheme 41). The major adduct was accompanied by small quantities of cycloadduct *198*, a *syn*-isomer of compound *197*, as well as *199*, a Diels-Alder dimer of diene *195*.

In studies towards oroidin derived alkaloids natural products Romo [63] overcome the problem of long reaction times for several of the Diels-Alder reactions in classical conditions by the use of microwave conditions. The Diels-Alder reaction of Tsv-diene *201* with dienophile *200* under microwave conditions, employing 1 equiv of diene relative to dienophile in aqueous benzene afforded initial cycloadduct *202*, isomerized cycloadduct *203*, and unreacted dienophile *200* (0.07:0.18:1 ratio, 150°C, 1 h). Addition of $LiClO_4$ and 3.0 equiv of diene *201* using only water as the heating medium resulted in much shorter reaction times possibly due to hydrophobic effects, and the initial cycloadduct *202* was not observed (Scheme 42).

Under optimal microwave conditions, the Diels-Alder reaction of Tsv-diene *201* gave a maximum overall yield of 48% after TBDPS deprotection. The major side product was determined to be the competing dimerization of *201* requiring excess diene to reach reaction completion which limits the use of this substrate. The cycloaddition of the Tse/DMB diene *204* in THF gave the initial Diels-Alder adduct *205* and regioisomer *206*. Conducting the reaction of *204* in benzene in the presence of the weak Lewis acid, resulted in the formation of isomerized adduct *207* as the major product along with regioisomer *208* using only 1.1 equiv of diene (Scheme 43). This contrasts to conventional heating, which required 2-2.5 equiv of diene to achieve similar conversions.

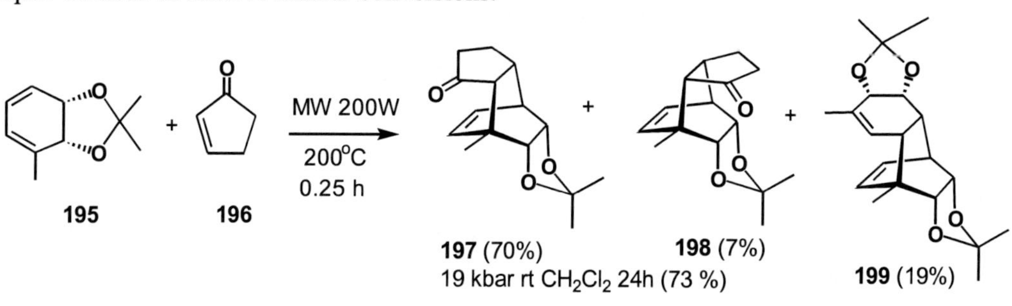

Scheme 41.

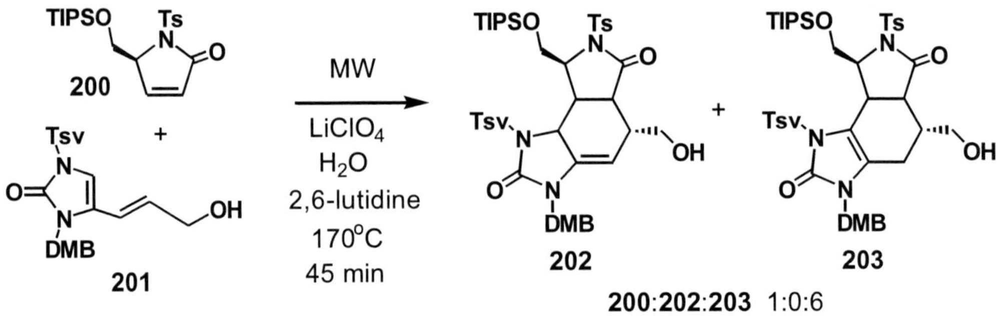

Scheme 42.

Four novel *murrapanine* analogues were synthesized by Guo [64] from 3-dehydroprenylindole *209* as a diene, via a Diels-Alder [4+2] reaction (Scheme 44). In the cycloaddition reaction with *213*, initially formed Diels-Alder product was oxidized to a more stable aromatized product *217* by 1,4-benzoquinone. The important feature of the Diels-Alder reaction was high stereoselectivity, and the original steric structure of diene and dienophile was not changed. Since the substituent groups of dienophiles (chloranil, naphthoquinone, maleic anhydride and 1,4-benzoquinone) have *cis* structure, these substituent groups in products obtained *cis*-configuration.

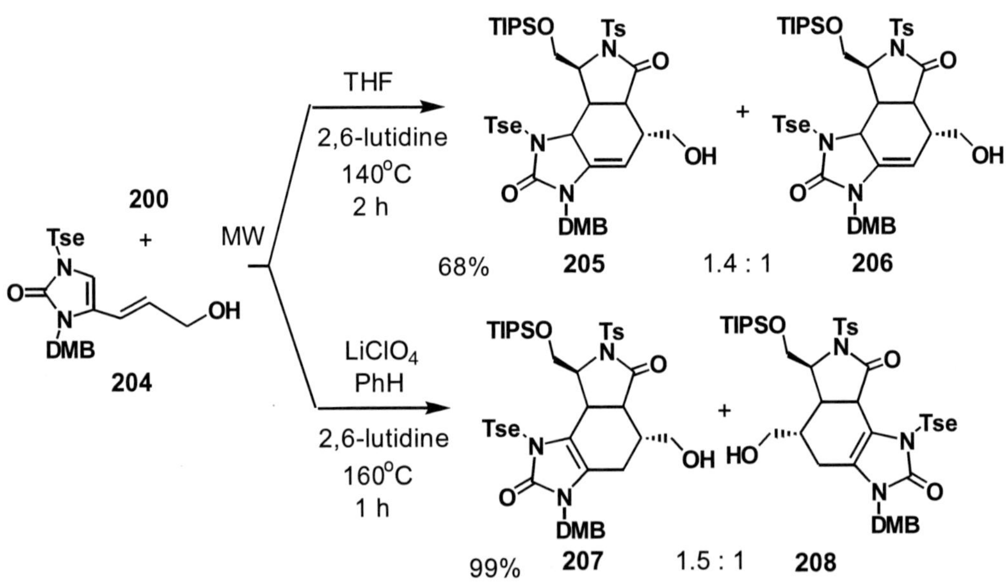

Scheme 43.

New cytotoxic diterpenyl naphthohydroquinone derivatives were obtained by del Corral [65] from a natural diterpenoid myrcecommunic acid *218* under microwave irradiation with *p*-benzoquinone. Reactants were deposited on solid support and dried material was subjected to microwave irradiation. The solvent-free reactions gave cycloadducts *220-222*, in ratios depending on the conditions and solid supports applied (Scheme 45). Interestingly, the change of irradiation power from 500W to 750W (Al_2O_3 solid support, 4 min), induced the change of products ratio *220*:*221*:*222* from 2:2:1 to 2:3:0, while the use of K-10 clay yielded cycloadduct *220* exclusively, no autooxidation products *221* and *222* were detected. Similarly, classical heating in diethyl ether with $BF_3 \cdot Et_2O$ catalyst (24 h) gave *220* without the aromatized products. This is one example of reactions where microwave conditions were not found to be better than conventional heating. Although the microwave irradiation has the advantage of shortening the reaction time, this procedure needs an excess of *219* that difficults the purification of the final products, while the aromatised products also appear with variable proportions in the reaction mixture.

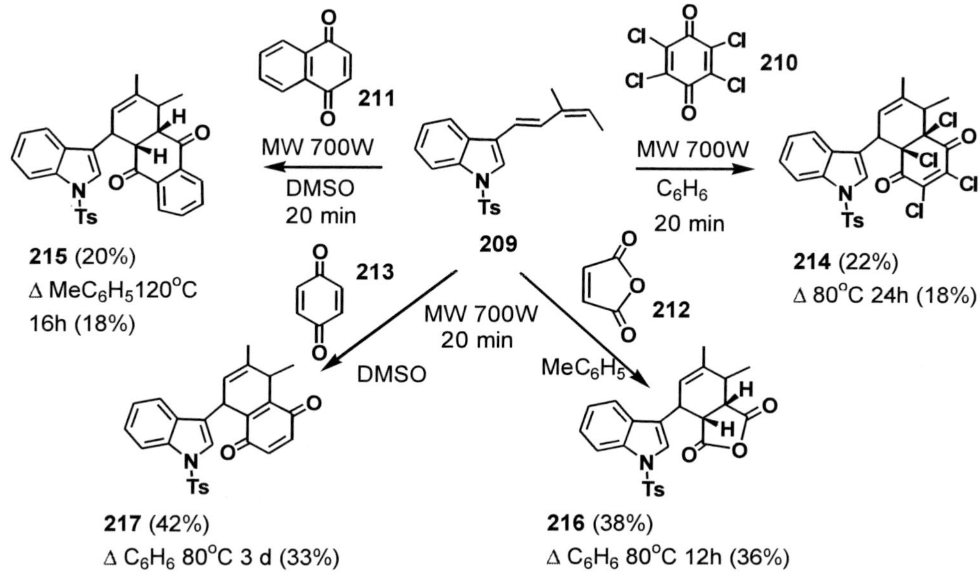

Scheme 44.

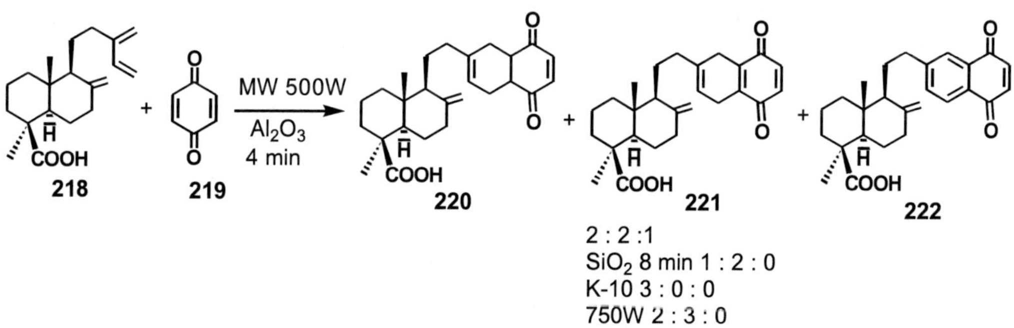

Scheme 45.

REFERENCES

[1] Fringuelli, F.; Taticchi, A. *The Diels-Alder Reaction: Selected Practical Methods*, John Wiley & Sons, Chichester, *2002*.

[2] Sauer, J. *Angew. Chem. Int. Ed. 1967*, *6*, 16-33.

[3] Boger, D. L. in *Comprehensive Organic Syntheses*, Vol. 7, Trost, B. M. (ed.), Pergamon, Oxford, *1991*.

[4] Boger, D. L.; Weinreb, S. M. *Hetero Diels-Alder Methodology in Organic Syntheses*, Academic Press, San Diego, *1987*.

[5] Giguere, R. J.; Bray, T. L.; Duncan, S. M.; Majetich, G. *Tetrahedron Lett. 1986*, *27*, 4945-4948.

[6] Díaz-Ortiz, A.; Carrillo, J. R.; Díez-Barra', E.; de la Hoz, A.; Gómez-Escalonilla, M. J.; Moreno, A.; Langa, F. *Tetrahedron 1996*, *52*, 9237-9248.

[7] Lacova, M.; Gasparova, R.; Loos, D.; Liptay, T.; Pronayova, N. *Molecules 2000*, *5*, 167-178.

[8] Bose, A. K.; Manhas, M. S.; Ghosh, M.; Shah, M.; Raju, V. S.; Bari, S. S.; Newaz, S. N.; Banik, B. K.; Chaudhary, A. G.; Barakat, K. J. *J. Org. Chem. 1991*, *56*, 6968-6970.

[9] Phutdhawong, W.; Buddhasukh, D.; Pyne, S. G.; Rujiwatra, A.; Pakawatchai, C. *Synth. Commun. 2006*, *36*, 881-883.

[10] Shore, G.; Organ, M. G. *Chem. Commun. 2008*, 838-840.

[11] Li, Z.; Chen, J.; Xu, H.; Hu, S.; Shen, S. *PIERS Proc. Hangzhou 2008*, 544-547.

[12] Safaei-Gomi, J.; Tajbakhsh, M.; Kazemi-Ani, Z. *Acta. Chim. Slov. 2004*, *51*, 545-550.

[13] Raner, K. D.; Strauss, C. R.; Vyskoc, F.; Mokbel, L. *J. Org. Chem. 1993*, *58*, 950-953.

[14] Pei, B.-J.; Lee, A. W. M. *Tetrahedron Lett. 2010*, *51*, 4519-4522.

[15] Polshettiwar, V.; Varma, R. S. (eds.), *Aqueous Microwave Assisted Chemistry - Synthesis and Catalysis*, RSC Green Chemistry No. 7, RSC, Cambridge, *2010*.

[16] Werner, S.; Curran, D. P. *Org. Lett. 2003*, *5*, 3293-3296.

[17] Tomé, J. P. C.; Cho, D.-G.; Sessler, J. L.; Neves, M. G. P. M. S.; Tomé, A. C.; Silva, A. M. S.; Cavaleiro, J. A. S. *Tetrahedron Lett. 2006*, *47*, 3131-3134.

[18] Sasaki, S.; Ishibashi, N.; Kuwamura, T.; Sano, H.; Matoba, M.; Nishikawa, T.; Maeda, M. *Bioorg. Med. Chem. Lett. 1998*, *8*, 2983-2986.

[19] Berlan, J.; Giboreau, P.; Lefeuvre, S.; Marchand, C. *Tetrahedron Lett. 1991*, *32*, 2363-2366.

[20] Sridhar, M.; Krishna, K. L.; Srinivas, K.; Rao, J. M. *Tetrahedron Lett. 1998*, *39*, 6529-6532.

[21] Sridhar, M.; Krishna, K. L.; Rao, J. R.*Tetrahedron 2000*, *56*, 3539-3545.

[22] Avalos, M.; Babiano, R.; Bravo, J. L.; Cintas, P.; Jiménez, J. L.; Palacios, J. C. *Tetrahedron Lett. 1998*, *39*, 9301-9304.

[23] de la Hoz, A.; Díaz-Ortiz, A.; Fraile, J. M.; Gómez, M. V.; Mayoral, J. A.; Moreno, A., Vázquez, E. *Synlett 2001*, 753-755.

[24] Fraile, J. M.; García, J. I.; Gómez, M. A.; de la Hoz, A.; Mayoral, J. A.; Moreno, A.; Prieto, P.; Salvatella, L.; Vázquez, E. *Eur. J. Org. Chem. 2001*, 2891-2899.

[25] Moreno, A.; Gómez, M. V.; Vázquez, E.; de la Hoz, A.; Díaz-Ortiz, A.; Prieto, P.; Mayoral, J. A.; Pires, E. *Synlett 2004*, 1259-1263.

[26] Karthikeyan, M.; Kamakshi, R.; Sridar, V.; Reddy, B. S. R. *Synth. Commun. 2003*, *33*, 4199-4204.

[27] Essers, M.; Mück-Lichtenfeld, C.; Haufe, G. *J. Org. Chem. 2002*, *67*, 4715-4721.

[28] Cativiela, C.; García, J. I.; Mayoral, J. A.; Pires, E.; Royo, A. J. *Appl. Catal. A 1995*, *131*, 159-166.

[29] Chen, I.-H.; Young, J.-N.; Yu, S. J. *Tetrahedron 2004*, *60*, 11903-11909.

[30] Elder, J. W. *J. Chem. Ed. 1994*, A142-A144.

[31] Nagarajan, S.; Barthes, C.; Gourdon, A. *Tetrahedron 2009*, *65*, 3767-3772.
[32] Trost, B. M.; Crawley, M. L. *J. Am. Chem. Soc. 2002*, *124*, 9328-9329.
[33] Loupy, A.; Maurel, F.; Sabatié-Gogova, A. *Tetrahedron 2004*, *60*, 1683-1691.
[34] Teyssot, M.-L.; Anh-Tuan Lormier, A.-T.; Chataigner, I.; Piettre, S. R. *J. Org. Chem. 2007*, *72*, 2364-2373.
[35] Kranjc, K.; Kočevar, M. *Tetrahedron 2008*, *64*, 45-52.
[36] Kranjc, K.; Kočevar, M.; Iosif, F.; Coman, S. M.; Parvulescu, V. I.; Genin, E.; Genêt, J.-P.; Michelet, V. *Synlett 2006*, 1075-1079.
[37] Kranjc, K.; Kočevar, M. *Bull. Chem. Soc. Jpn. 2007*, *80*, 2001-2007.
[38] Kranjc, K.; Kočevar, M. *Collect. Czech. Chem. Commun. 2006*, *71*, 667-678.
[39] Hohmann, E.; Keglevich, G.; Greiner, I. *Phosphorus Sulfur Silicon 2007*, *182*, 2351-2357.
[40] Oskooie, H. A. *Phosphorus Sulfur Silicon 2004*, *179*, 1165-1167.
[41] Monbaliu, J.-C.; Robiette, R.; Peeters, D.; Marchand-Brynaert, J. *Tetrahedron Lett. 2009*, *50*, 1314-1317.
[42] Diaz-Ortiz, A.; Carrillo, J.; Cossio, F.; Gomez-Escalonilla, M.; de la Hoz, A.; Moreno, A.; Prieto, P. *Tetrahedron 2000*, *56*, 1569-1577; Langa, F.; de la Cruz, P.; de la Hoz, A.; Espildora, E.; Cossio, F.; Lecea, B. *J. Org. Chem. 2000*, *65*, 2499-2507.
[43] Choshi, T.; Kumemura, T.; Nobuhiro, J.; Hibino, S. *Tetrahedron Lett. 2008*, *49*, 3725-3728.
[44] Zhu, J.; Bienaymé, H. Multicomponent reactions, Wiley-VCH, Weinheim, *2005*.
[45] Strübing, D.; Neumann, H.; von Wangelin, A. J.; Klaus, S.; Hübner, S.; Beller, M. *Tetrahedron 2006*, *62*, 10962-10967.
[46] Díaz-Ortiz, A.; Herrero, M. A.; de la Hoz, A.; Moreno, A.; Carrillo, J. R. *Synlett 2006*, 579-582.
[47] Díaz-Ortiz, A.; de la Hoz, A.; Moreno, A.; Prieto, P.; León, R.; Herrero, M. A. *Synlett 2002*, 2037-2038.
[48] Pinto, G. A.; Silva, A. M. S.; Almeida, L. M. P. M.; Carrillo, J. R; Díaz-Ortiz, A.; de la Hoz, A.; Cavaleiro, J. A. S. *Synlett 2003,* 1415-1418.
[49] Pinto, D. C. G. A.; Silva, A. M. S.; Brito, C. M.; Sandulache, A.; Carrillo, J. R.; Prieto, P.; Angel Díaz-Ortiz, A.; de la Hoz, A.; Cavaleiro, J. A. S. *Eur. J. Org. Chem. 2005*, 2973-2986.
[50] Silva, V. L. M.; Silva, A. M. S.; Pinto, D. C. G. A.; Cavaleiro, J. A. S. *Synlett 2006*, 1369-1373.
[51] Silva, V. L. M.; Silva, A. M. S.; Pinto, D. C. G. A.; Elguero, J.; Cavaleiro, J. A. S. *Eur. J. Org. Chem. 2009*, 4468-4479.
[52] Skoda-Földes, R.; Pfeiffer, P.; Horváth, J.; Zoltán Tuba, Z.; Kollár, L. *Steroids 2002*, *67*, 709-713.
[53] de la Hoz, A, Díaz-Ortiz, A.; Moreno, A.; Langa, F. *Eur. J. Org. Chem. 2000*, 3659-73.
[54] Patrick, T. B.; Gorerell, K.; Rogers, J. *J. Fluor. Chem. 2007*, *128*, 710-713.
[55] Jankowski, C. K.; LeClair, G.; Bélanger, J. M. R.; Pare, R. J.; VanCalsteren, M.-R. *Can. J. Chem. 2001*, *79*, 1906-1909.
[56] Stadler, A.; Henrik von Schenck, H.; Vallin, K. S. A.; Larhed, M.; Hallberg, A. *Adv. Synth. Catal. 2004*, *346*, 1773-1781.
[57] Songis, O.; Géant, P. Y.; Sautrey, G.; Martinez, J.; Calmès, M. *Eur. J. Org. Chem. 2008*, 308-318.

[58] Songis, O.; Didierjean, C.; Laurent, C.; Martinez, J.; Calmès, M. *Eur. J. Org. Chem. 2007*, 3166-3172.
[59] Clément, R.; Grisé, C. M.; Barriault, L. *Chem. Commun. 2008*, 3004-3006.
[60] Linders, J. T. M.; Kokje, J. P., Overhand, M.; Lie, T. S.; Maat, L. *Recl. Trav. Chim. Pays-Bas 1988*, *107*, 449-454.
[61] Linders, J. T. M.; Briel, P.; Fog, E.; Lie, T. S.; Maat, L. *Recl. Trav. Chim. Pays-Bas 1989*, *108*, 268-274.
[62] Reekie, T. A.; Austin, K. A. B.; Banwell, M. G.; Willis, A. C. *Aust. J. Chem. 2008*, *61*, 94-106.
[63] Dransfield, P. J.; Dilley, A. S.; Wang, S.; Romo, D. *Tetrahedron 2006*, *62*, 5223-5247.
[64] Guo, X.; Hu, W.; Cheng, S.; Wang, L.; Chang, J. *Synth. Commun. 2006*, *36*, 781-788.
[65] del Corral, J. M. M.; Angeles Castro, M.; Lucena Rodríguez, M.; Chamorro, P.; Cuevas, C.; San Feliciano, A. *Bioorg. Med. Chem. 2007*, *15*, 5760-5774.

1.2. Hetero Diels-Alder Reactions

Introduction

Hetero Diels-Alder reactions offer an efficient synthetic route to variety of important heterocyclic compounds. [1, 2] Different examples of hetero-Diels-Alder reactions have been carried out under microwave conditions, including both 1- and 2-azadienes, while cycloadditions involving heteroatoms other than nitrogen are fairly rare. As the environmentally friendly approach, microwave-assisted solvent-free heterocyclic synthesis is particularly important and therefore covered in review article. [3]

1.2.1. Nitrogen Dienes and Dienophiles

1.2.1.1. Acyclic Azadienes

Moody [4] used hetero-Diels-Alder methodology for synthesis of highly-functionalised pyridines. Reaction of 3-siloxy-1-aza-1,3-butadienes *223* with electron deficient acetylenes such as DMAD at 150 ^{o}C under microwave irradiation proceeded smoothly to afford the pyridines *224* in moderate yields after only a few hours (Scheme 46). Reaction times are significantly shorter than in classical heating reactions (several days). Increase of the temperature to 180°C shortens the reaction time even further. Similarly, irradiation of equimolar amounts of 1-azadiene *225a* and DMAD afforded pyridine product, still TBDMS protected and in poor yield, due to the competing Michael addition between the dienophile and liberated dimethylamine. Thus, 2 equiv of the dienophile were necessary to achieve complete consumption of *225a*. Cycloaddition of triethylsilyl (TES) analogue *225a* with DMAD gave, after acidic work-up, the deprotected pyridine *226a* in 34% yield, while the *tert*-butyldiphenylsilyl (TBDPS) analogue could not be induced to undergo cycloaddition, even at elevated temperatures. The derivatives *225b* and *225c* possessed similar reactivity to *225a*, including the formation of unwanted Michael adducts. However, the introduction of a second electron withdrawing group in hydrazone *225d*, completely suppressed Michael addition, leading to increased yield, although higher temperature is necessary to effect the cyclisation (180°C, 3h, 54%).

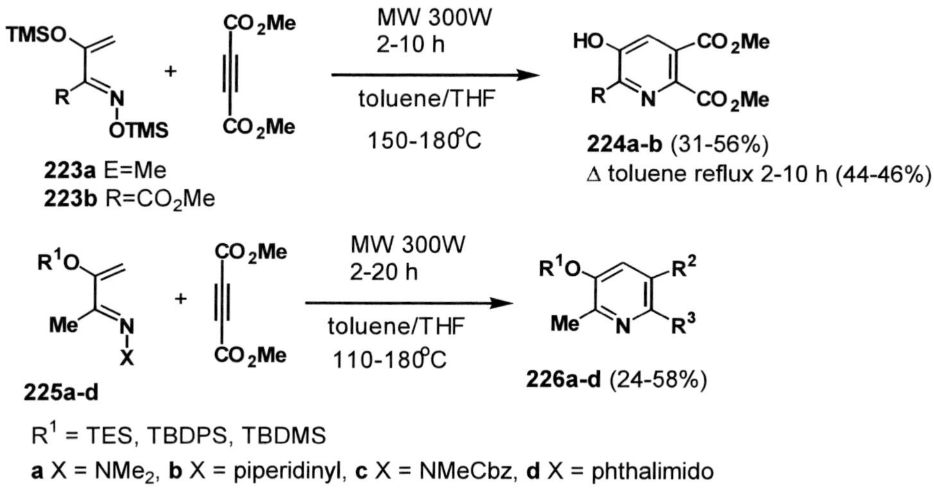

Scheme 46.

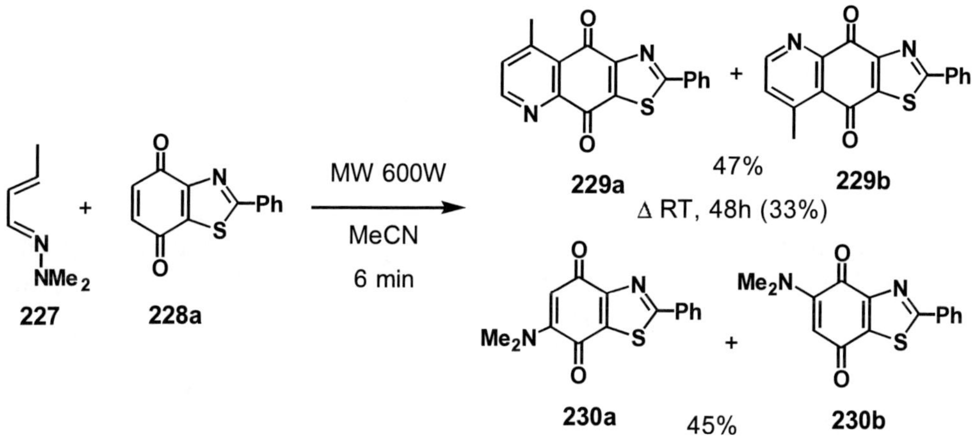

Scheme 47.

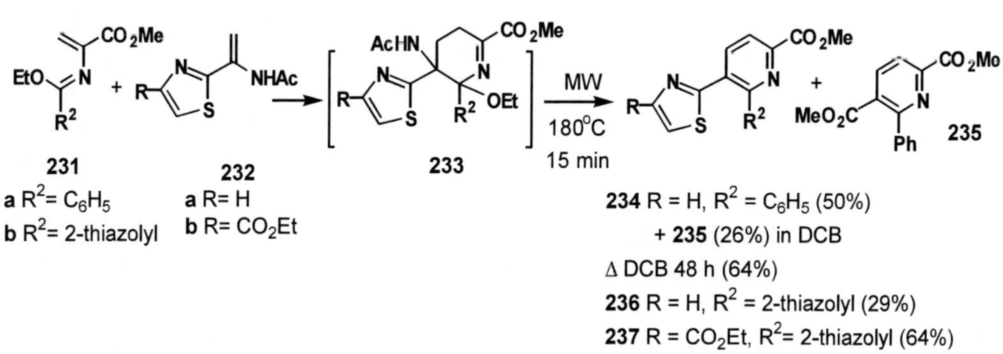

Scheme 48.

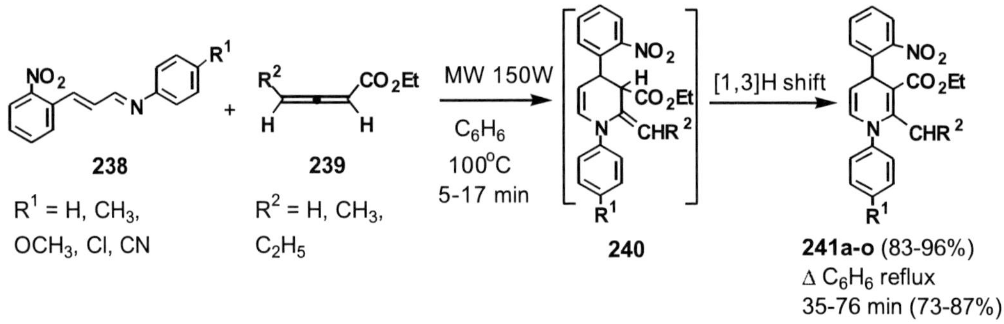

Scheme 49.

Running the reaction of the azadiene *227* with the thiazolobenzoquinone *228a* under microwave conditions (acetonitrile at 600 W for up to 10 minutes), regiosimeric adducts *229a* and *229b* Jackson [5] obtained in 47% yield as the fully aromatized Diels-Alder adducts (Scheme 47). In the same reaction, products *230a* and *230b* were obtained, resulting from addition of dimethylamine to the benzoquinone, in 45% yield. Other solvents were explored, affording lower conversion (toluene 10 min, *229a,b*:*230a,b* 43:21, *p*-xylene 10 min, 43:21, methanol 1 min, 15:19). In classical conditions in acetonitrile, a mixture of *229a,b* and *230a,b* (3:2 mixture) was obtained in 33% yield after 48 h.

Cycloaddition route to 2,3,6-trisubstituted pyridines, involving the Diels-Alder reaction of 1-alkoxy-2-azadienes *231* fixed in the required 'enol' form with *N*-acetyldehydroalanine derivatives *232* as dienophile was studied by Moody. [6] These dienes gave the corresponding pyridines *234* upon heating in Diels-Alder/aromatisation sequence (Scheme 48). The Diels-Alder reaction of 2-thiazolyl dienophile *232a* with *231a* under conventional or microwave heating in DCB gave the pyridine *234*, together with pyridine *235* formed by competing 'dimerisation' of the azadiene. Similarly, microwave reaction of thiazole containing 2-azadiene *231b* with *232a* gave the 2,3-dithiazolylpyridine *236* in 29% yield. Reaction of the corresponding 4-ethoxycarbonylthiazole dienophile *232b* gave the 2,3-dithiazolylpyridine *237* containing an additional ester group for further elaboration. These results establish the viability for the proposal of biosynthesis of the pyridine ring as a formal cycloaddition, followed by aromatisation of the intermediate tetrahydropyridine.

Singh Ishar [7] has found that thermal and microwave assisted [4+2] cycloadditions of 1,4-diaryl-1-aza-1,3-butadienes *238* with allenic esters *239* as dienophiles lead to cycloadducts *240*. After a [1,3]H shift these afford variedly substituted unsymmetrical 2-alkyl-1,4-diaryl-3-ethoxycarbonyl-1,4-dihydropyridines *241a-o* regioselectively and in high yields (Scheme 49). Reactions carried out under microwave irradiation are cleaner and give higher yields with shortened reaction times.

Panunzio [8] used catalytic amounts of lanthanide Lewis acids in conjuction with microwave heating to promote hetero Diels-Alder reactions. In the reaction of azadiene *242* and benzaldehyde *243*, the best conversion was achieved in chlorobenzene as solvent and EuFOD as lanthanide catalyst (Scheme 50). In microwave and classical conditions applied, no significant change in diastereomeric ratios of *cis/trans* hetero Diels-Alder cycloadducts *244a* and *244b* was observed (dr 72:22). Different aromatic substituents on diene and aldehyde were used with optimized method to obtain corresponding cycloadducts in moderate to high yields. Differences in reaction conditions could change the preference of intramolecular [2+2] electrocyclization versus [4+2] cycloaddition. For instance, microwave assisted Lewis catalyzed ($ZnCl_2$) hetero Diels-Alder reaction of azadiene *245* and cyclohexanone *246* afforded β-lactam *248*, while spiro-cycloadduct *247* was not isolated. When EuFOD catalyst was applied, *247* was formed in trace amounts, while *248* was obtained in 55% yield. On the other hand, in low-temperature conditions using BF_3 etherate as catalyst, product *247* was obtained exclusively.

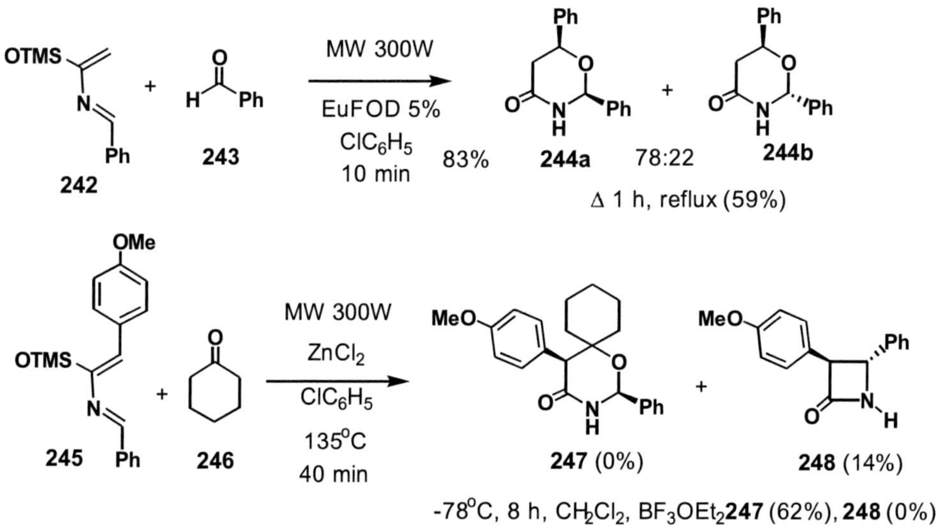

Scheme 50.

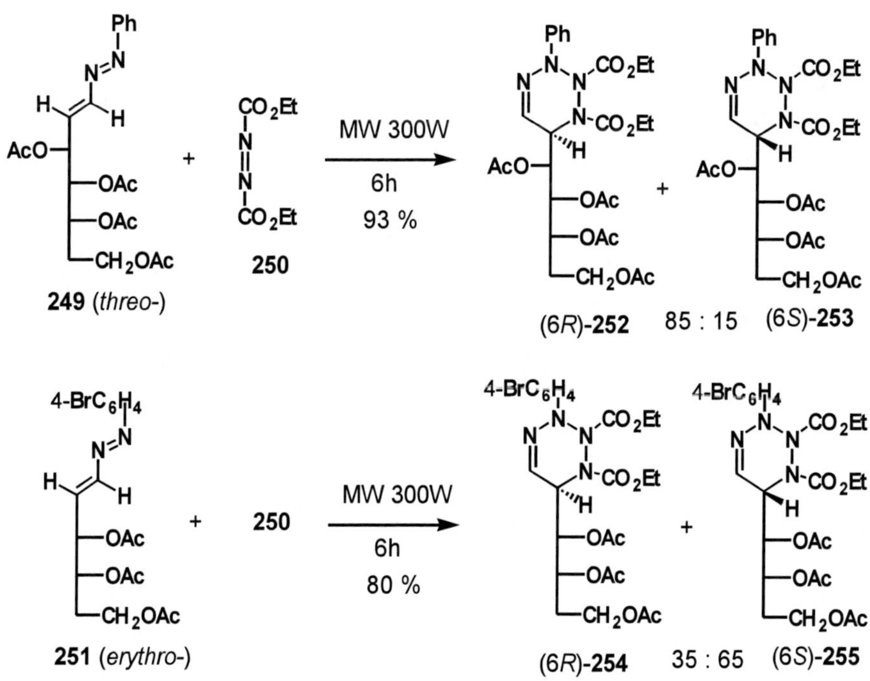

Scheme 51.

The stereoselective normal electron demand Diels-Alder reactions of chiral 1,2-diaza-1,3-butadienes, derived from acyclic carbohydrates, with diethyl azodicarboxylate *250* proceed slowly in benzene solution at room temperature. Jiménez [10] has found that reaction is greatly accelerated by microwave irradiation to form the corresponding functionalized 1,2,3,6-tetrahydro-1,2,3,4-tetrazines in good yields (Scheme 51). The observed stereoselectivity is dependent on the relative stereochemistry at C-1',2'. Thus, 1,2-diazoalkenes derived from per-*O*-acylated sugars having *threo* configuration at C-1',2' give tetrazines with a high facial selectivity, whereas those having *erythro* configuration at C-1',2' afford products in lower diastereomeric ratios. For instance, (1*E*,3*E*)-4-(1',2',3',4'-tetra-*O*-acetyl-D-*arabino*-tetritol-1'-yl)-1-phenyl-1,2-diaza-1,3-butadiene *249*, in microwave reaction with *250* without the presence of solvent affords cycloadducts (6*R*)- and (6*S*)-6-(1',2',3',4'-tetra-*O*-acetyl-D-*arabino*tetritol-1'-yl)-1,2-bis(ethoxycarbonyl)-3-phenyl-1,2,3,6-tetrahydro-1,2,3,4-tetrazines *252* and *253* in 85:15 ratio, while *251* produces *erythro*-tetrazines *254* and *255* in 36:65 ratio.

In solvent free conditions the azadienes *256* reacted with maleimides *257* with the aid of microwave irradiation to form pyrazolo[3,4-*b*]pyrrolo[3,4-*d*]pyridine derivatives *259* in various yields. Throughout the experiments, Fraga [11] used the microwave tube which was exposed to atmospheric oxygen to favor the oxidative aromatization step after initial Diels-Alder adduct *258* formation (Scheme 52).

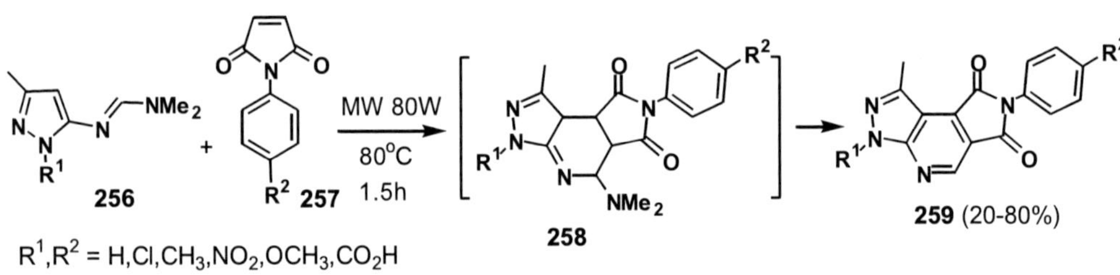

Scheme 52.

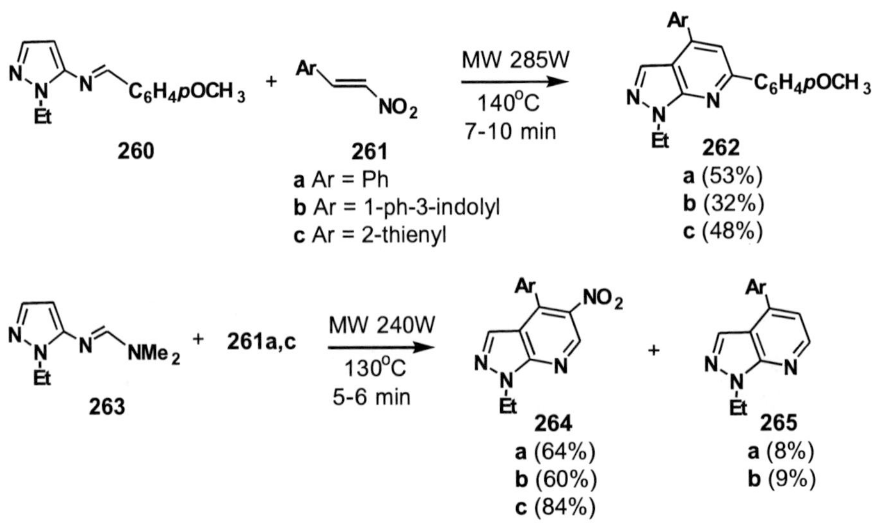

Scheme 53.

Díaz-Ortiz and de la Hoz [12] carried out Diels-Alder cycloaddition of pyrazolyl imines *260,263* with aromatic and aliphatic nitroalkenes (electron poor dienophiles) *261a-c* under microwave irradiation in the absence of solvent. This is the first example of a [4+2] cycloaddition of a 2-azadiene involving a pyrazole ring to give the corresponding pyrazolo[3,4-*b*]pyridines *262* (Scheme 53). Reactions are highly regioselective, in all the reactions with unsymmetrical dienophiles *261a-c* only one regioisomer was obtained. Reactions of 2-azadiene *263* and aliphatic nitroalkenes took place with loss of $HNMe_2$ and oxidation to yield the aromatic heterocycles. However, with diene *260* aromatization is possible only through loss of HNO_2. This is versatile synthetic procedure that permits the synthesis of several pyrazolopyridines by the introduction of different substituents in the heterocyclic ring. Shortening of heating time to 5-10 min reduces the decomposition or polymerization of the dienic pyrazole. When phenyl substituent is present in *260* instead of *p*-methoxybenzene, corresponding 2,4-diphenylpyrazolo[3,4-*b*]pyridine was obtained within 7 min of irradiation (285W, 140°C, 53%). By classical heating this [4+2] reaction does not occur or affords just traces of product.

Prajapati [14] used of 6-[(dimethylamino)methylene]amino-1,3-dimethyl uracil *266* as diene in several hetero Diels-Alder reactions. Firstly, three-component one-pot reaction, a mixture of 6-[(dimethylamino)methylene]amino-1,3-dimethyl uracil *266* with ethyl glyoxylate *267* and aniline in solvent free microwave irradiation conditions gave after elimination of dimethylamine from the 1:1 cycloadduct and tautomerism, the pyrimido[4,5-*d*]pyrimidine *268a* (Scheme 54). The reaction is less effective when carried out in DMF under reflux. Other pyrimido[4,5-*d*]pyrimidines *268* were prepared in 83-95% yields, while the reaction is less reactive and takes 10-15 min in 60% conversion when glyoxylate imines were reacted directly with *266* in a two-component system. There was no formation of Michael type products, which were the only products reported earlier. Various diaryl nitrones *269* reacted with amidine *266* under microwave irradiation in the solid state to afford the corresponding pyrimido[4,5-d]pyrimidines *270*, after elimination of 1,3-dimethylamine from the 1:1 cycloadduct and aromatization (in 80-90% yields). In addition, the treatment of *266* with the *N*-(arenesulfonyl) benzaldimines *271* in identical reaction conditions gave the pyrimido[4,5-*d*]pyrimidine *272* as the only product in 98% yield, whereas the use of substituted coumarins and quinones as dienophiles led to the formation of corresponding pyrido[2,3-*d*]pyrimidine derivatives *274* in an one-pot synthesis, after 1,3-dimethylamine elimination and aromatization of initially formed Diels-Alder cycloadduct.

Three-component aza-Diels-Alder reaction was used by Dai [16] to synthetise 1,2,3,4-tetrahydroquinolines via the CF_3CO_2H mediated aza-Diels-Alder reactions of 2-aminophenol with benzaldehyde and 3,4-dihydro-2*H*-pyran *277a* in MeCN at room temperature for 12-24 h to afford the pyrano[3,2-*c*]quinolines *278a* and *279a* (R^1=R^2=R^3=H) in 60% yield and in a 35:65 isomer ratio. Similar reaction using 2,3-dihydrofuran *277b* as the alkene component furnished the corresponding furo[3,2-*c*]quinolines in 67% yield and in a 48:52 isomer ratio of *278b* and *279b* (R^1=R^2=R^3=H). Optimization of reaction conditions under microwave heating: with the reactant ratio of amine:aldehyde:cyclic alkene to 1.2:1.2:1, the adducts *278a* and *279a* (R^1=R^2=R^3=H) were produced in 59% yield and in 18:82 ratio with microwave irradiation at 60°C for 15 min in the presence of 0.13 equiv of CF_3CO_2H, while other catalyst such as $Yb(OTf)_3$ and variation of conditions gave inferior yields and isomer ratios (Scheme 55). The reactions of *277a* generally gave higher yields (48-59%) and stereoselectivities

(24:76-16:84) of *278a/279a* as compared to those of 2,3-dihydrofuran *277b*. The latter afforded the *278a/279a* products in 39-41% yields and in 36:64-25:75 isomer ratios.

Yadav [17] used Montmorillonite clay impregnated with copper(I)bromide (30 mol%) as effective catalyst of Diels-Alder reactions in solvent free conditions. Aromatic imine *283* prepared *in situ* from aryl aldehyde *280* and aryl amine *281* undergoes hetero Diels-Alder cycloaddition with alkynes *282* on the surface of Montmorillonite under microwave irradiation to produce 2,4-disubstituted quinoline *284* in high yields and with good regioselectivity (Scheme 56). These reaction conditions are advantageous over conventional heating in terms of reduction of reaction times from 3-6 hours to 3-5 minutes, although obtained yields were not improved.

Moreno [18] reported that under microwave irradiation cycloaddition reactions of nitroindoles and nitropyrroles in solvent-free conditions give good yields of the aromatic carbazoles and indoles, through elimination of the nitro group and subsequent aromatization. The mechanism and stereoselectivity of the cycloaddition of nitroindole *205* with cyclopentadiene was determined by computational methods which indicated that the reaction takes place through a tandem hetero [4+2] cycloaddition/[3,3] sigmatropic shift (Claisen rearrangement). This mechanism gives stereoselectively the *endo* stereoisomer *287* (Scheme 57).

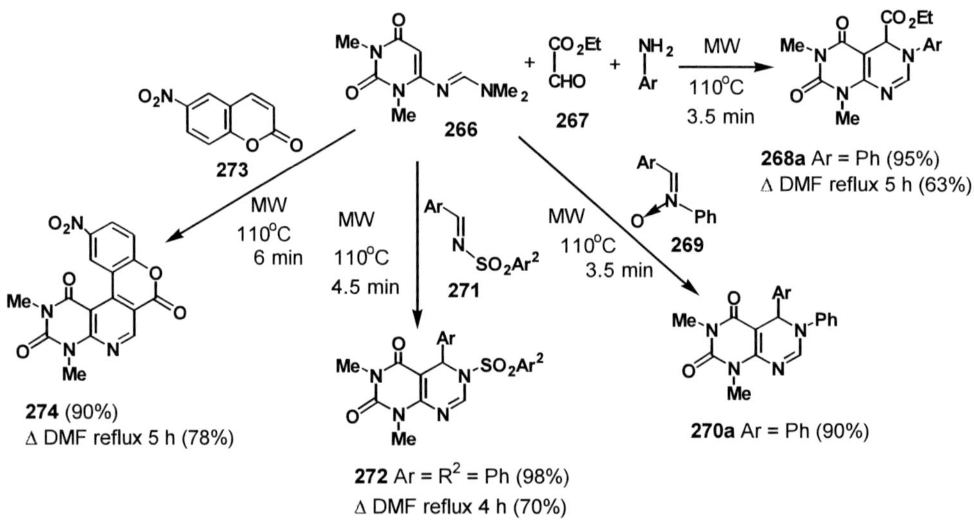

Scheme 54.

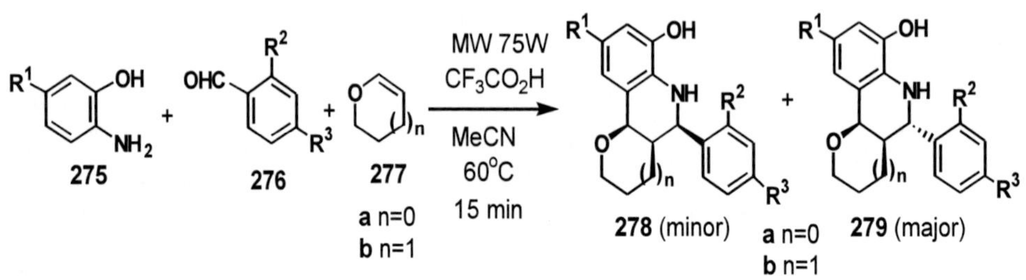

Scheme 55.

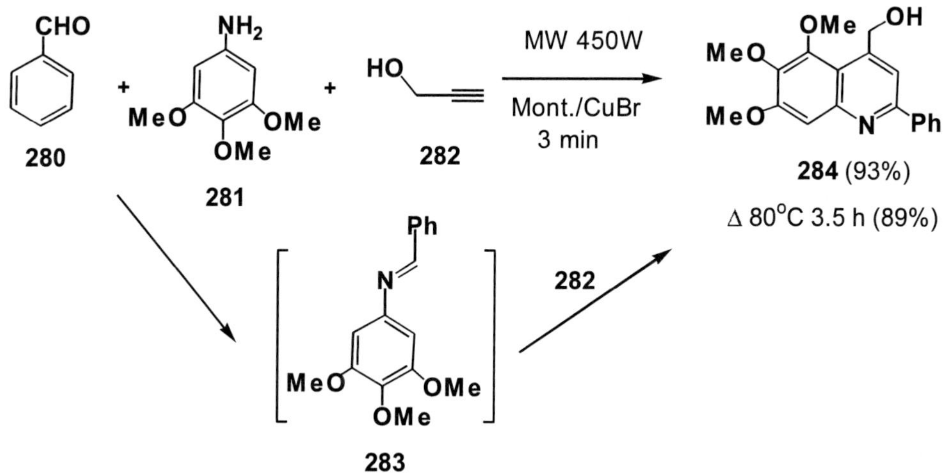

Scheme 56.

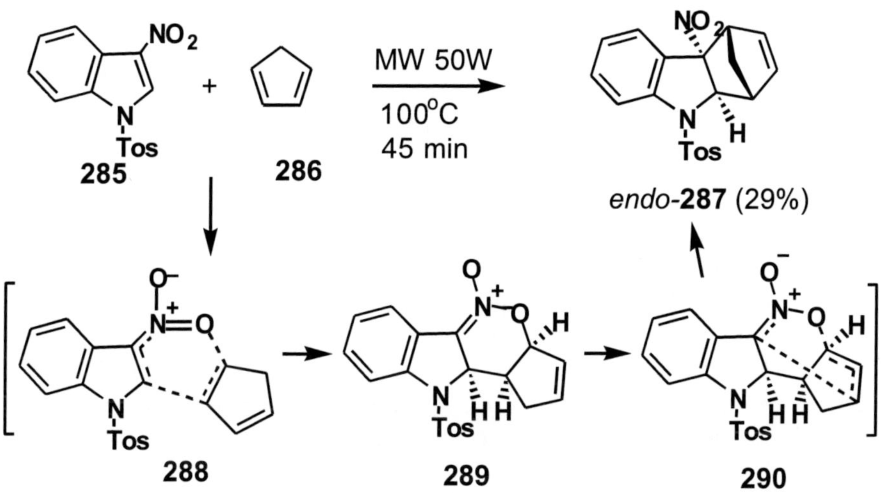

Scheme 57.

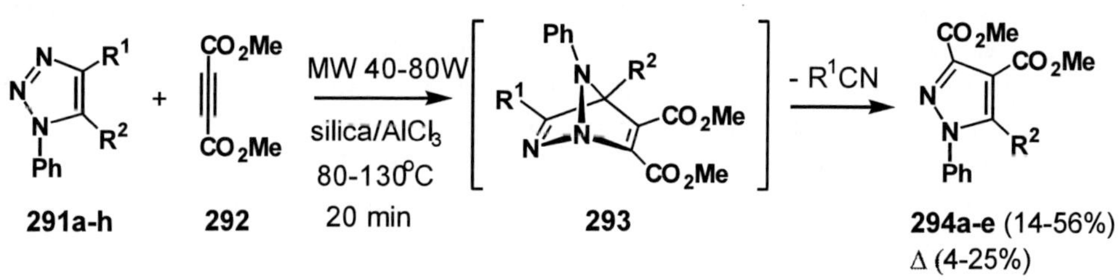

Scheme 58.

1.2.1.2. Cyclic Azadienes

The first example of a Diels-Alder cycloaddition involving a 1,2,3-triazole ring, a very unreactive and scarcely studied heterocyclic system was published by Prieto. [19] Substituted 1,2,3-triazoles *291a-h* undergo a Diels-Alder cycloaddition with DMAD under solvent-free conditions and microwave irradiation within 20 min to afford pyrazole-3,4-dicarboxylates

294a-e after extrusion of the substituent on position 4 of the triazole as a nitrile from the intermediate bicyclic cycloadduct *293* (Scheme 58). Yields can be increased significantly by using silica-bound $AlCl_3$ as a catalyst (0.1 mol %). These are the environmentally friendly and sustainable reaction conditions, that is, absence of solvent, microwave irradiation and recyclable supported catalyst.

Eycken [20] studied hetero Diels-Alder reaction of the triazole-functionalized 2(1*H*)-pyrazinones *295a-h* with DMAD. Intermolecular cycloaddition-elimination reaction of the 2-azadiene system of the 2(1*H*)-pyrazinones with acetylenes, generates pyridines and pyridinones. Reaction of triazole-functionalized 2(1*H*)-pyrazinones *295a-d* with dimethyl acetylenedicarboxylate at 180°C after spontaneous elimination of cyanogen chloride or isocyanate (*i.e.* retro Diels-Alder reaction), produced the pyridinones *297a-d* and pyridines *298a-d* in good yields (Scheme 59).

Analogous intermolecular Diels-Alder cycloaddition-elimination reaction of the 2-azadiene system of the 2(1*H*)-pyrazinones *299a-c* with DMAD takes place at higher irradiation power and temperature. This resulted in the exclusive formation of pyridines *300a-c*, whereas no traces of the corresponding pyridinones were detected (Scheme 60).

Díaz-Ortiz and de la Hoz [21] reported that Diels-Alder cycloaddition of 4,6-dimethyl-1,2,3-triazine *301* with enamines *302* (or enamines prepared *in situ* from their precursors amines and cyclic ketones) under microwave irradition produced fused pyridines such as *303* (Scheme 61). Reaction involves formation of an initial [4+2] cycloadduct *304*, nitrogen and pyrrolidine elimination and subsequent aromatization. Inverse electron demand reaction of triazine with electron-rich dienophiles under solvent free microwave conditions provides milder and shorter reaction conditions than required for classical heating.

In a similar manner as 1,2,3-triazines, the 1,2,4,5-tetrazines participate in inverse electron demand hetero Diels-Alder reactions. Schubert [22] carried out cycloaddition of 1-hexyne *305* to 3,6-di(pyridin-2-yl)-1,2,4,5-tetrazine *304* under microwave irradiation. After heating to 225°C the *n*-butyl 1,2-pyridazine *307* was isolated (Scheme 62). Low yield is due to the high reaction temperature, which induces decomposition of the tetrazine resulting in formation of 2-cyanopyridine and 3,6-di(pyridin-2-yl)-dihydro-1,2,4,5-tetrazine *308*. The reaction in less harsh conditions afforded better yield of cycloadduct *307*. Besides the acceleration, the use of CH_2Cl_2 also simplifies the workup procedure that includes evaporation of the solvent. Surprisingly, when microwave cycloaddition of 1-hexyne to *304* was carried out in acetone, besides product *307*, the 3,6-di(pyridin-2-yl)-4-methylpyridazine *311* was formed. Heating a solution of *304* alone in acetone also resulted in the formation of *311*, which demonstrates that acetone (its enol tautomer, propen-2-ol) participates in the inverse-electron-demand Diels-Alder reaction, which is facilitated by the electron-rich hydroxyl group of the propen-2-ol. Mechanism involves reaction of propen-2-ol with *304* forming the unstable intermediate *309* that spontaneously converts into the second intermediate *310* by elimination of nitrogen. Under the applied reaction conditions, rapid elimination of water and subsequent aromatization occurs resulting in *311*. The reaction between *304* and 3-methyl-2-butanone *312* provided a sufficient basis for the proposed reaction mechanism. Reaction of the enol tautomers with *304* resulted in pyridazine *313* and 4-hydroxy-4,5,5-trimethyldihydropyridazine *314* (which cannot aromatize by elimination of water).

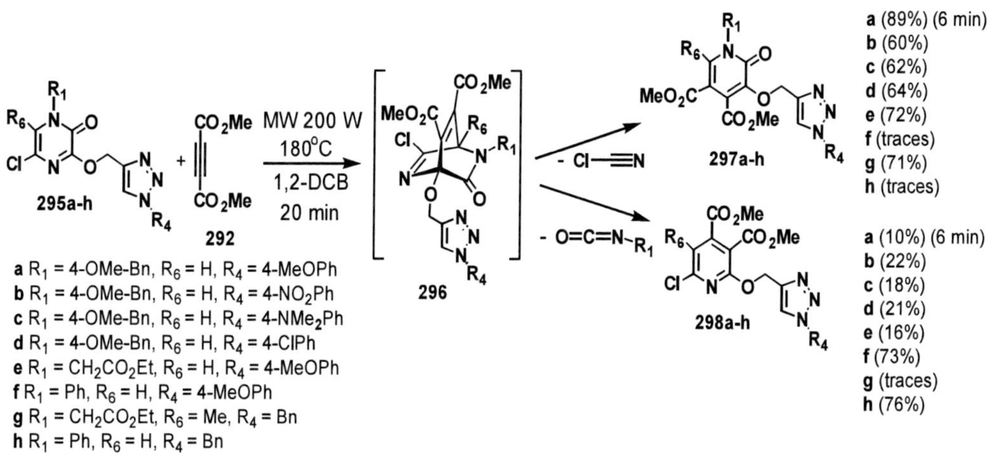

Scheme 59.

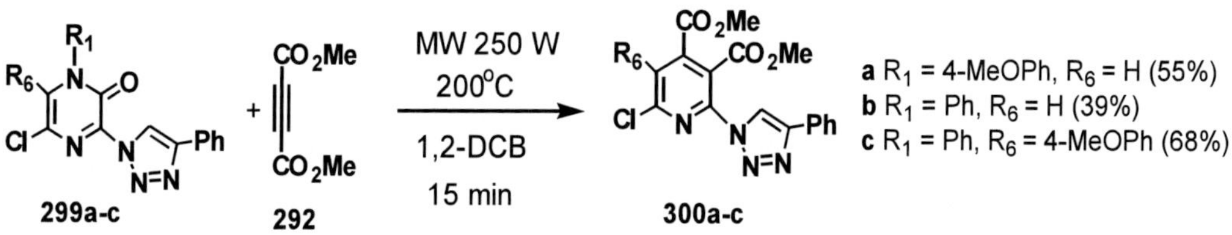

Scheme 60.

Based on these findings Schubert [23] performed the direct synthesis of poly(ε-caprolactone) functionalized 1,2-pyridazine *316* via the inverse-electron-demand Diels-Alder reaction between tetrazine *304* and the acetylene terminated pεCL *315* at 150 °C under microwave irradiation (Scheme 63).

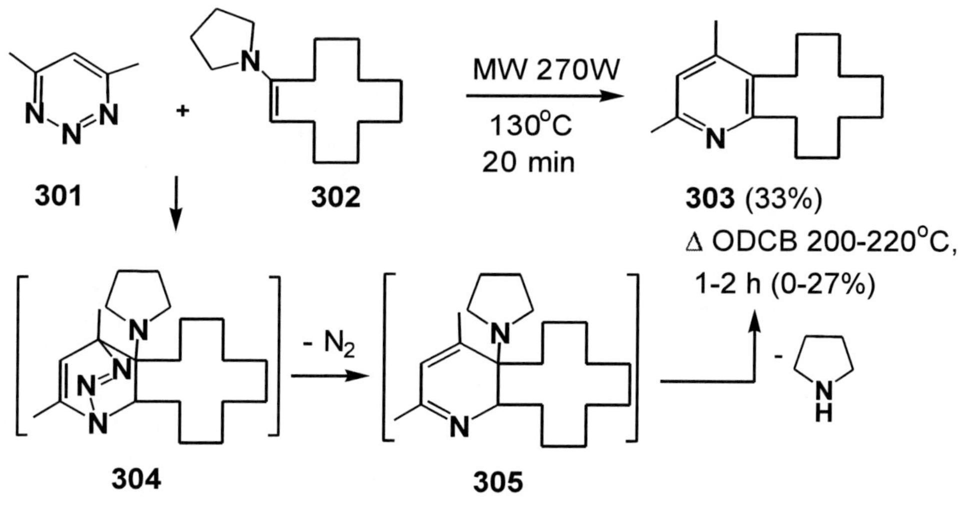

Scheme 61.

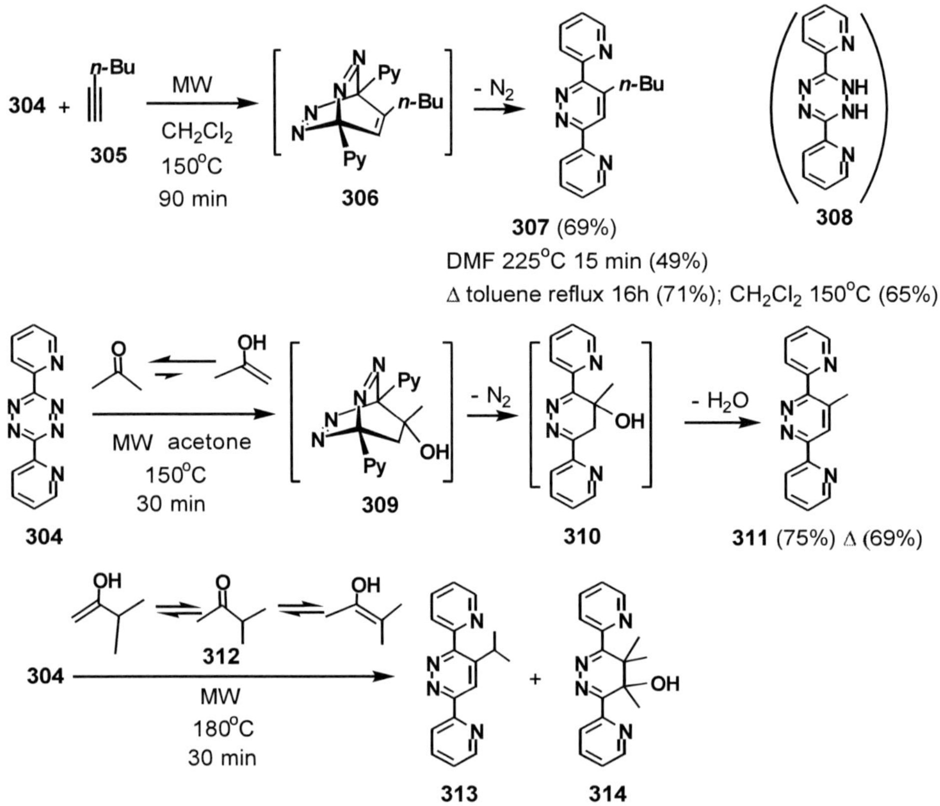

Scheme 62.

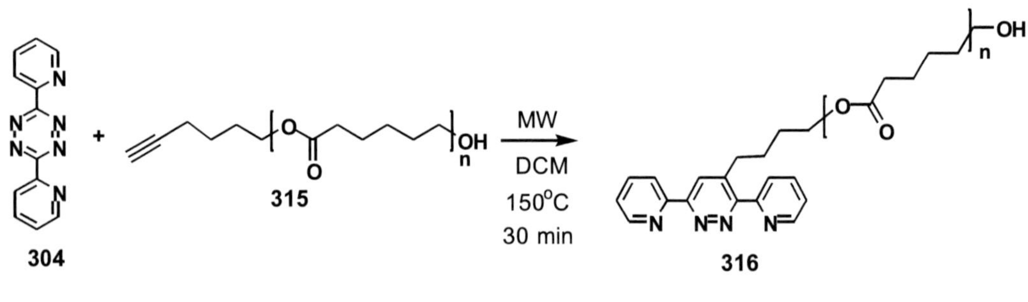

Scheme 63.

When the reactions of tetrazine *304* with norbornenes were conducted in the presence of triethylamine (to supress facile prototropic 1,3-shift of intermediate cycloadduct), tandem HDA[4+2]/retroDA[4+2]/HDA[4+2] reaction takes place, whereas two norbornene moieties are joined by new diazabicyclo[2.2.2]bridge at the junction. Construction of polarofacial molecules containing O and N atoms on the one side of the polycyclic molecular frame was achieved by coupling of diene *317* with tetrazine *304* (Scheme 64). [24] When coupling reaction was conducted under microwave irradiation, pyridazine *321* was obtained exclusively. Under high pressure (8 kbar), polarofacial O^2NO^2 product *318* was obtained, however in moderate yield, along with significant amounts of dihydropyridazines *319* and *320* and pyridazine *321*. When dienophile *317* was replaced by norbornene under microwave conditions (150W, Et_3N, 1h, 50-100°C, CH_2Cl_2), mixtures contained mainly of dihydropyridazine (analogue of *320*), pyridazine (analogue of *321*) and coupled product (analogue of *318*, in low yield ~5%).

Dubac [25] employed microwave assisted, solvent free organic reaction on graphite support to enhance [4+2] cycloadditions. Azadiene *323* does not react wih DMAD in conventional heating, while on graphite support, under microwave irradiation, aromatized cycloadduct *324* was obtained in high yield (Scheme 65). In similar solvent free conditions, 3,6-diphenyl-1,2,4,5-tetrazine *325* reacts with styrene to afford pyridazine *327*, while classical heating affords dihydropyridazine *328*. In microwave conditions, facile oxidation of *328* to aromatic pyridazine takes place. Furthermore, carbonyl group of ethyl mesoxalate *330* acts as dienophile in hetero-Diels-Alder reaction with 2,3-dimethyl-1,3-butadiene giving adduct *331* in 75% yield. [26] On the other hand, weaker dienophile ethyl glyoxylate required the addition of $ZnCl_2$ catalyst to get microwave reaction to proceed.

1.2.2. Dienes and Dienophiles with Heteroatoms other than Nitrogen

It could be seen from the Scheme 65 that several reactions involving heterodienes or dienophiles possessing heteroatoms other than nitrogen benefited from microwave irradiation. The oldest microwave paper on hetero Diels-Alder reaction published by Chastrette [27] deals with reaction of 2-methylpenta-1,3-diene *332* and 2,2-dimethoxyethanal *333* as dienophile. Microwave conditions, without the presence of solvent were found to obtain best yield, in the presence of catalytic amount of hydroquinone as antipolymerisant. The *cis/trans* ratio of cycloadducts *334a* and *334b* is 30/70 (Scheme 66). Reaction conditions are the improvement of classical reaction, which takes significantly longer, with lower yield.

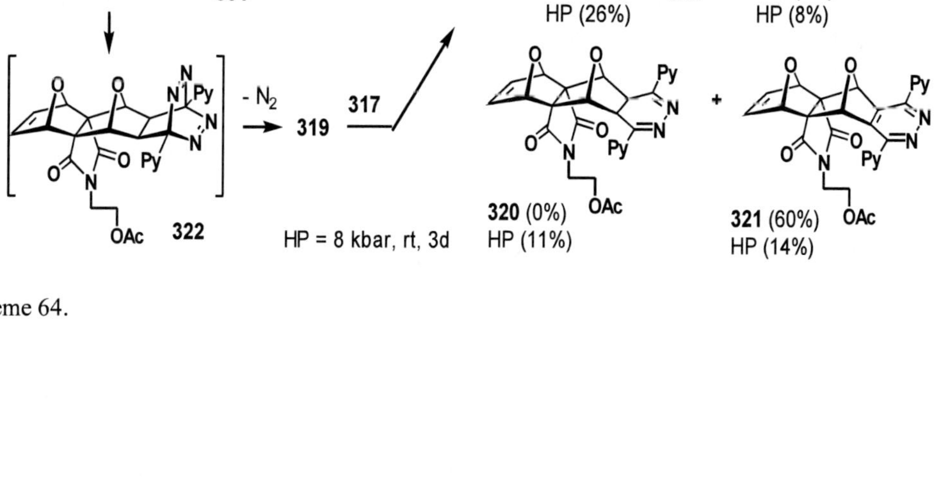

Scheme 64.

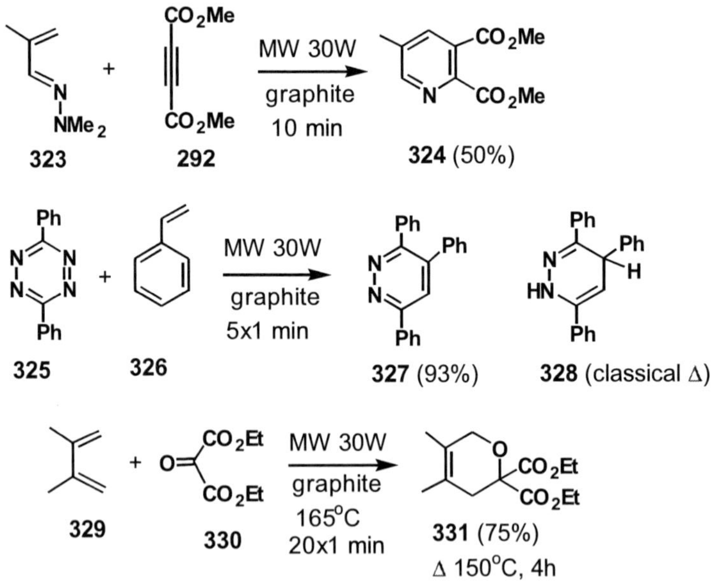

Scheme 65.

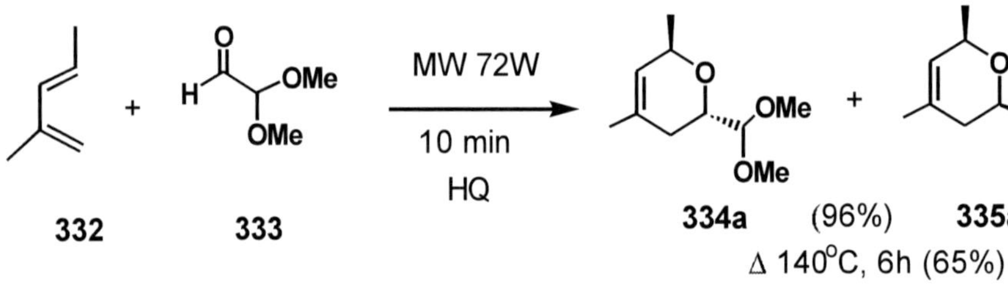

Scheme 66.

Hetero Diels-Alder cycloaddition of dienes *335a,b* to ketene acetals *336* and *338a,b* were carried out by Díaz-Ortiz [28] under microwave irradiation in solvent-free conditions proceed in high yield to obtain spiro cycloadducts *337* and *339a-c* (Scheme 67). When *338c* was used as dienophile, complex mixture of stereoisomers was obtained, which was studied in detail. [29] It was found that the diastereomeric ratios were *340a/340b* : 43/37, *340c/340d* : 14/6, with facial selectivity of 4:1.

Botta [30] applied multicomponent cross methathesis/hetero Diels-Alder (CM/HDA) methodology in synthesis of dihydropyrans. Different alkynes *341a-h* were mixed together with ethyl vinyl ether *342* and ethyl glyoxalate *343* in toluene and in the presence of 2nd generation Grubbs' catalyst *344*. Irradiation of mixture under microwaves afforded 2,3-dihydropyrans *345* as a mixture of *cis/trans* diastereoisomers in high yield in 2:1 *trans/cis* ratio. The first step of the reaction, the cross metathesis of alkyne *341* with *342*, led to a mixture of *E/Z*-diene *346* in a 2:1 ratio (Scheme 68).

Mechanistically, if an *endo* attack happened on both *E/Z*-dienes *346*, a 2:1 cis/trans mixture of products *345* should have been expected. On the contrary, only an *exo* attack can explain the observed 2:1 *trans/cis* regioselectivity. Optimized reaction conditions were then applied to the synthesis of 1,3-C-C-linked furanose-pyranose disaccharides. The reaction key step is multicomponent CM/HDA reaction of precursor C-ethynylribofuranose with *342* and

343 in the presence of *344* under microwave irradiation affording C-linked furanose-dihydropyran as a mixture of four diastereoisomers.

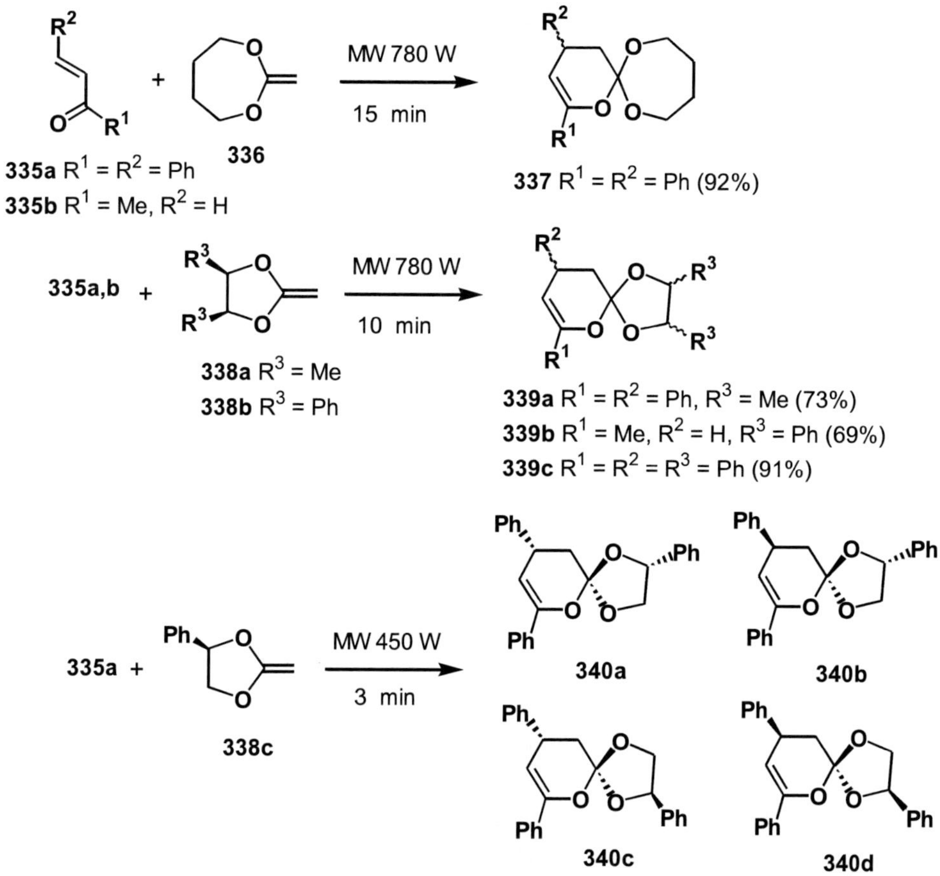

Scheme 67.

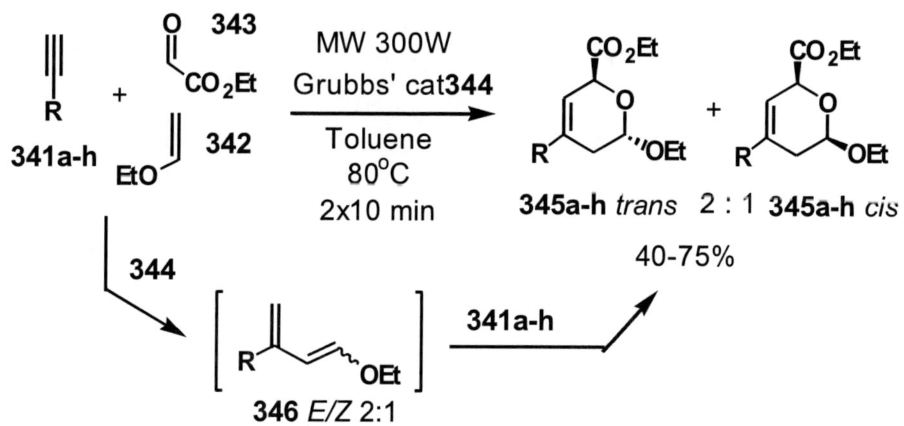

Scheme 68.

Using the domino Knoevenagel/hetero Diels-Alder reactions under microwave irradiation of a heterodiene (prepared from unsymmetrical 2,5-dihydroxy-1,4-benzoquinone embelin *346*), Estévez-Braun [31] obtained complex polycyclic structures. The reaction of *346*, paraformaldehyde, and alkenes is regioselective, obtaining only 1,4-benzoquinone derivatives from the more electron-poor *ortho* heterodiene *347*, in an inverse electron demand hetero Diels-Alder reaction. Optimized synthesis of the pyrano-1,4-benzoquinone derivative *348* from *346*, requires paraformaldehyde, and ethyl vinyl ether (Scheme 69). The reaction requires large excess of paraformaldehyde (8 equiv) in EtOH as a choice solvent. Under classical heating conditions, these reactions require long reaction times, high temperatures, and/or Lewis acid catalyst, which may result in decomposition of sensitive compounds. The short reaction time associated with microwave activation avoids decomposition of reagents and products, and prevents polymerization of the diene or dienophiles. The optimized conditions were applied to prepare a set of pyrano-1,4-benzoquinones with structural complexity. Tricyclic and tetracyclic adducts were obtained with cyclic alkenes such as 3,4-dihydro-2*H*-pyran or (±)-α-pinene to obtain cycloadducts *349* and *350* in high yield.

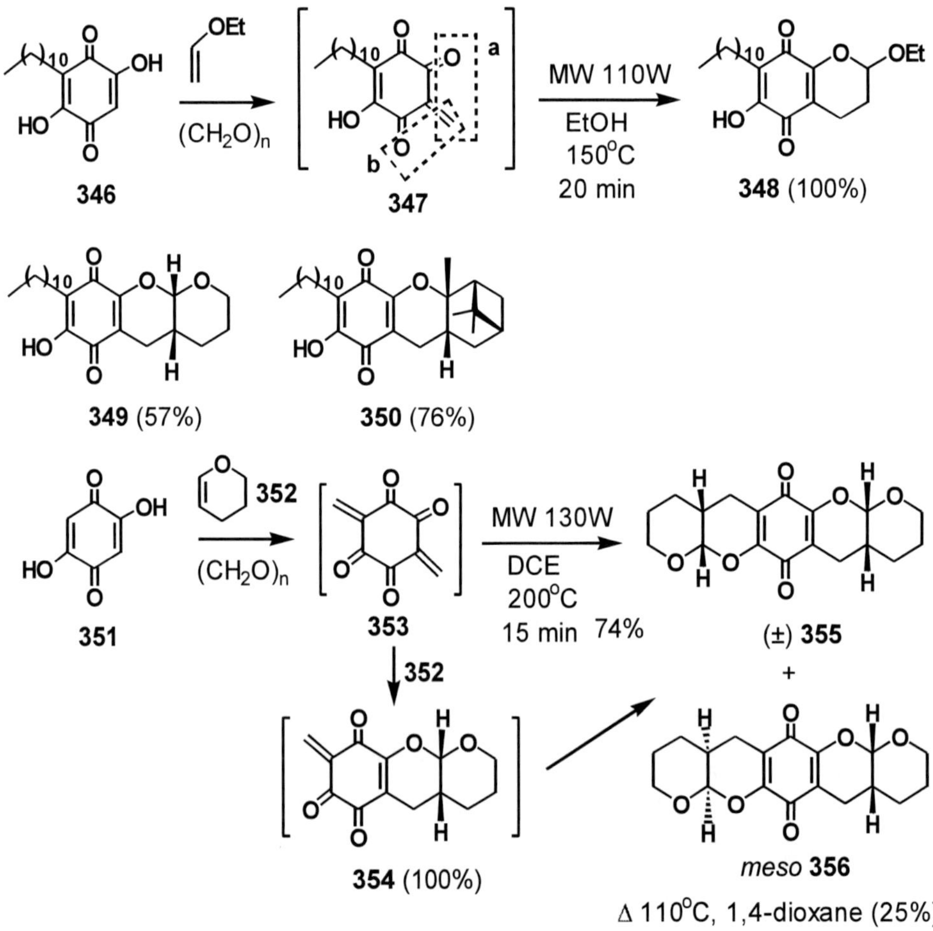

Scheme 69.

Double domino Knoevenagel/hetero Diels-Alder strategy was efficient towards bis-pyrano-1,4-benzoquinones formation through cycloaddition of 3,6-dimethylene-cyclohexane-1,2,4,5-tetraone intermediate *353*. [32] The 2,5-dihydroxy-1,4-benzoquinone *351* is synthetic equivalent to 1,3-dicarbonyl compound. In this case, the Knoevenagel condensation of 2,5-dihydroxy-1,4-benzoquinone and paraformaldehyde leads to a reactive intermediate *353*, which undergoes a double hetero Diels–Alder reaction with diverse electron-rich alkenes as dienophiles, yielding the corresponding bis-pyrano-1,4-benzoquinone derivatives *355,356* in one-pot reaction. The formation of the two pyran rings may occur simultaneously starting from the intermediate *353* or, alternatively, one of the two pyran rings is formed first to give a second intermediate *354*, which then undergoes the second hetero Diels-Alder reaction.

Bansal [33] successfully extended the application of microwave technique for the first time to the synthesis of a representative class of azaphospholes, starting with 1,3-bis(alkoxycarbonyl)-1,3-azaphospholo[5,1-*a*]-isoquinoline *357* (Scheme 70). Reaction of 1,3-bis(ethoxylcarbonyl)-1,3-azaphospholo[5,1-*a*]-isoquinoline *357*, 2,3-dimethyl-1,3-butadiene or isoprene (4 equiv.) and sulfur (1/8 equiv.) adsorbed on montmorillonite KSF upon microwave irradiation occurs rapidly giving higher yields of the cycloadducts *359*. The reactions with 2,3-dimethyl-1,3-butadiene and isoprene are regiospecific and only one regioisomer is formed in each case. When chlorobenzene was used as energy transfer medium cycloadducts *359a* and *359b* were obtained in comparable yields. Test reaction by classical heating needs much longer time to complete (12 and 28 days).

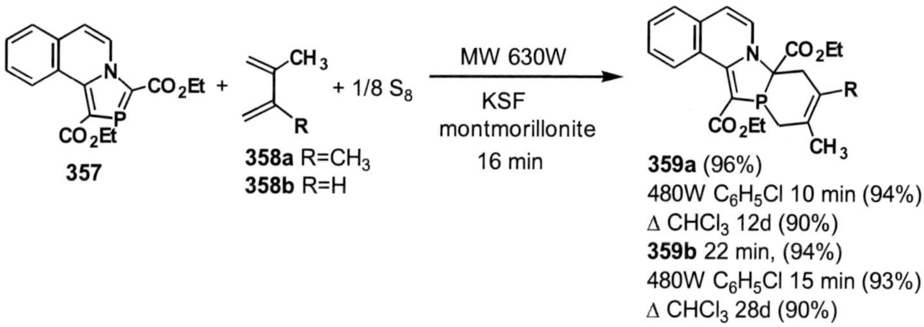

Scheme 70.

Addition of perfluoroalkyl substituents to a thiocarbonyl group substantially rises its dienophilic character in hetero Diels-Alder reactions with electron-rich 1,3-dienes. However, Shermolovich [34] has found that it was still insufficient to achieve addition of 2,3-dimethylbutadiene *358a* to thioamide *360a* in various conditions (thermal or high pressure) (Scheme 71). Synthesis of 2-aminosubstituted-2-perfluoroalkyl-3,6-dihydro-2*H*-thiopyran *360* by hetero Diels-Alder reactions of fluorinated thioamide *360* was achieved only under microwave heating in *N*-methylpyrrolidone in the presence of Weflon (Teflon filled with graphite). This material is heated by the microwave field and transfers heat to the reaction mixture (the same heating effect may be obtained with graphite but in the case of Weflon dispersion of the powder and dangerous hot-spots are suppressed). The choice of NMP as the solvent was justified by its efficiency to transform electromagnetic energy into thermal energy. [35] This polar aprotic solvent allows to work up to 200°C without trouble with vapor and high pressure in the tubes. It is now a common replacement for more sensitive solvents (e.g., DMSO and DMF) in pharmaceutical applications under microwaves and it may be eliminated by simple extraction with water during the work up. Longer heating (>3 h) of the reaction mixture decreased the yield of cycloadduct *361* and several minor by-products appeared, possibly by retro Diels-Alder reaction. The best yields were obtained when *358a* was added in three portions, and reaction heated for 30 min after each addition (2.5 h in total).

Wolff rearrangement of α-diazosulfoxides *362* has been induced by microwave irradiation forming α-oxosulfines *363* (as mixture of *E*- and *Z*-species). This methodology allowed Maguire [36] generation of synthetically useful α-oxosulfines in the absence of a transition-metal catalyst. After generation, α-oxosulfines *363* were trapped *in situ* with 2,3-dimethyl-1,3-butadiene (10 equiv) via hetero Diels-Alder reaction to obtain two diastereomeric cycloadducts *364a* and *365a* (Scheme 72). The outcome of the transformation was similar to the outcome under thermal conditions (for the rhodium(II)-catalysed reactions, the cycloadduct *364a* derived from the *Z*-stereoisomer of the intermediate *363a* is also the predominant product of the microwave-induced Wolff rearrangement reaction and subsequent cycloaddition). In the case of the methyl substituted derivative ratio of stereoisomeric cycloadducts vary with reaction conditions employed (MW, $Rh_2(OAc)_4$, Δ *364b*:*365b* 1:15.5; 1:24 and 1:11, respectively).

Marchand-Brynaert [37] has found that 1-(diethylphosphono)-1,3-butadiene *366* reacted smoothly with *t*-butyl azodicarboxylate *367* under classical thermal conditions to furnish 1,2-(di-*t*-butyloxycarbonyl)-3-diethoxyphosphono-3,6-dihydro-1,2-pyridazine *338*, while microwave activation provided a significant increase in yields and reaction rate (Scheme 73). Diene *366* reacted more easily with 2-nitrosotoluene *369* in refluxing dichloroethane to afford dihydro-1,2-oxazine cycloadducts *370* as a 10:1 mixture of proximal and distal regioisomers. Interestingly, when microwave activation was applied, the regioselectivity became complete in favour of the proximal isomer and the reaction time was shortened. This observation was explained by the greater synchronicity of the distal transition state compared to the proximal one.

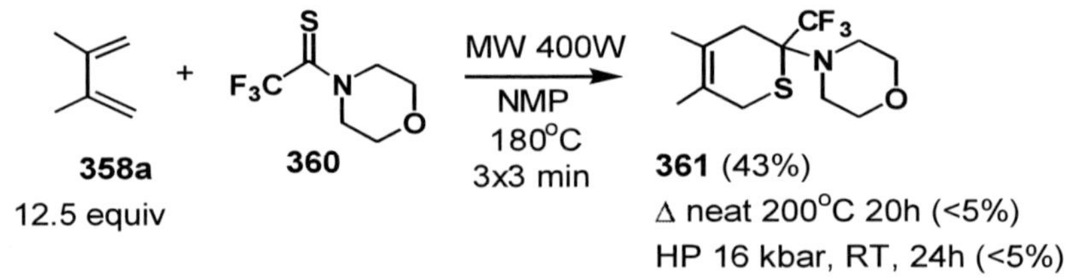

Scheme 71.

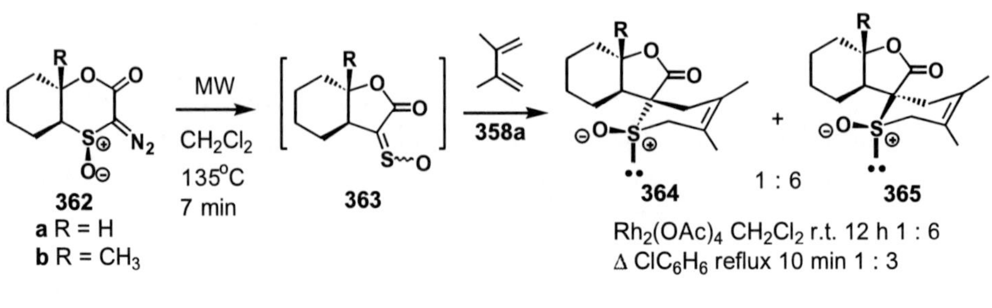

Scheme 72.

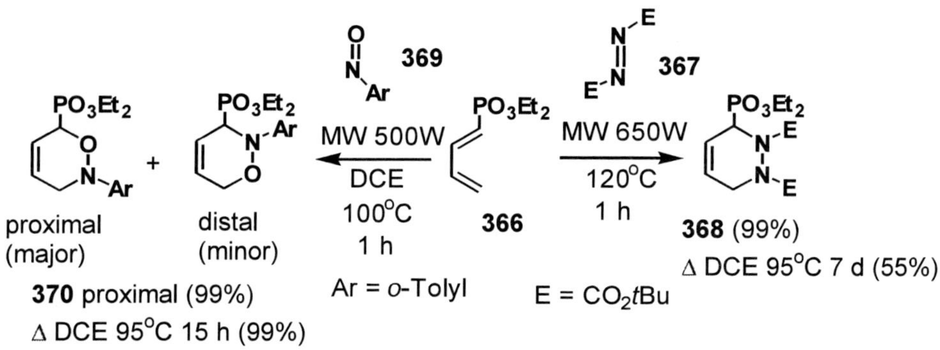

Scheme 73.

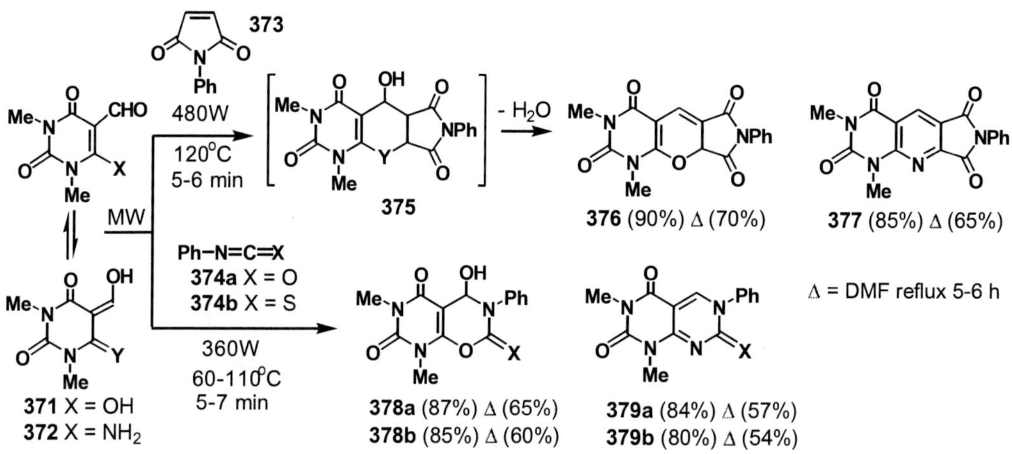

Scheme 74.

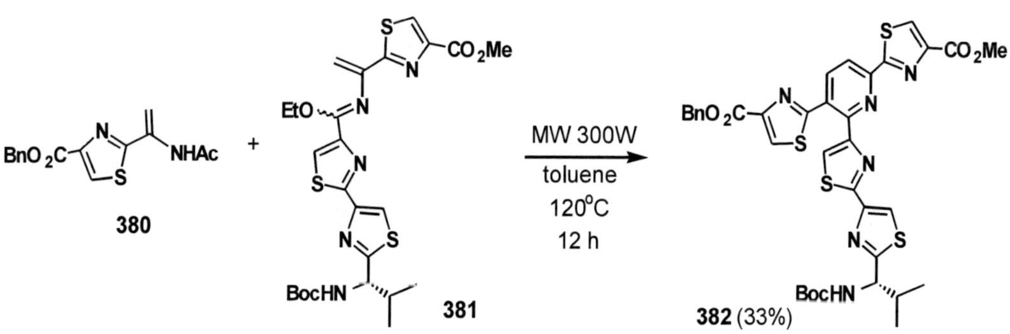

Scheme 75.

Hetero Diels-Alder reactions of heterodienes where heteroatoms are positioned at the termini of diene moiety were studied by Bhuyan. [38] In these reactions, *N,N*-dimethyl-5-formylbarbituric acid *371* reacts with maleimide *373* and phenyl isocyanate/phenyl isothiocyanate *374a/b* under microwave-assisted conditions in the solid phase to afford pyrano[2,3-*d*]pyrimidines *376* and oxazino[4,5-*d*]pyrimidines *378* in excellent yields. Under identical reaction conditions, *N,N*-dimethyl-6-amino-5-formyluracil *372* reacts with *373* and *374* to give pyrido[2,3-*d*]pyrimidine derivative *377* and pyrimido[4,5-d]pyrimidines *378* in high yields (Scheme 74). Products *376* and *377* were formed via hydroxy intermediate of type *375*, which spontaneously eliminates water in reaction conditions employed and when X=N further aromatizes. Cycloaddition reaction is highly regioselective when unsymmetrical dienophiles *374* were used. These reaction conditions are superior to conventional heating under reflux in chloroform (both in reduced time and in higher yields).

1.2.3. Natural Product Synthesis

The key step in total synthesis of thiopeptide antibiotic Amythiamicin D was Diels-Alder reaction. Moody [39] used heating of the 2-azadiene *381* (containing three thiazole rings) with enamide thiazole dienophile *380* (containing an orthogonally protected carbonyl) under microwave irradiation in toluene at 120 °C for 12 h to obtain the required 2,3,6-tris(thiazolyl)pyridine *382*, albeit in a modest 33% yield (Scheme 75). [40] Given that pyridine *382* forms the core of the natural product, its stereochemical integrity was established confirming that no racemization had occurred during the synthesis of *382*.

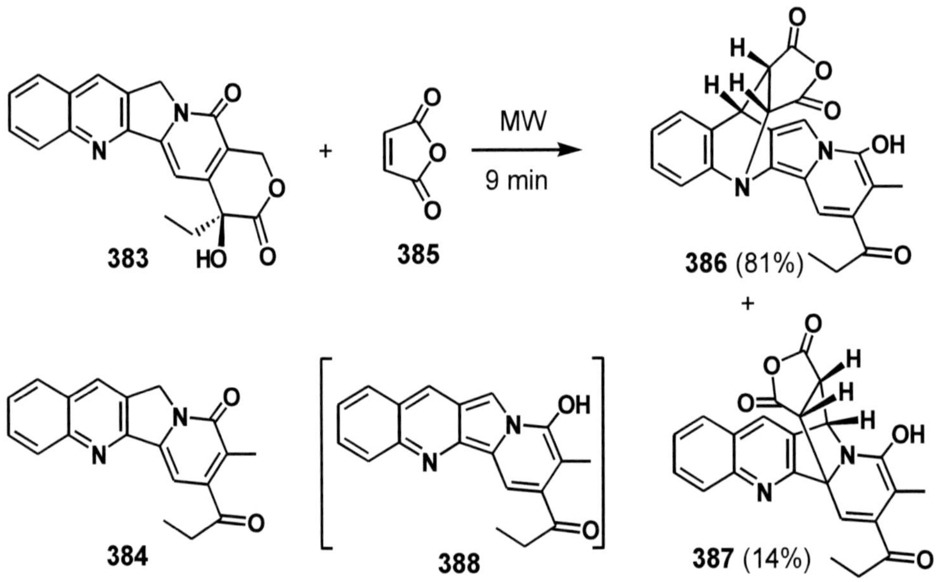

Scheme 76.

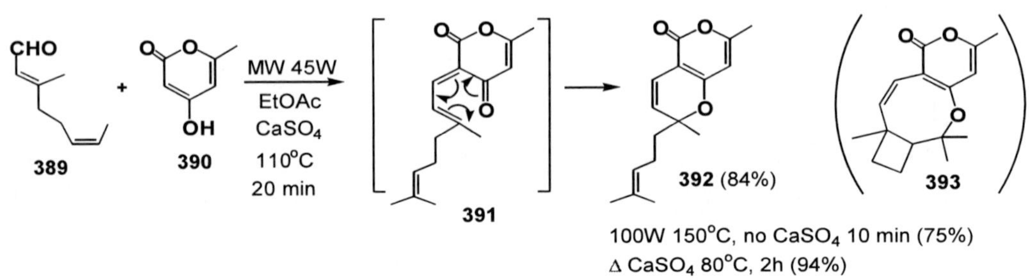

Scheme 77.

Diels-Alder methodology was used by Das [41] exploiting naturally occurring alkaloids, camptochecin *383* and mappicine ketone *384* as dienes. In reaction with maleic anhydride *385* in solvent-free conditions under microwave irradiation in a commercial domestic oven, two cycloadducts *386* and *387*, were obtained (Scheme 76). The first adduct was formed by involvement of B-ring of camptochecin with maleic anhydride, while the second was formed by involvement of the C-ring of the alkaloid with the dienophile *385*. Mappicine ketone *384* underwent similar reaction with maleic anhydride under microwave irradiation (9 min) to produce the same adducts *386* and *387*, probably through a Diels-Alder reaction of intermediate *388*.

The use of microwave radiation has thus far shown beneficial effect on Diels-Alder reactions, while Narsaiah reported that some fluorinated 2(1*H*)pyridones would not undergo inverse electron demand cyclization with DMAD in microwave, rather migration of functional group takes place. [42] In addition, Beifuss [43] has found that the variation of the aldehyde component in domino Knoevenagel condensation/hetero Diels-Alder cycloaddition reaction could reverse the preference for formation of 6π-electrocyclisation over cycloaddition products. For instance, microwave assisted reaction of aldehyde *389* with 6-methyl-4-hydroxy-2*H*-pyran-2-one *390* affords 6π-electrocyclisation product *392* (2*H*,5*H*-pyrano[4,3-*b*]pyran-5-one), while cycloaddition product *393* was not observed (Scheme 77). Although the formation of pyranone moiety in *393* is a driving force for cycloaddition, the presence of strained cyclobutane ring in final product is counterproductive.

REFERENCES

[1] Tietze, L. F.; Kettschau, G., *Hetero Diels-Alder reactions in organic chemistry, Stereoselective Heterocyclic Synthesis I*, Topics in Current Chemistry 189, Springer, Berlin, *1997*.

[2] Boger, D., Weinreb, S. Hetero Diels-Alder Methodology in Organic Synthesis, *Academic Press, San Diego,* 1987.

[3] Bougrin, K.; Loupy, A.; Soufiaoui, M. *J. Photochem. Photobiol. C 2005*, *6*, 139-167.

[4] Fletcher, M. D.; Hurst, T. E.; Miles, T. J.; Moody, C. J. *Tetrahedron 2006*, *62*, 5454-5463.

[5] Lyon, M. A.; Lawrence, S.; Williams, D. J.; Jackson, Y. A. *J. Chem. Soc.*, *Perkin Trans. 1 1999,* 437-442.

[6] Moody, C. J.; Hughes, R. A.; Thompson, S. P.; Alcaraz, L. *Chem. Commun. 2002*, 1760-1762.

[7] Singh, L.; Singh Ishar, M. P.; Elango, M.; Subramanian, V.; Gupta, V.; Kanwa, P. *J. Org. Chem. 2008*, *73*, 2224-2233.

[8] Panunzio, M.; Bandini, E.; D'Aurizio, A. D.; Millemaggi, A.; Xia, X. *Synthesis 2007*, 2060-2062.

[9] Panunzio, M.; Bandini, E.; D'Aurizio, A. D.; Xia, X.; Mu, X. *Synthesis 2008*, 1753-1756.

[10] Avalos, M.; Babiano, R.; Cintas, P.; Clemente, F. R.; Jiménez, J. L.; Palacios, J. C.; Sánchez, J. B. *J. Org. Chem. 1999, 64,* 6297-6305.

[11] Nascimento-Júnior, N. M.; Mendes, T. C. F.; Leal, D. M.; Corrêa, C. M. N.; Sudo, R. T.; Zapata-Sudo, G.; Barreiro, E. J.; Fraga, C. A. M. *Bioorg. Med. Chem. Lett. 2010, 20*, 74-77.

[12] Díaz-Ortiz, A.; Carrillo, J. R.; Cossío, F. P.; Gómez-Escalonilla, M. J.; de la Hoz, A.; Moreno A.; Prieto, P. *Tetrahedron 2000*, *56*, 1569-1577.

[13] Díaz-Ortiz, A.; Carrillo, J. R.; Gómez-Escalonilla, M. J.; de la Hoz, A.; Moreno, A.; Prieto, P. *Synlett 1998*, 1069-1070.

[14] Prajapati, D.; Gohain, M.; Thakur, A. J. *Bioorg. Med. Chem. Lett. 2006*, *16*, 3537-3540.

[15] Gohain, M.; Prajapati, D.; Gogoi, B. J.; Sandhu, J. S. *Synlett 2004*, 1179-1182.

[16] Xing, X.; Wu, J.; Dai, W.-M. *Tetrahedron 2006*, *62*, 11200-11206.

[17] Yadav, J. S.; Reddy, B. V. S.; Rao, R. S.; Naveenkumar, V.; Nagaiah, K. *Synthesis 2003*, 1610-1614.

[18] Victoria Gómez, M.; Aranda, A. I.; Moreno, A.; Cossío, F. P.; de Cózar, A.; Díaz-Ortiz, A.; de la Hoz, A.; Prieto, P. *Tetrahedron 2009*, *65*, 5328-5336.

[19] Díaz-Ortiz, A.; de Cózar, A.; Prieto, P.; de la Hoz, A.; Moreno, A. *Tetrahedron Lett. 2006*, *47*, 8761-8764.

[20] Kaval, N; Ermolat'ev, D.; Appukkuttan, P.; Dehaen, W.; Kappe, C. O.; Van der Eycken, E. *J. Comb. Chem. 2005, 7,* 490-502.

[21] Díaz-Ortiz, A.; de la Hoz, A.; Prieto, P.; Carrillo, J. R.; Moreno, A., Heunhoeffer, H. *Synlett 2001*, 236-237.

[22] Hoogenboom, R.; Moore, B. C.; Schubert, U. S. *J. Org. Chem. 2006*, *71*, 4903-4909.

[23] Hoogenboom, R.; Brian C. Moore, B. C.; Schubert, U. S. *Chem. Commun. 2006*, 4010-4012.

[24] Margetić, D. *Trends in Heterocycl. Chem. 2009*, *14*, 79-89.

[25] Garrigues, B.; Laporte, C.; Laurent, R.; Laporterie, A.; Dubac, J. *Liebigs Ann. Chem. 1996*, 739-741.

[26] Garrigues, B.; Laurent, R.; Laporte, C., Laporterie, A.; Dubac, J. *Liebigs Ann. 1996*, 743-744.

[27] Stambouli, A.; Chastrette, M.; Soufiaoui, M. *Tetrahedron Lett. 1991*, *32*, 1723-1724.

[28] Díaz-Ortiz, A.; Díez-Barra, E.; de la Hoz, A.; Moreno, A.; Prieto, P. *J. Chem. Soc. Perkin Trans. 1 1994*, 3595-3598.

[29] Díaz-Ortiz, A.; Díez-Barra, E.; de la Hoz, A.; Prieto, P.; Moreno, A.; Langa, F.; Prangé, T.; Neuman, A. *J. Org. Chem. 1995*, *60*, 4160-4165.

[30] Castagnolo, D.; Botta, L.; Botta, M. *Tetrahedron Lett. 2009*, *50*, 1526-1528.

[31] Jiménez-Alonso, S.; Chávez, H.; Estévez-Braun, A.; Ravelo, A. G.; Feresin, G.; Tapia, A. *Tetrahedron 2008*, *64*, 8938-8942.

[32] Jiménez-Alonso, S.; Estévez-Braun, A.; Ravelo, A. G.; Zárate, R.; López, M. *Tetrahedron 2007*, *63*, 3066-3074.

[33] Bansal, R. K.; Dandia, A.; Gupta, N.; Jain, D. *Heteroat. Chem. 2003*, *14*, 560-563.

[34] Mikhailichenko, S.; S.; Bouillon, J. P.; Besson, T.; Shermolovich, Y. B. *Tetrahedron Lett. 2010*, *51*, 990-993.

[35] Lamazzi, C.; Dreau, A.; Bufferne, C.; Flouzat, C.; Patrick Carlier, P.; ter Halle, R.; Besson, T. *Tetrahedron Lett. 2009*, *50*, 5402-5405; Pereira, M.; De, F.; Thiéry, V.; Besson, T. *Tetrahedron Lett. 2007*, *48*, 7657-7659.

[36] O'Sullivan, O. C. M.; Collins, S. G.; Maguire, A. R. *Synlett 2008*, 659-662.

[37] Monbaliu, J.-C.; Marchand-Brynaert, J. *Tetrahedron Lett. 2008*, *49*, 1839-1842.
[38] Devi, I.; Borah, H. N.; Bhuyan, P. J. *Tetrahedron Lett. 2004*, *45*, 2405-2408.
[39] Hughes, R. A.; Thompson, S. P.; Alcaraz, L.; Moody, C. J. *J. Am. Chem. Soc. 2005*, *127*, 15644-15651.
[40] Hughes, R. A.; Thompson, S. P.; Alcaraz, L.; Moody, C. J. *Chem. Commun. 2004*, 946-948.
[41] Das, B.; Madhusudhan, P.; Kashinatham, A. *Indian J. Chem. B 2000*, *39*, 326-328; *ChemInform 2001*, *32(3)*, DOI: 10.1002/chin.200103200
[42] Krishnaiah, A.; Narsaiah, B. *J. Fluor. Chem. 2002*, *113*, 133-137.
[43] Leutbecher, H.; Conrad, J.; Klaiber, I., Beifuss, U. *QSAR Comb. Sci. 2004*, *23*, 895-898.

1.3. Intramolecular Diels-Alder Reactions

Introduction

Intramolecular cycloaddition reactions often benefit from higher reactivity and greater control of stereoselectivity relative to their intermolecular counterparts. Specifically, connecting the two π-fragments via a "tether" generally facilitates the rate of the [4+2] cycloaddition reaction. [1] With the application of microwave heating, intramolecular Diels-Alder reactions (IMDA) could be even more expedited.

The first account on intramolecular and also retro-Diels-Alder reactions conducted in microwave conditions was published by Takano. [2] This is also the first example of microwave cycloaddition reaction conducted in water. While the (-)-form of Vincadifformine alkaloid is unnecessary in the synthesis, its (+)-form shows significant biological activity, which could not be obtained in sufficient quantitites from natural sources. The optically active alkaloid (-)-*394* racemized through the triene intermediate *395*, by the concurrent Diels-Alder cycloreversion and cycloaddition (Scheme 78). Solvent choice was found to be the most important in this process, since the solvents other than DMF yielded much less product.

1.3.1. All-carbon Intramolecular Diels-Alder Reactions

Mihovilovic [3] published an example of IMDA reaction which highly benefited from employment of microwave irradiation. Intramolecular Diels-Alder reactions of dihydroxylated benzoic acid derivative amide *396* gave tricyclic product *397* in microwave conditions, while in thermal conditions the reaction did not proceed (Scheme 79). The formation of dimeric by-products such as *398* was also observed under microwave conditions, arising from intermolecular Diels-Alder reactions. In a similar manner, the IMDA cyclization of allyl ether *399* led to bicyclic compound *401* as the only reaction product under thermal and microwave conditions in almost identical yields. Reaction of the ester *400* was much slower, indicating higher activation energy needed for the process. Again, the formation of dimeric products *405* was observed, and the IMDA reaction competes against a dimerization reaction of the precursors. In addition, it is apparent that the dienophile of the allyl ether derivative *399* represents the least activated double bond of all precursors, however its IMDA reaction gave almost quantitative yields. These results were explained by the fact that the side chain of allyl ether *399* represents the most flexible system, which plays a significant role to activation energy in this reaction.

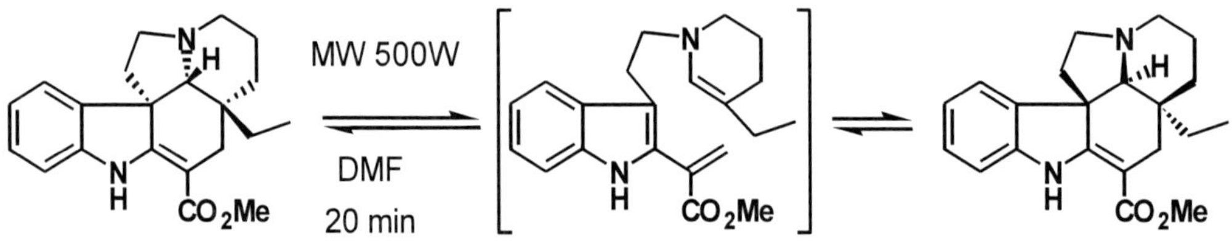

(-)-394 **395** **(+)-394** (92%)

Scheme 78.

The cyclization reaction with substituted allyl ether *403* under microwave irradiation led to polycyclic compound *404a* possessing the methyl group in the *anti* position relative to the oxygen-containing bridge, and to the unexpected product *404b* with the methyl adopting a *syn* configuration. [4] The presence of the isomer *404b* may indicate the possibility of a cyclization process, which does not exclusively follow the classical concerted Diels-Alder pathway but rather proceeds via a stepwise mechanism.

In an early account on the microwave assisted IMDA reactions, Roskamp [5] has developed microwave synthesis employing silicagel/water as reaction medium. Furan hemiacetal *406* reacted in microwave reactor, when adsorbed on silica gel and saturated with water to afford two diastereomeric Diels-Alder adducts *407* in 1:1 ratio (Scheme 80). Addition of water to the silica was crucial to the success of cycloaddition reaction, where water may perform three functions during heating: efficient generation of heat during the microwave process, acceleration of Diels-Alder reaction via the hydrophobic effect, and facilitation of the hemiacetal-hydroxy ketone equilibrium.

Using a domestic microwave oven, Demircan [6] carried out an intramolecular, solvent free reaction of bromofurylalkenes *408a,b* to afford fused tetracyclic bromides *409a,b*. Obtained yields were moderate, since further irradiation decomposes the furans (Scheme 81). Although these yields are slightly lower than obtained in the thermal process, the considerable time saving make this procedure as a synthetic alternative.

Jakopčić [7] has carried out the intramolecular Diels-Alder reaction of the *N*-(5-methyl-2-furfuryl)-*N*-prop-2-ynyl-*p*-toluidine *410*. The IMDAF reaction was carried out by heating a benzene solution of the amine *410*. Under these conditions cycloaddition product, *N*-*p*-tolyl-5-methyl-5,7a-dihydro-5,7a-epoxyisoindoline *411* was obtained (Scheme 82). The best results by classical heating were accomplished in toluene heating for 72 h at 80°C, with the ratio *410*:*411* in crude product was established at 80%:20%. Under the microwave irradiation conditions was much better yield (46%) and with much less decomposition.

Indoles were synthetised by Wipf [8] *via* the intramolecular Diels-Alder cascade from homoallylic furanyl amines. For instance, when alcohol *412* was heated in *o*-dichlorobenzene under microwave irradiation for 20 min at 180°C, complete consumption of the starting material was observed, and 4-phenylindole *416* was isolated in 79% yield, presumably according to the mechanism shown in (Scheme 83). This process proceeded concomitantly with thermal Boc deprotection. In contrast, when *412* was heated in toluene at reflux, after 48 h gradual decomposition of the starting material was observed. This annulation strategy is quite tolerant of functional groups, with variation the 4-aryl substituents from electron-rich (R=Tol, PMB) to electron-deficient arenes (R=*p*-F-C_6H_4), observing a slight increase in yield in the case of the 4-fluoro aryl substituent (the highest yield obtained was 84%).

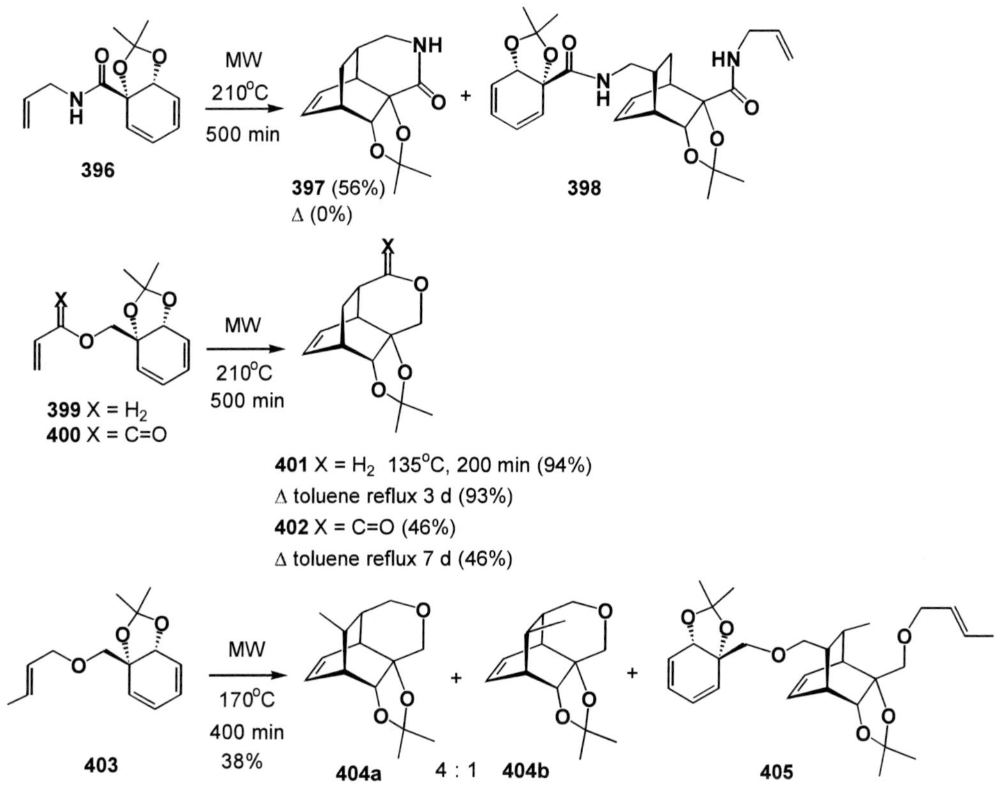

Scheme 79.

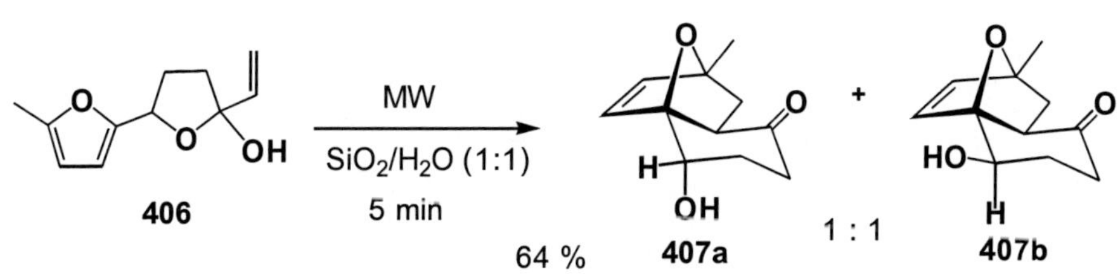

Scheme 80.

Unique reaction sequence was developed by Coquerel and Rodriguez [9] featuring a three-component domino Wolff rearrangement/ intermolecular hetero Diels-Alder/intramolecular Diels-Alder reaction. Three-component domino reaction of 1:1:1 mixture of a 2-diazo-1,3-diketone *417*, an aldehyde *418* and a primary amine *419* produced pentacyclic oxazinone derivatives *423a* and *423b* (Scheme 84). Reaction takes place by hetero-Diels-Alder reaction of acylketenes *420* (generated by Wolff rearrangement *in situ*) with *in situ* generated C=N double bonds (from aldehyde and amine), to form bicyclic 1,3-

oxazin-4-ones derivatives *422* in an intermolecular hetero Diels-Alder reaction. The kinetics of the imine formation is superior to the kinetics of the acylketene formation by Wolff rearrangement thus avoiding competitive nucleophilic addition of the amine to the acylketene. Intermediate *422* further reacts in a domino manner by subsequent intramolecular Diels-Alder reaction of furan moiety with alkene bond to afford *423a* and *423b* stereoselectively. With enantiopure primary branched amines a modest chiral induction was observed (d.r. 2.2:1). The reaction with the six- and seven-membered ring diazo precursors also provides oxazinones and variety of amines could be used.

Interesting example of how solvent effects influence bond formation under microwave irradiation to generate molecular diversity in a single-pot operation arising from a multicomponent coupling reaction was provided by Andreana. [10] Heating of mixture of aromatic aldehyde, aromatic amine, fumaric acid monoethyl ester *425*, and alkyl isocyanide. In protic solvents (such as water) unexpected products 2,5-diketopiperazines *430* from an aza-Michael reaction, and 2-azaspiro[4.5]deca-6,9-diene-3,8-diones *431* from a 5-*exo* Michael cyclization of the intermediate acyclic Ugi product *428* were obtained. However, in aprotic solvent unique intramolecular thiophene Diels-Alder-derived tricyclic lactams *429a,b* were obtained, by heating of mixture of thiophene aldehyde *427*, aromatic amine *424*, ester *425*, and benzyl isocyanide (Scheme 85). These products were formed by tandem Ugi/intramolecular Diels-Alder reaction cascade, in low yield, possibly due to reversibility, but also due to low reactivity of thiophene.

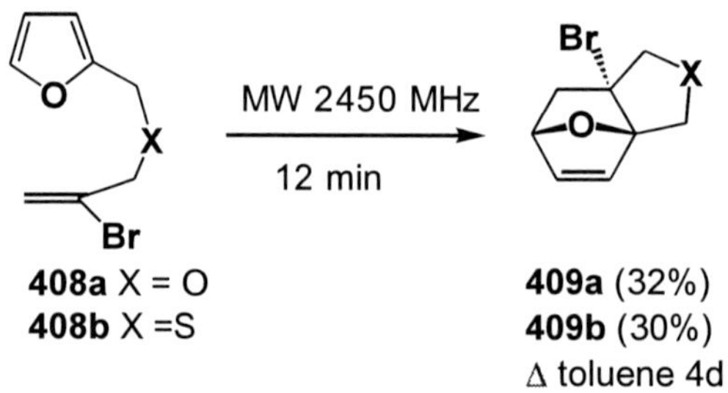

Scheme 81.

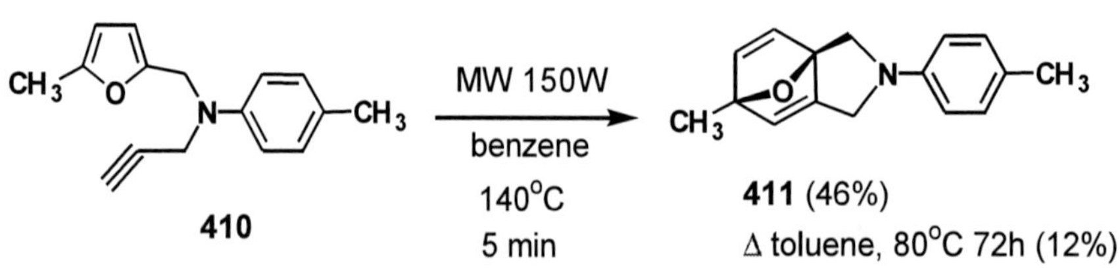

Scheme 82.

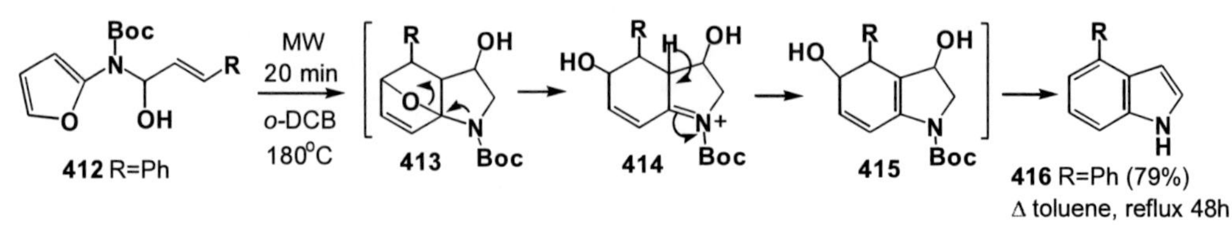

Scheme 83.

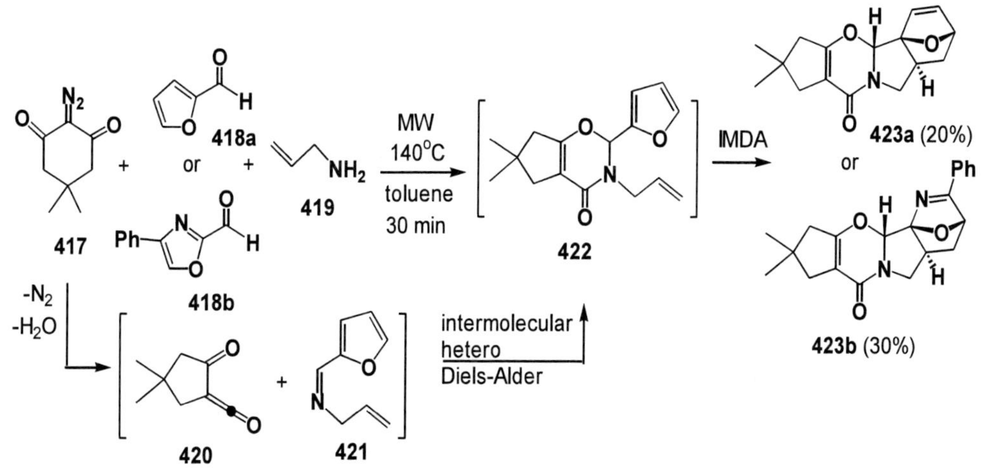

Scheme 84.

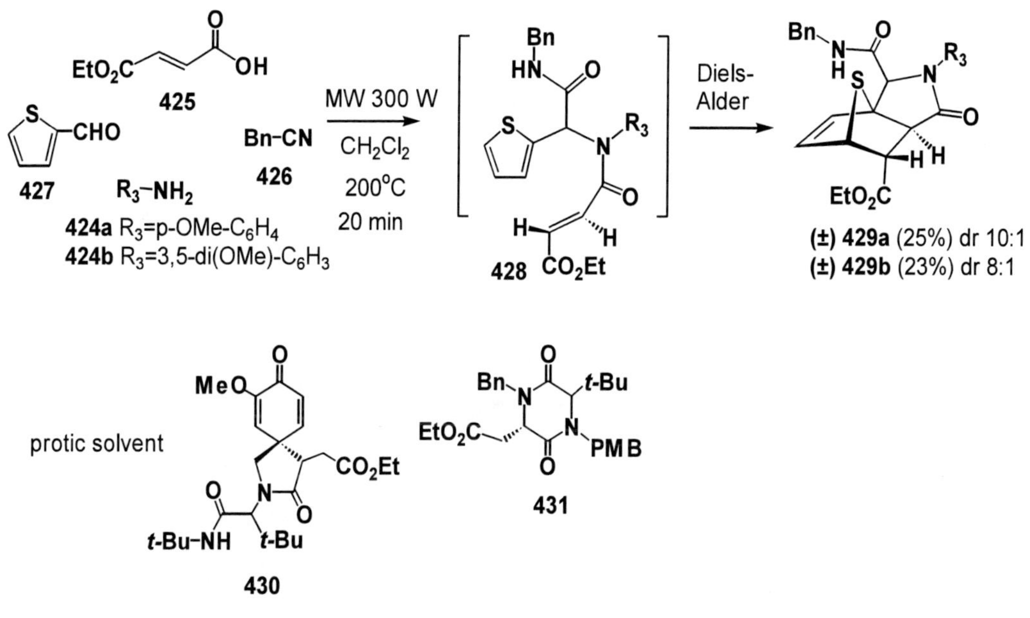

Scheme 85.

The IMDA cycloadditions of series of ester-tethered 1,3,8-nonatrienes have been carried out by Dai [11] under microwave heating, giving high yields of bicyclic lactones *433-434*. The stereochemical outcome of the reaction favors the formation of *exo* adducts, regardless of the *E*/*Z* configuration of the dienophilic moiety in *431* (*432*) (Scheme 86). The reactivity of the 9-benzoyl-substituted triene *435* parallels with the polarity of the solvent in the order of PhMe < MeCN and the conversions increase with the polarity, which is consisted with the rate enhancement observed in polar solvents. Both solvents favored the *endo-436b* isomer. A microwave-assisted tandem Wittig/IMDA cycloaddition protocol has been established for a modular synthesis of the bicyclic lactones starting from α-bromoacetates of 2,4-pentadien-1-ols and α-oxo carbonyl compounds. The reaction sequence consists of alkylation of PPh_3 with the bromides *437*, deprotonation of the phosphonium salts with 2,6-lutidine, olefination of the ylides with the α-oxo carbonyl *438*, and finally the IMDA cycloadditions. The overall yields of a single operation tandem reactions are 68-80% and the stereoselectivity of the major adducts formed from *E*-substituted 1,3,8-nonatriene is the same as that observed for the isolated 1,3,8-nonatrienes.

Dai [12] has also found that IMDA reaction of sorbate-related 1,3,8-nonatrienes *439a,b* under microwave heating yielded partial epimerization of the major adduct *441* (Scheme 87). In the case of microwave heating of *439a* at 180°C, three bicyclic γ-butyrolactones *440a-442a* were produced. Treatment of the reaction mixture with silicagel produced two *cis*- adducts. These results indicate that partial epimerization occured during the IMDA reaction at the elevated temperature. When *439b* was subjected to IMDA reaction, the initially formed *441b* is largerly epimerized to *442b*. The major adducts formed from *439a* and *439b* should be *trans-441a* and *trans-441b*, which epimerized to the thermodynamically more stable *cis* isomers presumably through the enol form of the lactone. Presence of the double bond at C6-C7 is a prerequisite for epimerization.

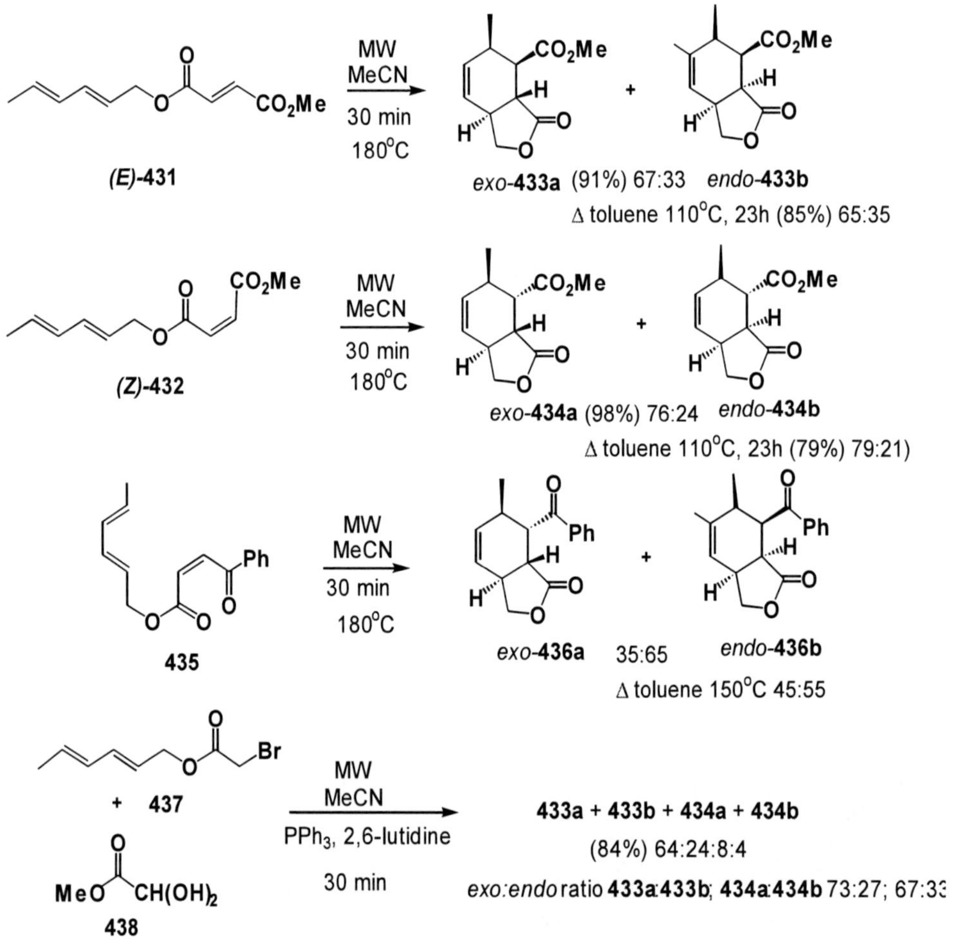

Scheme 86.

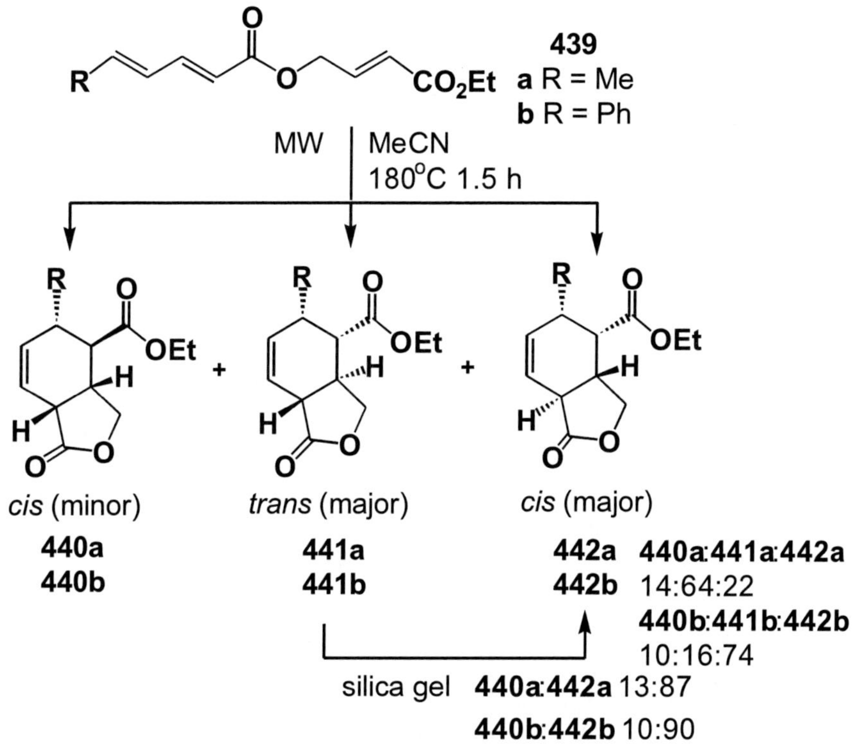

Scheme 87.

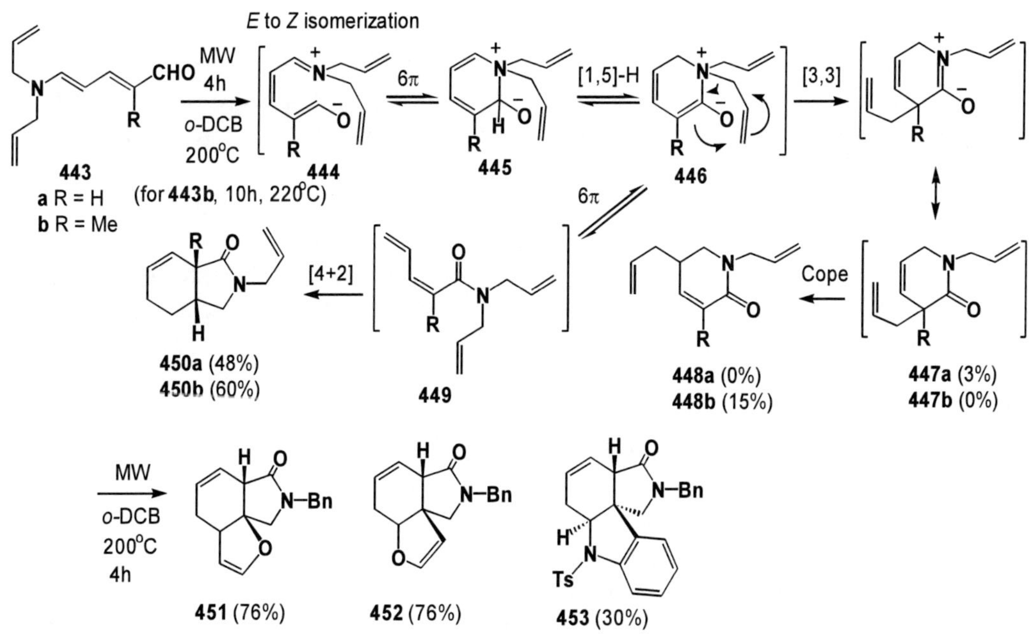

Scheme 88.

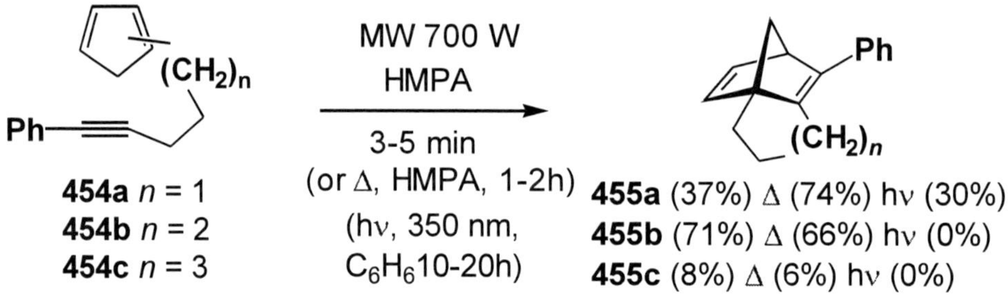

Scheme 89.

The pericyclic reaction cascade that results from heating Zincke aldehydes (derived from unsaturated amines) delivers rigid polycyclic lactam scaffolds of diverse structure. In this way, substantial complexity is obtained in a single step. Vanderwal [13] by heating of aldehyde *443a* at 200°C in a microwave reactor generated the dihydropyridone *447a* in trace amounts, while the bicyclic lactam *450a* was the major product. Methyl-substituted Zincke aldehyde *443b* similarly generated bicyclic lactam *450b* as the major product; but also dihydropyridone *448b* was isolated, in which a terminal Cope rearrangement has occurred (Scheme 88). The cascade reaction that affords bicyclic lactams *450a* and *450b,* starts with alkene isomerization, and is followed by 6π-electrocyclization, [1,5]-sigmatropic shift of hydrogen, 6π-electrocyclic ring-opening, and intramolecular Diels-Alder cycloaddition. Products *450a,b* are formed as single diastereomers with a *cis-* ring junction. A series of Zincke aldehydes with various unsaturated secondary amines upon microwave heating afforded the products of the rearrangement/Diels-Alder cascade. The polycyclic γ-lactams were all formed as single diastereomers. Interestingly, furans bearing aldehyde led to tricyclic dihydrofurans *451* and *452*, where furan had served as the 2π component in the [4+2] cycloaddition. Indole-containing substrate also behaved as a dienophile toward the *in situ*-generated diene, affording product *453* possessing a common tetracyclic core of many indole monoterpene alkaloids.

Chae [14] studied intramolecular Diels-Alder reaction of alkynylcyclopentadienes *454a-c* of various linker lengths upon microwave heating in HMPA solution which produce double bridged olefins (Scheme 89). Reversible thermal interconversion of cyclopentadienes yields a single cycloadduct *455* derived from 1-substituted isomer. Identical products were obtained by heating in HMPA, or irradiation in a DCA-saturated benzene solution at 320 nm.

1.3.2. Intramolecular Hetero Diels-Alder Reactions

In a series of papers Raghunathan [15] developed novel synthetic strategy to the pyrano fused heterocycles, which frequently occur in the natural products. High chemoselectivity was obtained in microwave accelerated domino Knoevenagel condensation /intramolecular hetero Diels-Alder reactions of various diones with aromatic aldehydes containing an internal dienophile. This procedure is efficient synthetic strategy to the pyrano [3-2c] fused coumarin and [2-3b]chromone frameworks, when hydroxy coumarins and O-prenylated benzaldehydes were used. The reaction of 4-hydroxy coumarin *456* with 2-(3-methyl-2-butenyloxy)benzaldehyde *457* in refluxing ethanol afforded pyrano [3-2c]coumarin and [2-3b]chromone derivatives *459* and *460* (Scheme 90). In this reaction, both the keto and lactone carbonyls could be involved in the cycloaddition reactions of intermediate *458* leading to *459* and *460*. Under microwave irradiation in ethanol, pronounced chemoselectivity was achieved with the considerable reduction of reaction time and increased yield. Similar chemoselectivity was obtained in the reaction of *456* with 2-(3-methyl-2-butenyloxy)naphthaldehyde *462* (i.e. benzo-annulated aldehyde *457*). Less pronounced chemoselectivity was achieved in the reactions of benzo analogues of 1-hydroxycoumarins with *457* and *462*. For instance, reaction of *452* with 4-hydroxy-α-naphthocoumarin *461* produced the mixture of polycyclic pyrano[3-2c]coumarin and pyrano[2-3b]chromone derivatives *463* and *464* (in 80:20 ratio, while reaction of *461* with *457* produced 57:43 ratio). When citronellal *465* was used in reaction with 4-hydroxy coumarin *456*, the reaction afforded the *trans* fused pyrano[3-2c]coumarin derivative *466* as a major product, with the formation of intramolecular ene product *467* in minor amounts. Again, the microwave enhanced reactions were more chemoselective and faster than classical reflux in ethanol.

Domino Knoevenagel condensation/ intramolecular hetero Diels-Alder reactions were further improved by change of conditions to solvent-free reaction with K-10 montmorillonite clay under microwave irradiation. [16] Further increase of yield and reduction of time were obtained. The synthesis of novel polycyclic thiopyrano coumarin/chromone frameworks *469/470* was achieved through reactions of 4-hydroxy coumarin and its benzo-analogues such as *461* with S-prenylated aromatic aldehydes *468*. A higher degree of chemoselectivity on coumarin products was achieved by the application of microwave irradiation and a solid support (Scheme 91). Change of methyl substituent in aldehyde *468* to phenyl does not change reactivity and stereoselectivity.

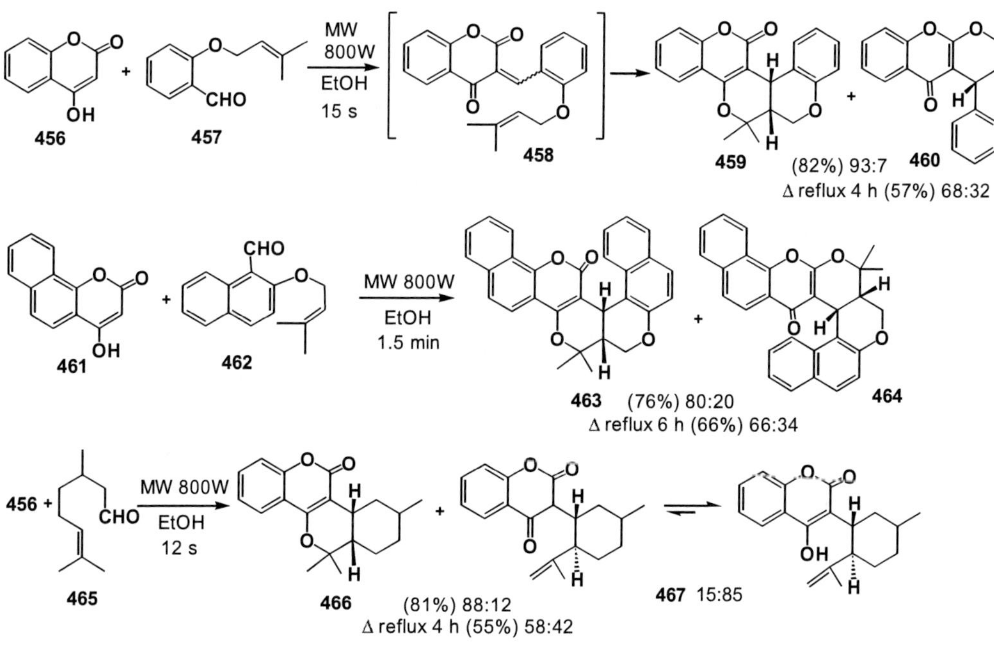

Scheme 90.

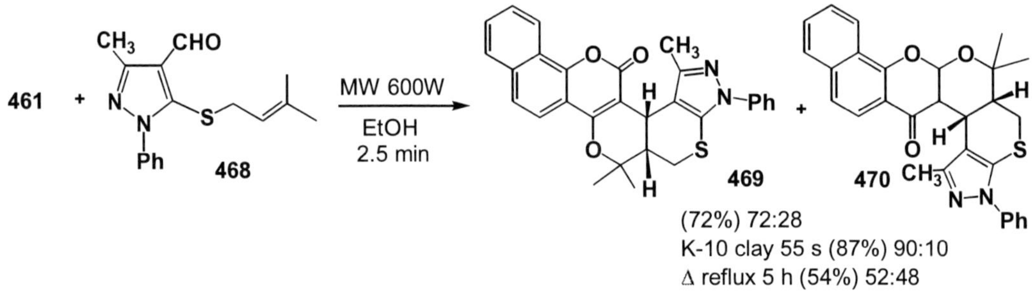

Scheme 91.

Raghunathan [17] used analogous reaction conditions in domino Knoevenagel condensation/ intramolecular hetero Diels-Alder reactions for the rapid access to indolo[2,1-a]pyrrolo [40,30:4,5]-pyrano[5,6-*c*]coumarin/[6,5-*c*]chromone derivatives. In this synthesis, reaction of indole-2-carbaldehydes containing an internal dienophile with α-naphtho-coumarins was carried out. Hence, reaction of *N*-alkenyl indole-2-carbaldehydes *471a,b* with coumarin *456* afforded stereo- and chemoselective mixtures of coumarin and chromone derivatives *472* and *473* (Scheme 92). Significant increase in chemoselectivity was obtained when instead of microwave heating in ethanol, microwave reaction was carried out in solvent-free conditions on montmorillonite K-10 clay. The corresponding polycyclic products *474-475* were obtained rapidly with α-naphthocoumarin *461*, with similar results and selectivities also obtained with β-naphthocoumarin.

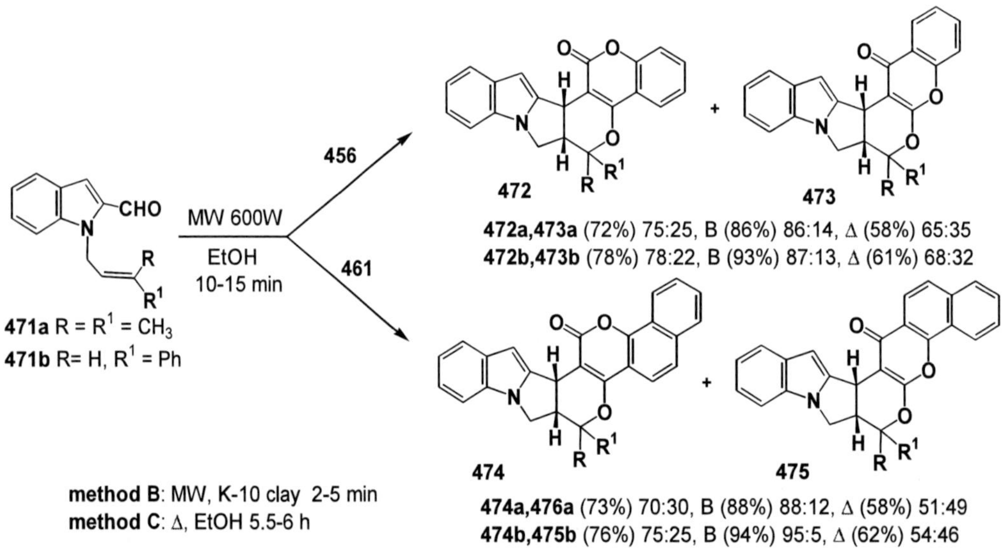

Scheme 92.

Identical conditions and synthetic protocols were used for the synthesis of indolo[2,1-*a*]pyrrolo[4',3':4,5]pyran ring systems (Scheme 93). [18] For instance, by classical heating of the aldehyde *471a* with pyrazolone *476* in toluene cycloadducts *478a* and *478b* were obtained, accompanied by the secondary tetrahydropyran product *470*, which lowered the isolated yield of the cycloaddition compounds. Product *480* arises from the 1,6-electrocyclization of the initially formed Knoevenagel adduct *477*, followed by a 1,3-

prototopic shift in intermediate *479*. The reaction in toluene under microwave irradiation provided *478-480* with an increase in the chemical yield and a slight improvement in the chemoselectivity. Solvent-free reaction with K-10 montmorillonite clay under microwave irradiation further increased yield to 84% without the formation of dihydropyran product.

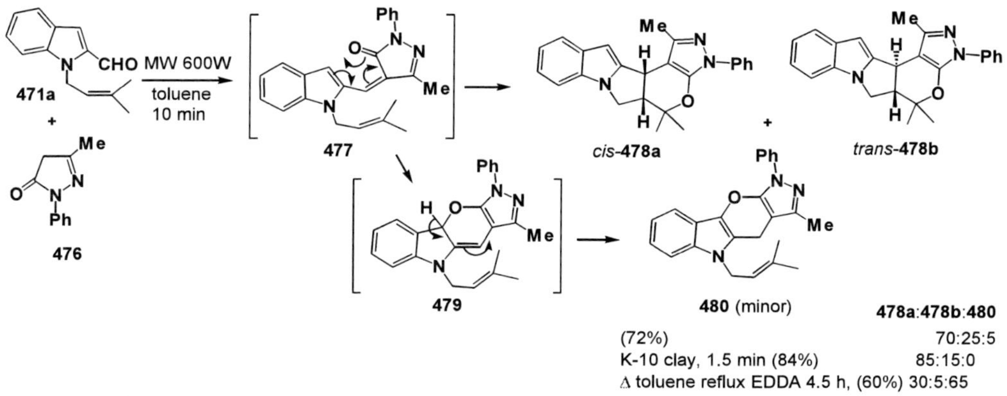

Scheme 93.

By employing other symmetrical 1,3-diones, *N*-acetyl oxindole, indane-1,3-dione, dimedone and meldrum's acid with aldehydes such as *471a*, related cycloadducts were obtained, whereas the *cis*- adduct are always dominant product in microwave enhanced reactions, but the dihydropyran is the major product in classical reflux reaction.

When domino Knoevenagel condensation/ intramolecular hetero Diels-Alder reactions were applied to reactions of uracil annulated pyrrole aldehydes with chromones, the uracil annulated pyrrolo[2,3-*d*]pyrimidine-pyrano[5,6-*c*]coumarin/[6,5-*c*]-chromone derivatives were obtained [19] Solvent-free approach on a solid support accelerated by microwaves proved to be synthetically useful in achieving high degrees of chemo- and stereoselectivity with a substantial reduction in reaction time. Treatment of 4-hydroxycoumarin *456* with *481a* in refluxing toluene resulted in the formation of the *cis*-fused coumarin *484a* and chromone *484b* with an overall yield of 47% in the ratio 58:42 (Scheme 94). The reaction proceeded via a domino Knoevenagel/intramolecular hetero Diels-Alder pathway in one-step process. When the same reaction was carried out under microwave irradiation for 5 min in toluene, the coumarin and chromone derivatives were obtained in the ratio 65:35 with an overall yield of 52%. The reaction under solvent-free conditions, simply by grinding the two components with K-10 montmorillonite clay and irradiating the mixture under microwave conditions afforded the cycloadducts in excellent yields with high chemoselectivity. The ratio of the coumarin and chromone derivatives increased to 74:26 with an improved overall yield of 74%. The reactions of 4-hydroxy-thiocoumarin *482*, α- and β-naphthocoumarins gave similar results in terms of yields and selectivity.

In an analogous reaction, heating of 2-(cyclohex-2-enyloxy)benzaldehydes *488a-c* with 1,3-diones such as pyrazolone, indane-1,3-dione and dimedone *489* produced mixtures of the pyrano[2,3,4-*kl*]xanthene *cis*- and *trans*- derivatives *491a-c* and *492a-c* (Scheme 95). [20] In a short duration of time, microwave irradiation proved to be highly beneficial for obtaining good yield of the products with high stereoselectivity. While in the reflux conditions *cis*-products *491a-c* were obtained in low stereoselectivity (50-53%), applying microwave conditions amount of the *cis-* isomers significantly increased to (74-83%). By applying microwave conditions on solid support, the overall yield increased, while ethylenediaminediacetate (EDDA) catalyst was not necessary. In addition, a high degree of chemoselectivity on pyrano[2,3,4-*kl*]xanthene products was achieved.

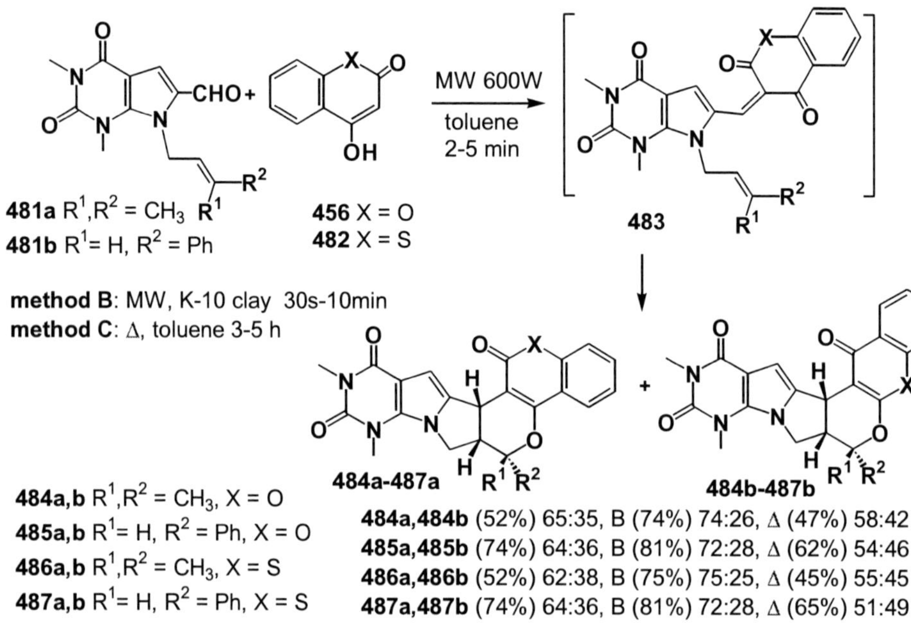

Scheme 94.

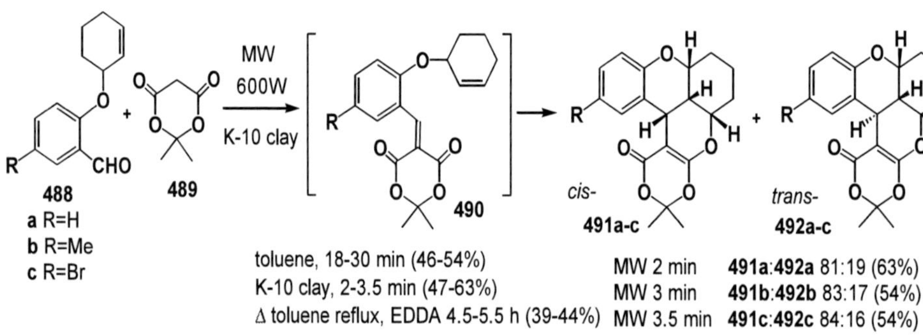

Scheme 95.

Further variation of reaction conditions of domino Knoevenagel condensation/ intramolecular hetero Diels-Alder reactions is indium chloride/silica gel supported synthesis of pyrano/thiopyranoquinolines. [21] $InCl_3$/silicagel supported conditions using microwave irradiation were applied to reaction of 3-formylquinoline-2-O-alkenyl (3-formylquinoline-2-S-alkenyl) aldehydes with aniline leading to efficient synthesis of tetrahydroquinoline[4,3:2,3]pyrano/thiopyrano[2,3-b]quinoline derivatives (Scheme 96). In conventional conditions, aldehyde *493* reacts with anilines such as *494* in the presence of $InCl_3$ in acetonitrile forming a mixture of *cis-* and *trans-* products *497* in the one pot-reaction. Imine product of Knoevenagel condensation *495* undergoes intramolecular hetero Diels-Alder reactions to afford cycloadducts *496*. Their aromatization led to final products, tetrahydroquinoline[4,3:2,3]thiopyrano[2,3-b]quinoline derivatives *497*. A dramatic increase in the overall yield of the products was achieved when the same reaction was carried out using $InCl_3$ impregnated silica gel as catalyst under microwave irradiation however, the ratio of the *cis*- and *trans*- adducts remained almost the same. Moreover, silica gel can function as a mild acidic catalyst, and another advantage of silica gel as effective solid support is the easy separation of the end products. The method avoids the use of a solvent and the reaction was complete within 1-3 min.

The domino Knoevenagel condensation/ intramolecular hetero Diels-Alder reaction was also used by Lenardão and Jacob [22] as a green and efficient method for the synthesis of 1,2,3,4,4a,9,9a,10-octahydroacridines *500*. It has been developed by a simple one-pot hetero-Diels-Alder reaction starting from (+)-citronellal *498* and *N*-arylamines *499a-f* in the presence of a solid supported catalyst ($SiO_2/ZnCl_2$, 10%), under microwave irradiation and without any solvent. In short reaction time, mixturs of *cis*- and *trans*- products *500* were obtained in high yields (Scheme 97). The *cis/trans* ratio depends on the substitution pattern, while yields obtained by classical heating were poor, and 80 min at 40-45°C were required to achieve same yields as in microwave.

When selenonium salts ionic liquid were used as green solvents and/or catalyst in the preparation of 1,2,3,4,4a,9,9a,10-octahydroacridines *8* and *9* by hetero-Diels-Alder reaction involving (*R*)-citronellal *6* and arylamines *7a,b*, octahydroacridines were obtained in good yields in short times. [23] Phenyl butyl ethyl selenonium tetrafluoroborate (acidic liquid at room temperature) was used as the catalyst in this reaction, in view of the high stability, facile preparation and low viscosity. In the course of reaction, the catalyst decomposes to phenyl butyl selenide that can be used to prepare new selenonium salt.

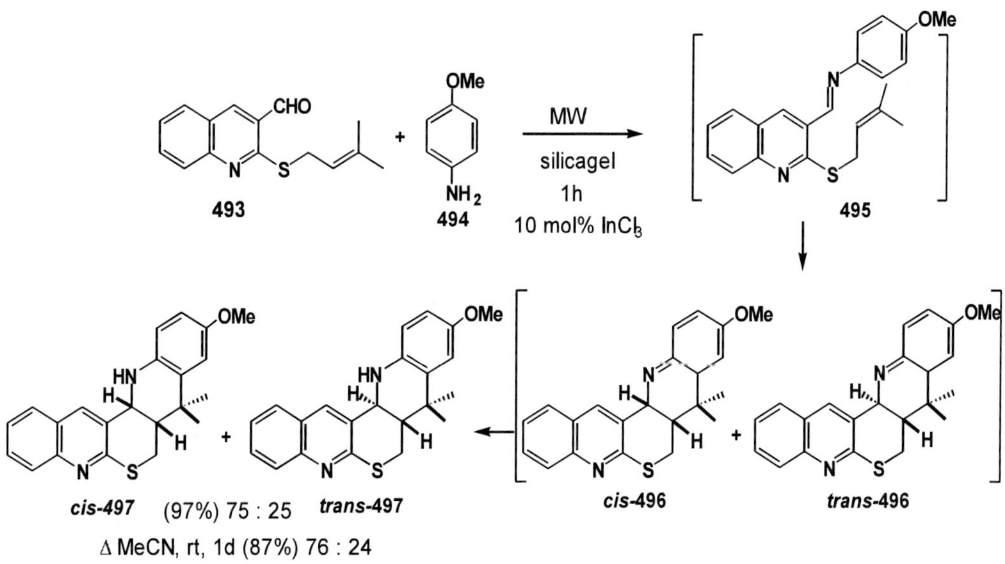

Scheme 96.

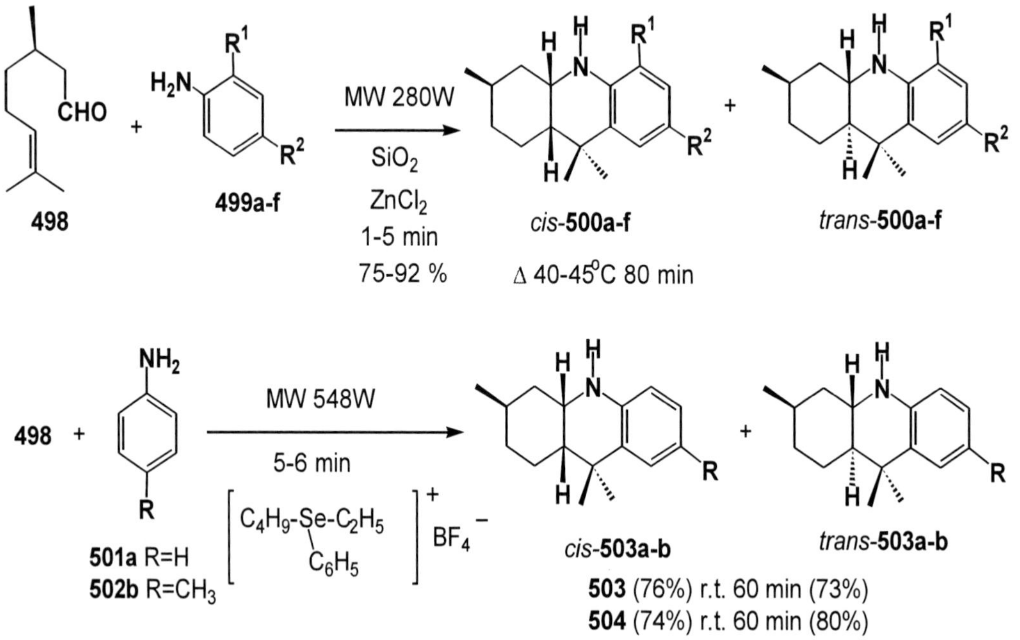

Scheme 97.

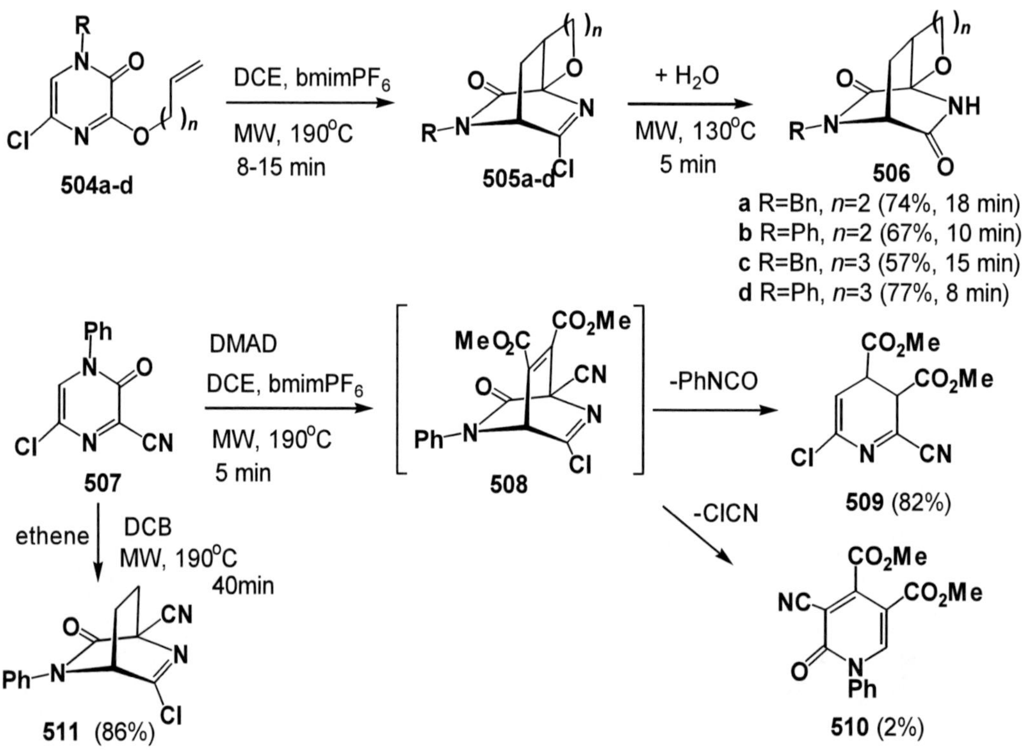

Scheme 98.

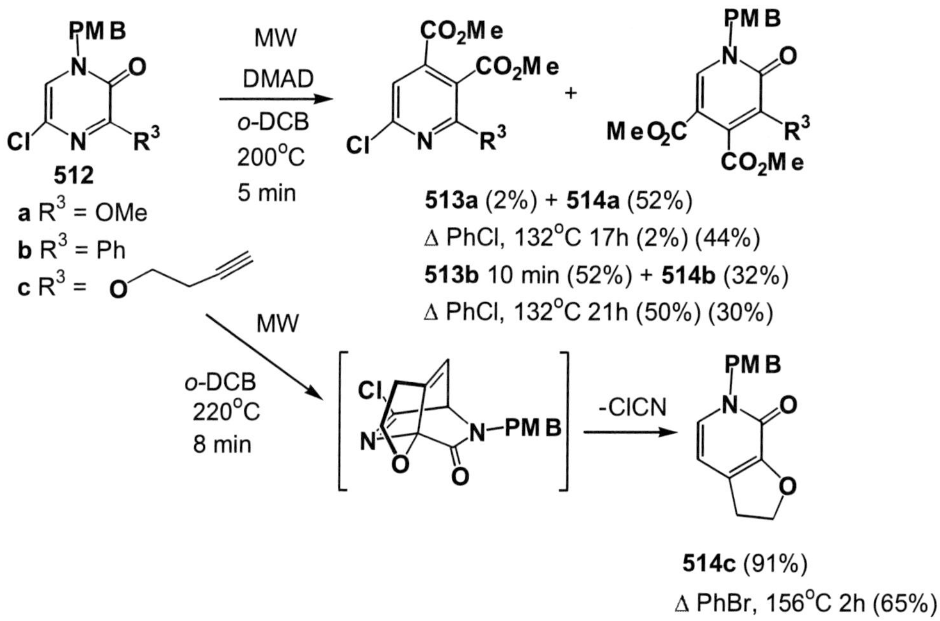

Scheme 99.

Another example on the use of ionic liquid for microwave assisted hetero intramolecular Diels-Alder reaction was given by Kappe. [24] The cycloaddition chemistry of 2(1*H*)-pyrazinones suffers from low reactivity, usually rather long reaction times at high temperatures and sometimes high pressures are required. Microwave/ionic liquid strategy of intramolecular Diels-Alder reactions was successfully applied to a series of alkenyl-tethered 2(1*H*)-pyrazinones *504a-d* to obtain products in shortened reaction times (Scheme 98). These yields are in the same range as for the conventional thermal protocols. However, the microwave-promoted transformations allow a more rapid and efficient one-pot access to cycloadducts of type *506*. Chlorobenzene solvent used under reflux conditions (b.p. 132 °C) was replaced by 1,2-dichloroethane (DCE), which has a considerably lower boiling point (83°C), but couples with microwaves reasonably well. To further enhance the maximum attainable reaction temperature (and therefore shorten the reaction time) the solvent was doped with ionic liquid, 1-butyl-3-methylimidazolium hexafluorophosphate (bmimPF_6) in order to increase the dielectric constant of solvent medium. In these reaction conditions, intermolecular hetero Diels-Alder cycloaddition of 2(1*H*)-pyrazinone *507* with acetylenic dienophile, DMAD afforded pyridine *509* and pyridone *510*. On the other hand, cycloaddition reaction of *507* with ethene, led to the bicyclic cycloadduct *511*. This is one rare example of the use of gaseous reagents in microwave oven reactions. To perform microwave-assisted protocols utilizing gaseous reagents, the solution of *507* in the microwave vial was saturated with ethene at atmospheric pressure, before sealing the vial. It was found that reaction does not require a significant ethene pressure (1-2 bar).

Van der Eycken [25] has shown that hetero Diels-Alder/retro Diels-Alder reactions of 2(1*H*)-pyrazinones *5a-b* with DMAD in 1,2-dichlorobenzene enhanced by microwave irradation gave mixture of pyridines *6a* or *6b* and pyridinones *7a* or *7b*, in moderate to good yields (Scheme 99). The intramolecular reaction of *5c* was performed at a slightly higher temperature (220 °C), and completion was reached after 8 min of irradiation, whereas only the dihydrofuropyridinone product *7c* is isolated. The yields obtained after microwave irradiation are comparable with those reached after thermal heating for compounds *6a,b* and *7a,b* (with a comparable product distribution), whereas for the dihydrofuropyridinone *7c*, the yield is markedly improved from 65 to 91%.

These Diels_Alder cycloadditions could be carried out with resin-bound 2(1*H*)-pyrazinones *515a,b-517a,b*. Microwave-assisted reaction of these substrates was performed in 1,2-dichlorobenzene at 220 °C in short reaction times (Scheme 100). Cleavage of compounds *518-520a,b* was then effected with neat TFA under reflux affording pyridinones *521a-b*. To drive the cycloaddition/retrocycloaddition reaction of the resin-bound pyrazinones *515c-517c* to completion in conventional conditions, an excess of the DMAD in chlorobenzene was necessary, and the reaction mixture was heated under reflux for 1-2 days. The product distribution and the yields of the reaction in solution phase and on solid support are comparable. Conventional cleavage of the resin-bound pyridinones *518a-c-520a-c* using different ratios of TFA gave unsatisfactory results. In contrast, a smooth release from the support could be performed upon microwave irradiation of a suspension of resin-bound products *518a,b-520a,b* and *518c-520c* in TFA/DCM (5:95) at 120 °C. Although the commercially available Wang resin is suited to perform the cycloaddition/ retrocycloaddition reactions of the coupled 2(1*H*)-pyrazinones, the required cleavage conditions are rather harsh. It was demonstrated that linker derived from Merrifield resin and syringaldehyde is well suited for the disclosed pyrazinone chemistry. It combines an appropriate chemical stability with very mild cleavage conditions.

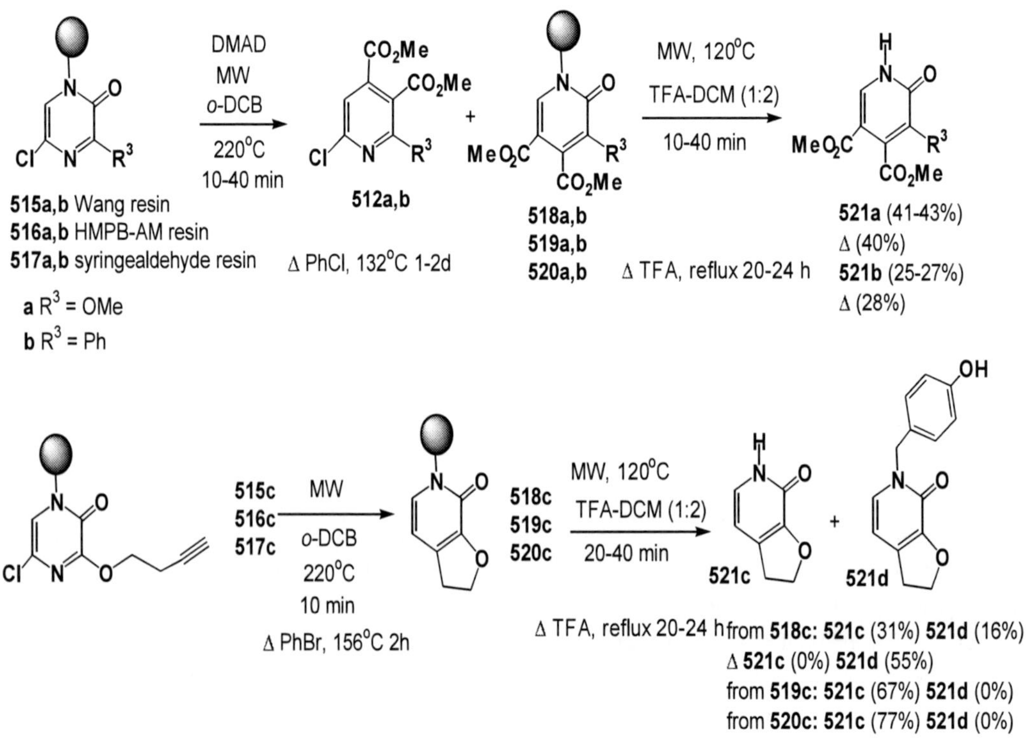

Scheme 100.

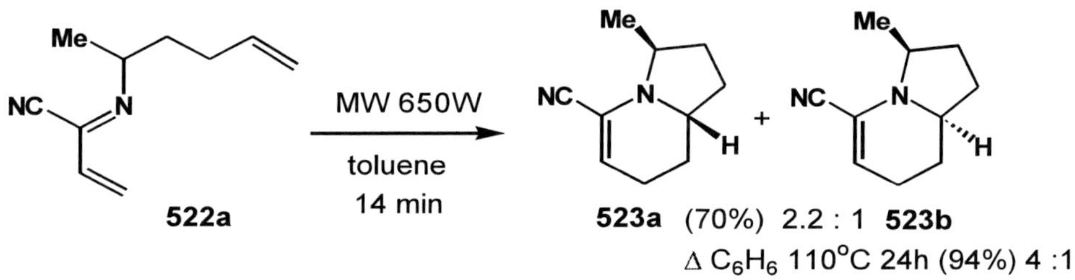

Scheme 101.

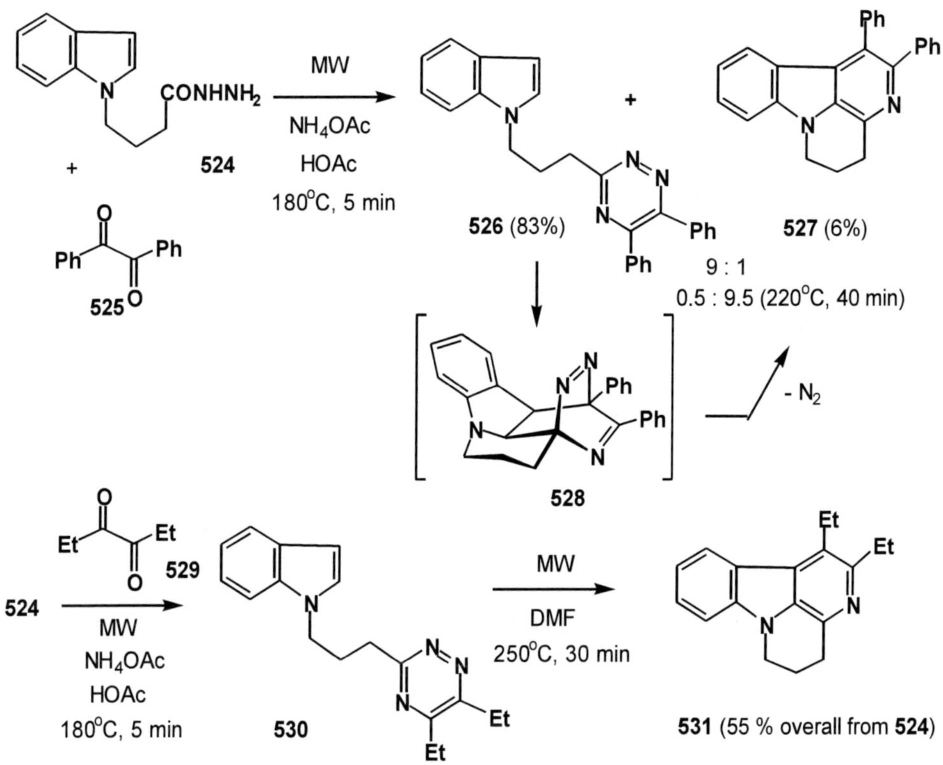

Scheme 102.

Grierson [26] carried out the conversion of azadiene *522a* (1-(1-methylpent-4-enyl)-2-cyano-1-aza-1,3-butadiene) to indolizidines *523a,b* (mixture of diastereomers) via intramolecular hetero Diels-Alder cyclization in microwave conditions. To complete, classical heating of *522a* in benzene has to be conducted overnight in a sealed tube at 110 °C. It was found that reaction time could be reduced to a mere 14 min when this Diels-Alder reaction was carried out in a microwave oven (Scheme 101).

Lindsley [27. 28] has found that acyl hydrazide *524* under microwave conditions in the presence of 1,2-diketone *525* and NH_4OAc delivers desired triazine *526*, as well as 1,2-diphenyl canthine derivative *527* (Scheme 102). This is three component condensaton, where triazine *526* is generated first, followed by the intramolecular hetero Diels-Alder / retro Diels-Alder reaction. Optimization of the reaction conditions led to reverse of selectivity and preferential formation of *527* (which was isolated in 80% yield). This reaction is general and variety of functionalized aromatic diketones could be used to prepare derivatives of *527* in moderate to excellent yields (30-80%). On the other hand, all attempts to obtain satisfactory yield in the case of C1/C2 dialkyl analogs failed, and the standard two step triazine protocol has to be employed. Reaction of *530* was heated at 250°C, which is almost 100°C above the boiling point of DMF, for 30 min. Other solvents (dichloroethane, dioxane, toluene and ethanol) were less satisfactory.

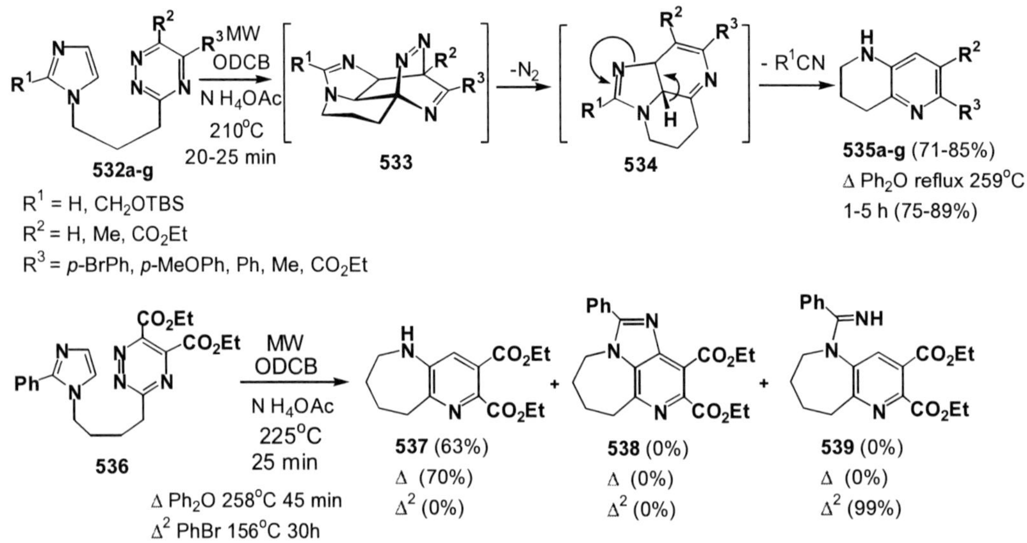

Scheme 103.

Another study of intramolecular hetero Diels-Alder cyclization of tethered 1,3,4-triazine was published by Snyder. [29] Microwave irradiation of the intramolecular inverse electron-demand Diels-Alder reaction between imidazoles and 1,2,4-triazines linked by a trimethylene tether from the imidazole N1 position to the triazine C3 *532o-u* proceed in excellent yields to produce 1,2,3,4-tetrahydro-1,5-naphthyridines *535o-u* (Scheme 103). The reaction proceeds by a cycloaddition to afford intermediate *533* with subsequent loss of nitrogen, forming intermediate *534*, followed by a stepwise loss of a nitrile. The best conditions for the cycloaddition were found to be microwave irradiation at 210°C, for 20 min, in *o*-dichlorobenzene in the presence of NH_4OAc (15 equiv) as an energy transfer agent. There conditions proved comparable to refluxing in diphenyl ether (at 259°C) in yield, but in considerably less time. The tether length could be increased, thus tetramethylene tethered imidazole/triazine pairs *536* in microwave conditions generate 2,3,4,5-tetrahydro-1*H*-pyrido[3,2-*b*]-azepines *537*. While the nitrile elimination products *537* were dominant, small amounts of the aromatized deazapurines *538* were also obtained when diphenyl substituted triazines were employed in classical heating. Classical reaction of the diester triazine *536* produced unstable amidine *539* in quantitative yield upon refluxing in bromobenzene. Brief refluxing of *536* in diphenyl ether or microwave promoted cycloaddition yielded the pyridoazepine *537* without any amidine precursor or deazapurine. This is one rare example where the temperature of microwave conducted reaction is actually lower than thermal (usually by 20°C higher for transfer of thermal to microwave conditions). When the tethered triazine preparation with subsequent cycloaddition was carried in a single step via the acyl hydrazide/α,β-diketone route in the microwave-promoted synthesis, tethered triazines *532* were major products, while naphthyridine cycloadducts *535* had been detected in small quantitites under these conditions (refluxing acetic acid) presumably due to the higher temperature required to effect the cycloaddition. The one-pot approach to the naphthyridines resulted in the production of *535* in considerably lower yield than resulted from the two-step preparation, wherein the tethered triazine was isolated and then subjected to microwave irradiation to promote the cycloaddition.

Pyrimidines could be also used as dienes in hetero Diels-Alder reactions. Shao [30] reported that formation of fused lactams and lactones by the microwave-assisted hetero-Diels–Alder cycloaddition reactions of series of acetylenic pyrimidines tend to be much cleaner than those using conventional thermal conditions, resulting in higher isolated yields of the desired product (Scheme 104). When pyrimidine/ester tethered acetylene *540* was subjected to microwave irradiation, benzo fused lactone *542* was obtained. In the reaction of pyrimidine/amide tethered acetylene *543,* the benzo fused lactam *544* was prepared. In addition, the oxidized cyclization products such as *545*, were observed. The extent of the formation of *545* decreased when the solution of the reactants in nitrobenzene was deoxygenated prior to heating. Under the same microwave irradiation conditions pyrimidine *543* produced a mixture of *544* and *545*. At 260°C for 2 min, the ratio of *544*:*545* became 2:1, while the additional heating at 280°C for 2 min changed the ratio to 1:2.

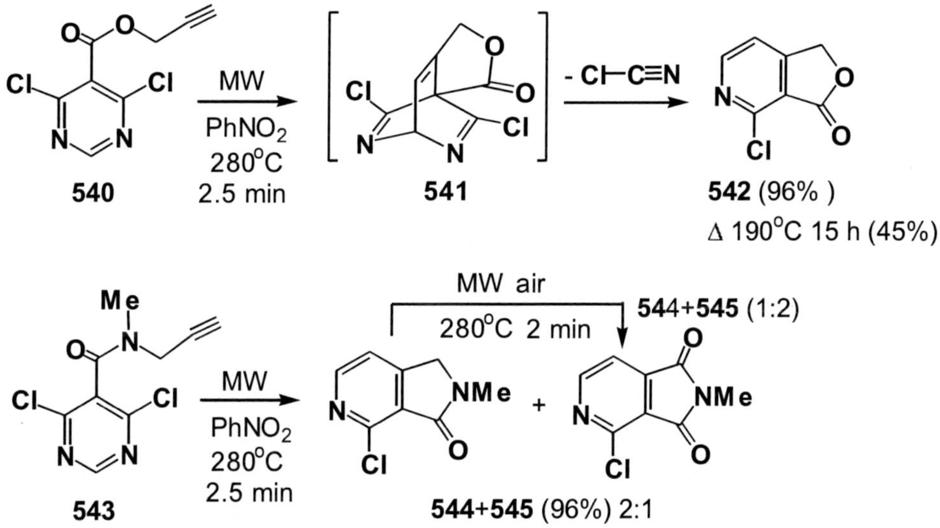

Scheme 104.

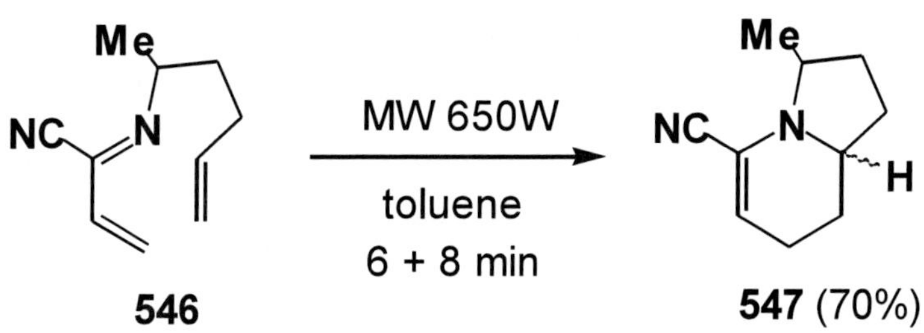

Scheme 105.

Although the conversion of the 2-cyano-substituted *N*-alkyl azadiene *546* to indolizidine *547* (mixture of diastereomers) had already been achieved by heating in benzene overnight in a sealed tube at 110°C, Grierson [31] has found that reaction time could be reduced to a mere 14 min when this intramolecular Diels-Alder reaction was carried out in a microwave oven (Scheme 105).

Analogous thermal intramolecular hetero-Diels-Alder reaction of 2-cyano-1-aza-1,3-butadiene *548* failed to produce the desired cycloadduct. Gratefully, Zhu [32] reported that the cycloaddition of *548* took place smoothly under microwave irradiation conditions to afford the indolizidine in 99% yield, as a mixture of two diastereomeric indolizidines *549* and *550* (ratio 1:1) (Scheme 106).

Guillaumet [33] has shown that intramolecular hetero Diels-Alder reactions of alkenyl tethered triazines *551*, enhanced by microwave irradiation gave corresponding furo- *552a* and pyranopyridines *552b* were formed in high yields (Scheme 104). For the six-membered ring formation, higher temperatures were required. Interestingly, Bu_3Sn substituent is capable to withstand the microwave conditions. Subsequent cycloaddition reaction of stannyl derivatives in identical reaction conditions afforded *554a* and *554b*, possessing functionality suitable for further synthetic elaboration.

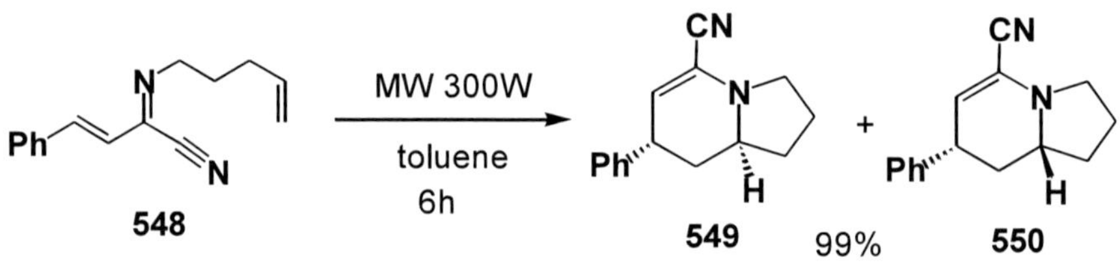

Scheme 106.

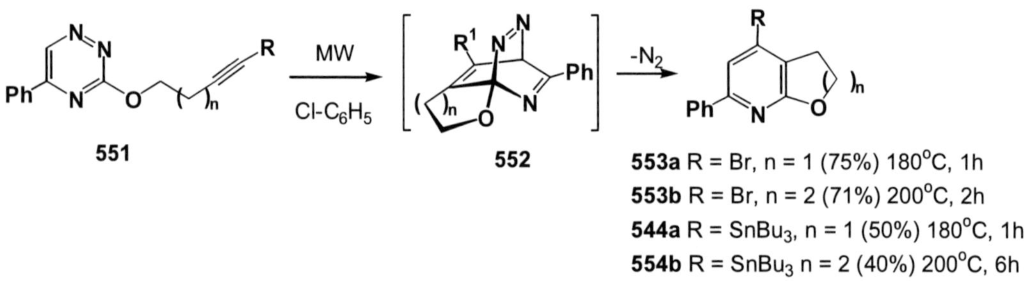

Scheme 107.

When the alkynol triazine *555* was engaged in an intramolecular cycloaddition in chlorobenzene at reflux for 12 h the 3-hydroxy-6-phenyl-2,3-dihydrofuro[2,3-*b*]-pyridine *558* was obtained in 93% yield. [34] The same yield was obtained within 45 min in microwave conditions (Scheme 108). More drastic cyclization conditions were needed for less reactive alkynol *556*, which was heated at 220°C under microwave irradiation to afford 4-hydroxy-7-phenyl-3,4-dihydro-2H-pyrano[2,3-*b*]pyridine *559*. By classical heating, *556* did not react in refluxing dichlorobenzene, but reaction takes place in a sealed tube in chlorobenzene at 220 °C (48 h, 90%). Even at extremely high pressure (dichloromethane, 11.3 kbar, 48 h, r.t.), *556* was recovered unchanged. In the case of several alkynol triazines possessing aromatic substituents (*560* and *561*), besides 3-hydroxy-6-phenyl-2,3-dihydrofuro[2,3-*b*]-pyridines *562* and 4-hydroxy-7-phenyl-3,4-dihydro-2H-pyrano[2,3-*b*]pyridines *563* also smaller amounts of dehydrated products *564,565* were obtained. Shorter reaction times applied for triazines *560,561* precluded the exposure of cycloadduct to conditions which would favor dehydration.

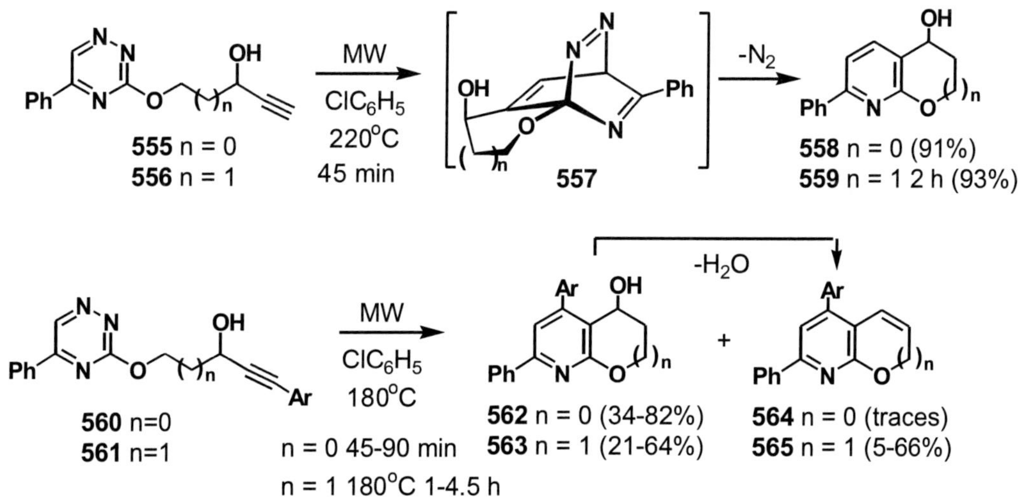

Scheme 108.

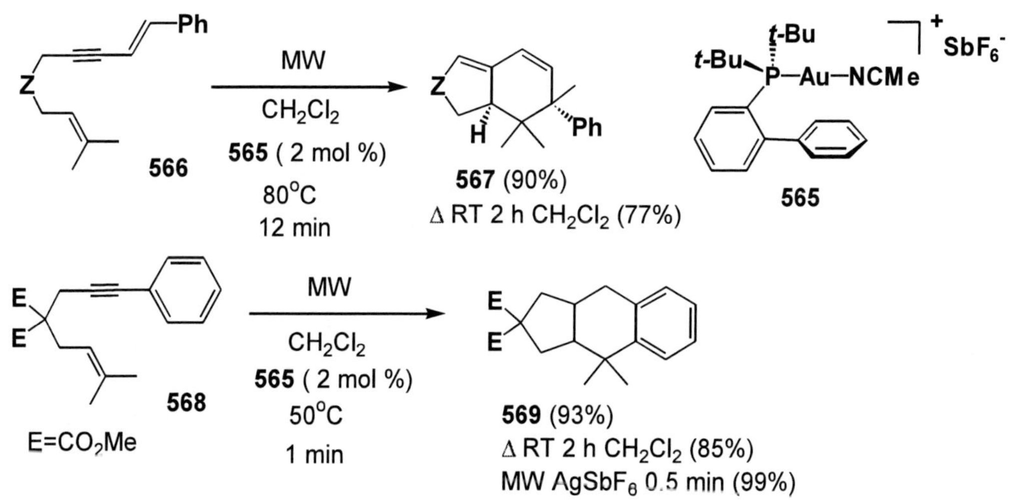

Figure 109.

1.3.3. Formal Intramolecular Diels-Alder Reactions

Gold(I) catalyzed formal intramolecular [4+2] cycloadditions of arylalkynes or 1,3-enynes with alkenes (dienynes) were studied by Echavarren. [35] Hydrindane *567* was obtained in short time and in good yield with catalyst *565* under microwave heating of 1,8-dien-3-yne *566*, which react by a 5-*exo-dig* route (Scheme 109). Analogous reaction of arylenyne with Au(I) catalyst gave the best results in CH_2Cl_2. The cationic catalyst generated from $AgSbF_6$ complex provided *569* instantaneously in quantitative yield.

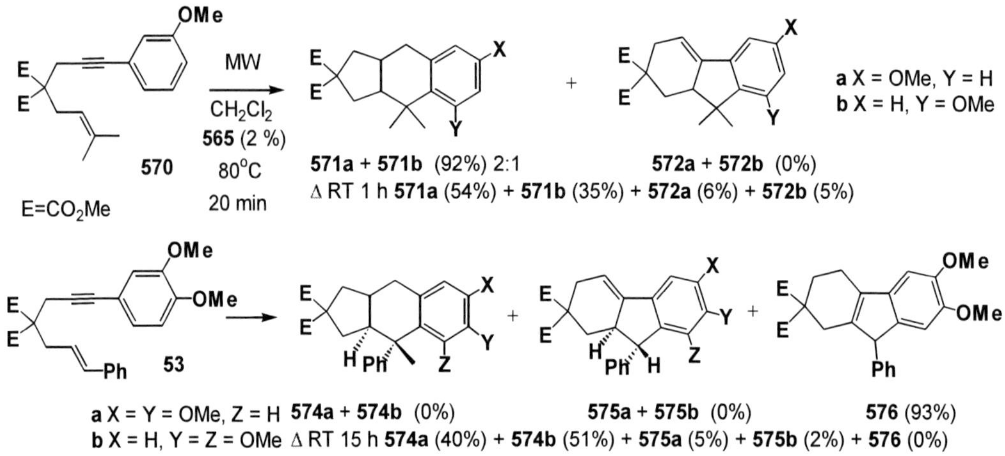

Scheme 110.

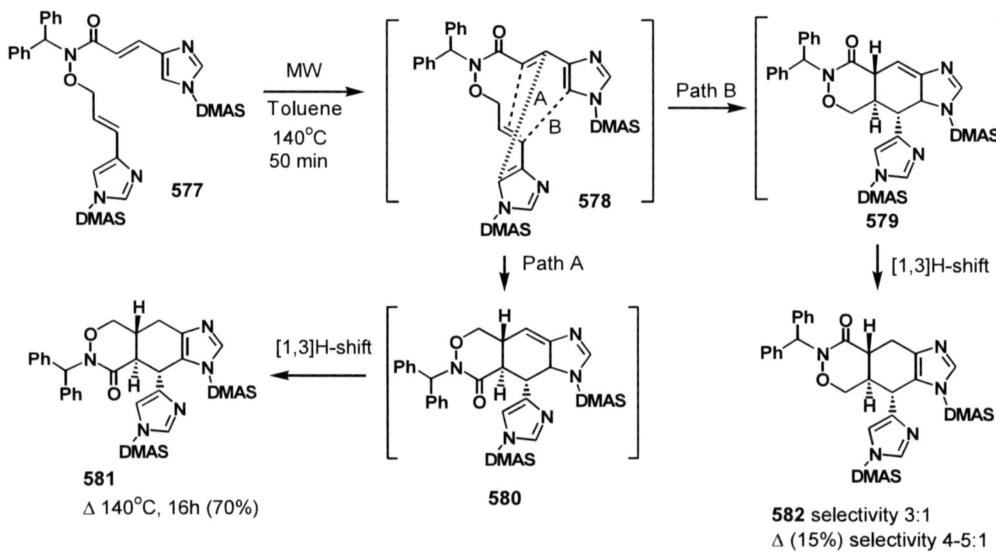

Scheme 111.

Substrates with *meta-* substituents *570* and *573* gave mixtures of regioisomeric cycloadducts. Thus, with *570* four products were obtained with catalyst *565* at room temperature (Scheme 110). However, under microwave heating with *565*, substrate *570* afforded only compounds *571a* and *571b*. The result with substrate *573* is noteworthy since the reaction with *565* at room temperature led to a mixture of four products, but in contrast reaction under microwave heating with *565* gave *576* as single regioisomer in excellent yield. Product *576* is the result of an initial *endo-dig* formal [4+2] cyclization, followed by the isomerization of the alkene to the internal position.

1.3.4. Intramolecular Diels-Alder Reactions in Natural Product Synthesis

Intramolecular cycloadditions of functionalized tethered bisimidazoles were used in synthesis of marine alkaloid ageliferin by Lovely. [36] Intramolecular Diels-Alder reaction of bisvinylimidazole substrate *577* gives two cycloadducts *581* and *582* (Scheme 111), but the major one *581* arises from the cycloaddition via Path A, where electron-defficient double bond acts as dienophile component. Diels-Alder reactions of this substrate carried out by conventional heating using toluene as a solvent at 140-150 °C for 16 h afforded moderate levels of chemoselectivity (4-5:1). It was found that this reaction would proceed under microwave irradiation, leading to significantly reduced reaction times (50 min), but there was a slight decrease in the selectivity (3:1). Both isomers possess the correct relative stereochemistry (all *trans*-diaxial relationship) for use in synthetic approach to ageliferin.

Fallis used IMDA synthetic approach to tricyclic taxoid nucleus.[37] Cyclization of ketone *583* was effected by heating in a sealed glass tube in a modified microwave oven [38] (0.05 M toluene solution, 1% mol equiv. hydroquinone) to stereoselectively afford the tricyclic system *585* (Scheme 112). The major adduct possessing a tricycle [9.3.1.0^{3,8}]pentadec-ene skeleton arose via the preferred *endo-* transition state in which the nonbonded interactions were minimized due to the alignment of the dienophile under the triene remote from the MOM substituent. Due to the ease with which the cyclohexene double bond migrated into conjugation with the acetylenic ketone accounted for the uncyclized material in the thermal reaction.

Fallis [39] reported another intramolecular Diels-Alder strategy to tricyclo [5.4.0^{1,7}.0^{2,9}]undecanes for the direct total synthesis of sesquiterpene (+)-longifolene. This [4+2] approach to the tricyclo[5.4.0^{1,7}.0^{2,9}] undecene skeleton with considerable synthetic appeal proved challenging and only a 10 % yield of the adduct *588* was obtained after 24 h reflux in toluene, while decomposition predominated at higher temperatures. On the other hand, microwave heating has been very useful technique when the triene *587* was heated in a sealed glass tube as a 0.05 M solution in toluene containing 1% molar equiv of hydroquinone for 2 h in a modified, household microwave oven to afford the adduct *588* in 92% yield (Scheme 113).

Kobayashi [40] described an interesting microwave-assisted biomimetic Diels-Alder reaction towards the total synthesis of symbioimine. The intramolecular Diels-Alder reaction of *trans*-enone *589* was efficiently accelerated by microwave heating, while any other conventional methods were not as effective (Scheme 114). It was found that a steric interaction between phenyl and isopropyl groups in the transition state of the Diels-Alder reaction would be the origin of the difficulty. The [4+2] cycloaddition of related allyl alcohol *591* was significantly facilitated compared with that of *trans*-enone *589*, suggesting steric hindrance as the determining factor of the reaction rate. The IBX oxidation of *591* by conventional heating at 80°C for 3 hours gave a mixture of the intermediate enone (which was readily converted into *592* (*endo*/*exo* = 4:1) by further heating at 80°C. This reaction proceeded very cleanly under microwave heating conditions (MeOH, 160°C, 4 h), although the *endo*/*exo* selectivity was moderate (3:1).

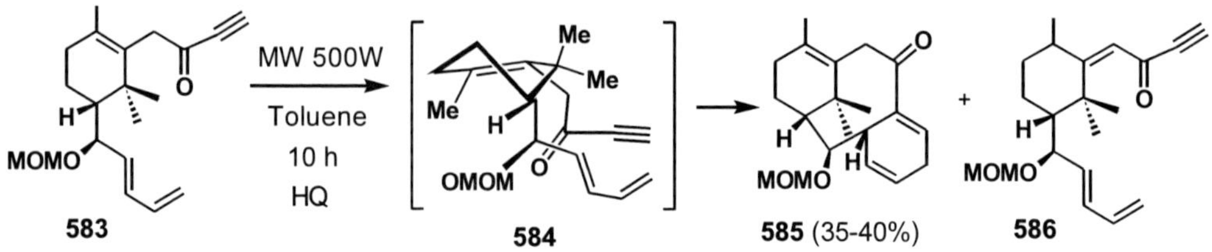

Scheme 112.

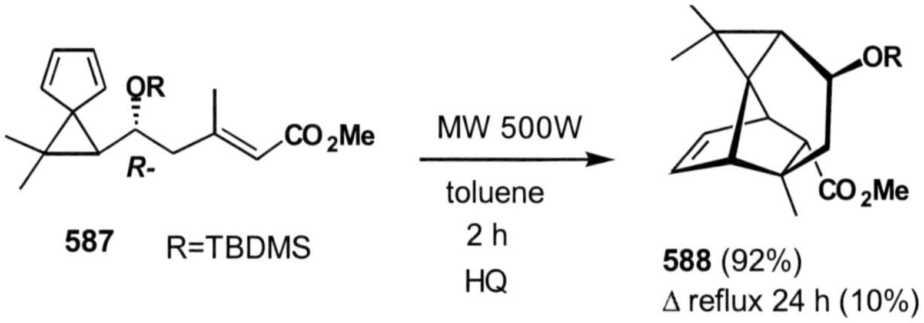

Scheme 113.

Kajiwara [41] used intramolecular 4+2 cycloaddition approach to synthetize compactin. IMDA of substrate *583a* was initially carried out by conventional heating at 130°C for 10 h in o-xylene (Scheme 115). The *exo-* adduct *584a* was obtained with high stereoselectivity, alongside small amounts of the *endo-* adduct *585a* and another diastereomer *585b*. The formation of *585b* occurs by isomerisation from *584a*, which was proven by heating of *584a* in xylene affording *585b*. In microwave conditions, cycloaddition of the malate *583a* (irradiated for 2x9 min) in an open vessel at temperatures below 130°C, was drastically accelerated. The carboxylic acid *583c* irradiated for 5x9 min give *exo-* adduct *584c* as the major product.

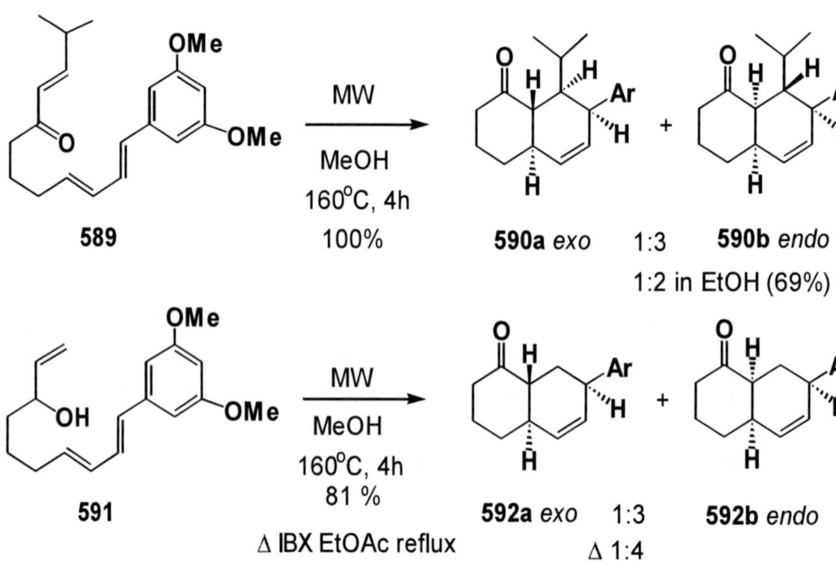

Scheme 114.

Total synthesis of (±)-strychnine carried out by Padwa [42] features intramolecular domino [4+2] cycloaddition/rearrangement cascade of 2-amidofurans. Furanylindole *597* with the presence of the large *o*-methylbenzyl group on the amido nitrogen atom causes the reactive *s-trans* rotamer to be more highly populated and promotes the [4+2] intramolecular cycloaddition. The cycloaddition/rearrangement cascade of *597* was remarkably efficient given that two heteroaromatic systems are compromised in the reaction. Thus, heating a sample of *597* at 150°C (toluene) in a microwave reactor for 30 min in the presence of catalytic amount of MgI_2 afforded aza-tetracycle *600* in 95% yield (Scheme 116). This reaction cascade can be rationalized by a nitrogen-assisted ring opening of the initially formed cycloadduct *598* to produce *N*-acyliminium ion *599*. A subsequent deprotonation followed by ketonization of the resulting enol accounts for the formation of *600*.

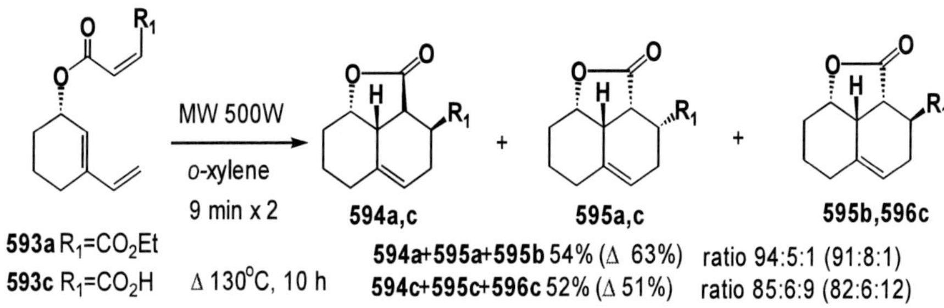

Scheme 115.

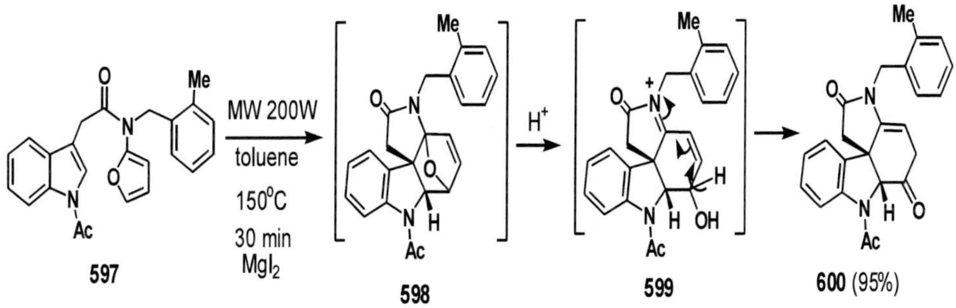

Scheme 116.

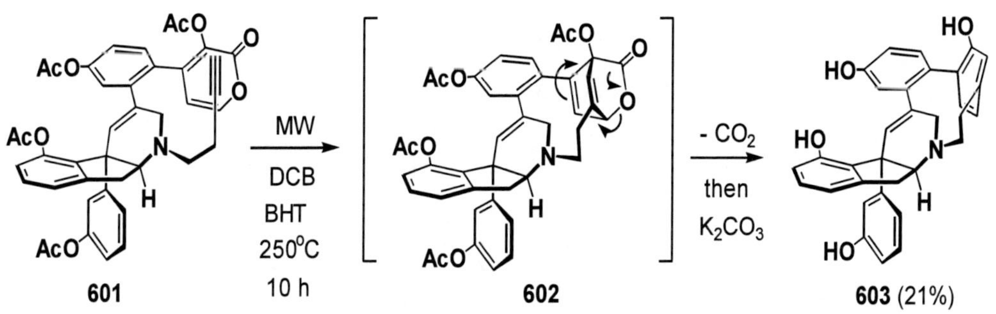

Scheme 117.

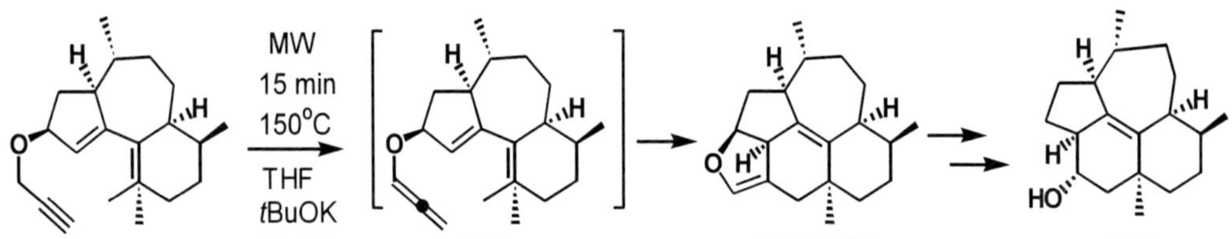

Sheme 118.

Intramolecular Diels-Alder reaction was employed as a key-step in total synthesis of alkaloid (±)-Haouamine A *603*. Baran [43] converted acetylene precursor *601* into *603* upon microwave irradiation in dichlorobenzene at 250°C, followed by basic acetate hydrolysis (Scheme 117). The use of fully acetylated precursor *601* was essential to minimize thermal decomposion of cycloadduct *602*. Remarkably, the macrocyclization event proceeded with high atropselectivity in favor of *602* (10:1). Synthetic (±)-haouamine A *603* was isolated in 21% yield.

Enantioselective synthesis of 4-desmethyl-3α-hydroxy-15-rippertene *607* was carried by Metz [44] *via* the intramolecular [4+2] cycloaddition of allene as dienophilic partner (Scheme 118). Construction of the tetracyclic rippertene core was achieved by an efficient intramolecular Diels-Alder reaction from the cyclization precursor *604*. Compound *604* firstly underwent isomerization under basic conditions to give the corresponding allenyl ether *605*, which then cyclized under microwave irradiation to yield the enol ether *606* with complete diastereoselectivity.

References

[1] Padwa, A.; Brodney, M. A.; Dimitroff, M. *J. Org. Chem. 1998*, *63*, 5304-5305.

[2] Takano, S.; Kijima, A.; Sugihara, T.; Satoh, S.; Ogasawara, K. *Chem. Lett. 1989*, 87-88.

[3] Mihovilovic, M. D.; Leisch, H. G.; Mereiter, K. *Tetrahedron Lett. 2004*, *45*, 7087-7090.

[4] Fischer, T. C. M.; Leisch, H. G.; Mihovilovic, M. D. *Monatsch. Chem. 2010*, *141*, 699-707.

[5] Wang, W.-B.; Roskamp, E. J. *Tetrahedron Lett. 1992*, *33*, 7631-7634.

[6] Karaarslan, M.; Gokturk, E.; Demircan, A. *J. Chem. Res. 2007*, 117-120.

[7] Mance, A. D.; Jakopčić, K. *Mol. Diversity 2005*, *9*, 229-232.

[8] Petronijevic, F.; Timmons, C.; Cuzzupe, A.; Wipf, P. *Chem. Commun. 2009*, 104-105.

[9] Presset, M.; Coquerel, Y.; Rodriguez, J. *Org. Lett. 2009*, *11*, 5706-5709.

[10] Santra, S.; Andreana, P. R. *Org. Lett. 2007*, *9*, 5035-5038.

[11] Wu, J.; Sun, L.; Dai, W.-M. *Tetrahedron 2006*, *62*, 8360-8372.

[12] Wu, J.; Yu, H.; Wang, Y.; Xing, X.; Dai, W.-M. *Tetrahedron Lett. 2007*, *48*, 6543-6547.

[13] Steinhardt, S.; Vanderwal, C. D. *J. Am. Chem. Soc. 2009*, *131*, 7546-7547.

[14] Yoon, H. J.; Lee, J. M.; Im, S. S.; Chae, W. K. *Bull. Korean Chem. Soc. 2001*, *22*, 543-544.

[15] Shanmugasundaram, M.; Manikandan, S.; Raghunathan, R. *Tetrahedron 2002*, *58*, 997-1003.

[16] Jayashankaran, J.; Manian, R. D. R. S.; Raghunathan, R. *Tetrahedron Lett. 2006*, *47*, 2265-2270.

[17] Manian, R. D. R. S.; Jayashankaran, J.; Raghunathan, R. *Tetrahedron Lett. 2007*, *48*, 1385-1389.

[18] Manian, R. D. R. S.; Jayashankaran, J.; Raghunathan, R. *Synlett 2007*, 874-880.

[19] Ramesh, E.; Raghunathan, R. *Tetrahedron Lett. 2008*, *49*, 1812-1817.

[20] Ramesh, E.; Raghunathan, R. *Synth. Commun. 2009*, *39*, 613-625.

[21] Ramesh, E.; Vidhya, T. K. S.; Raghunathan, R. *Tetrahedron Lett. 2008*, *49*, 2810-2814.

[22] Jacob, R. G.; Gelson Perin, G.; Botteselle, G. V.; Lenardão, E. J. *Tetrahedron Lett. 2003*, *44*, 6809-6812.

[23] Lenardão, E. J.; Mendes, S. R.; Ferreira, P. C.; Perin, G.; Silveira, C. C.; Jacob, R. G. *Tetrahedron Lett. 2006*, *47*, 7439-7442.

[24] Van der Eycken, E.; Appukkuttan, P.; De Borggraeve, W.; Dehaen, W.; Dallinger, D.; Kappe, C. O. *J. Org. Chem. 2002*, *67*, 7904-7907.

[25] Kaval, N; Van der Eycken, J.; Caroen, J.; Dehaen, W.; Strohmeier, G. A.; Kappe, C. O.; Van der Eycken, E. *J. Comb. Chem. 2003*, *5*, 560-568.

[26] Motorina, I. A.; Fowler, F. W.; Grierson, D. S. *J. Org. Chem. 1997, 62,* 2098-2105.

[27] Lindsley, C. W.; Wisnoski, D. D.; Wang, Y.; Leister, W. H.; Zhao, Z. *Tetrahedron Lett. 2003*, *44*, 4495-4498.

[28] Zhao, Z.; Leister, W. H.; Strauss, K. A.; Wisnoski, D. D.; Lindsley, C. W. *Tetrahedron Lett. 2003*, *44*, 1123-1127.

[29] Lahue, B. R.; Lo, S.-M.; Wan, Z.-K.; Woo, G. H. C.; Snyder, J. K. *J. Org. Chem. 2004*, *69*, 7171-7182.

[30] Shao, B. *Tetrahedron Lett. 2005*, *46*, 3423-3427.

[31] Motorina, I. A.; Fowler, F. W.; Grierson, D. S. *J. Org. Chem.* 1*997*, *62*, 2098-2105.

[32] Fontaine, P.; Chiaroni, A.; Masson, G.; Zhu, J. *Org. Lett. 2008*, *10*, 1509-1512.

[33] Hajbi, Y.; Suzenet, F.; Khouili, M.; Lazar, S.; Guillaumet, G. *Synlett 2009*, 92-96.

[34] Hajbi, Y.; Suzenet, F.; Khouili, M.; Lazar, S.; Guillaumet, G. *Tetrahedron 2007*, *63*, 8286-8297.

[35] Nieto-Oberhuber, C.; Pérez-Galán, P.; Herrero-Gómez, E.; Lauterbach, T.; Rodríguez, C.; López, S.; Bour, C.; Rosellón, A.; Cárdenas, D. J.; Echavarren, A. M. *J. Am. Chem. Soc. 2008*, *130*, 269-270.

[36] Du; H.; He, Y.; Sivappa, R.; Lovely, C. J. *Synlett 2006*, 965-992.

[37] Lu, Y.-F.; Fallis, A. G. *Tetrahedron Lett. 1993*, *34*, 3367-3370.

[38] Lei, B.; Fallis, A. G. *J. Am. Chem. Soc. 1990*, *112*, 4609-4610; Lei, B.; Fallis, A. G. *J. Org. Chem. 1993*, *58*, 2186-2195.

[39] Lei, B.; Fallis, A. G. *J. Org. Chem. 1993*, *58*, 2186-2195.

[40] Born, S.; Bacani, G.; Olson, E. E.; Kobayashi, Y. *Synlett 2008*, 2877-2881.

[41] Takatori, K.; Hasegawa, K.; Narai, S.; Kajiwara, M. *Heterocycles 1996*, *42*, 525-528.

[42] Zhang, H.; Boonsombat, J.; Padwa, A. *Org. Lett. 2007*, *9*, 279-282.

[43] Baran, P. S.; Burns, N. Z. *J. Am. Chem. Soc. 2006*, *128*, 3908-3909.

[44] Henning, R.; Metz, P. *Angew. Chem., Int. Ed. 2009*, *48*, 1157-1159.

Chapter 2

1,3-DIPOLAR CYCLOADDITIONS

INTRODUCTION

Importance of Huisgen 1,3-dipolar reactions [1,2] in organic synthesis is especially obvious for the effective construction of five-membered heterocycles. [3] In a simple reaction of dipole with dipolarophile, a several bonds undergo transformation during the reaction to obtain products which are considerably more complex than the reactants. By this process, it is possible to generate several contigous stereocenters in one synthetic operation. The asymmetric versions of 1,3-dipolar cycloadditions are also covered within this chapter, with the main body of work is dealing with cycloadditions of azides, [4] nitrones, [5] nitrile oxides, [6] nitrile imines, azomethine ylides, [7] carbonyl ylides and diazoalkanes, [8] with some examples of miscellaneous mesoionic heterocycles [9] such as münchnones (1,3-oxazolium-5-olates) and sydnones (1,2,3-oxadiazolium-5-olates). Microwave methodology was successfully applied to generate 1,3-dipoles and to promote the subsequent cycloaddition and usually avoids harsh reaction conditions. [10] This nonconventional energy source is able to reduce chemical reaction times and to increase yields and in some cases can lead to different outcomes from those obtained with conventional heating. [11] The chapter describes all developments in the area, from the early accounts, to the most recent results, with particular emphasis on aspects related to efficiency and to regio- and stereoselectivity and an eco-friendly approach to five-membered heterocycles.

2.1. AZIDES

Introduction

Reactions of azides and alkynes represents an important group of 1,3-dipolar cycloadditions, which is often called 'click' reaction. [12] In the course of 1,3-dipolar cycloadditions of azides and alkynes triazole rings were formed, and when nitriles are used as dipolarophiles, tetrazoles are obtained. [13] Apparently, there is only one literature example of 1,3-dipolar cycloadditions of azides with alkenes yielding triazoline rings as transient species, which may be due to thermal instability of triazolines. [14] Triazolines could be conventionally synthetized by reaction of azides and alkenes under high pressure at room temperature. [15] In 2002, Tornøe and Sharpless [16] independently reported copper-catalyzed versions of ‘click’ reaction that often result in quantitative yields under very mild reaction conditions. Sharpless method was subsequently transferred to microwave conditions with numerous variations. [17, 18] The body of the work carried out in microwave conditions is dealing with the synthesis of triazole rings, while the syntheses of tetrazoles are ocassionally reported. There are also several literature examples of the intramolecular version of ‘click’ reaction. This reaction was employed in synthesis of natural products, drugs, and peptidomimetics for use in biotechnology and material sciences.

2.1.1. Synthesis of Triazoles

First account on microwave assisted 1,3-dipolar cycloaddition of azides and alkynes was published by Loupy [19] in 1998. It was demonstrated that β-functionalized alkyltriazoles can be efficiently prepared under solvent-free conditions with microwave activation. In cycloadditions of azides and poorly activated acetylenes or enamines, regardless of conditions employed, mixtures of regioisomers *3* and *4* were obtained. The reaction time and yields were generally improved by the reaction carried out in microwave solvent-free conditions (Scheme 1). The specific effect of the irradiation is dependent on the dipolarophilic nature of reactants: it is absent for the most activated acetylene *2c*, and become very important for the less activated ones (*2a*, *2d* and *2e)*. This specific (non-purely thermal) microwave effect on reactivity is probably due to the polarity and reactivity of these dipolarophiles. Selectivities *3*:*4* are not affected by the change heat source. The reversibility of the cycloaddition was demonstrated when isomer *4* was heated (10 min, 160°C), obtaining a mixture of *3*:*4* (63:38). Cycloadditions of azides with substituted enamines *5a,b* as dipolarophiles are also effective, giving corresponding triazoles *3f* and *4f* after the *in-situ* elimination of tetrahydropyrrole.

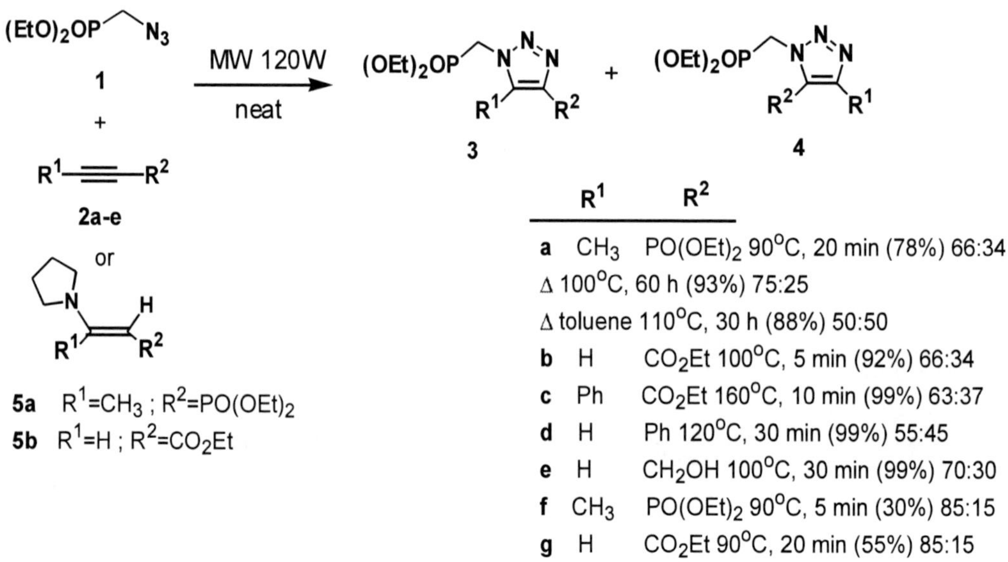

Scheme 1.

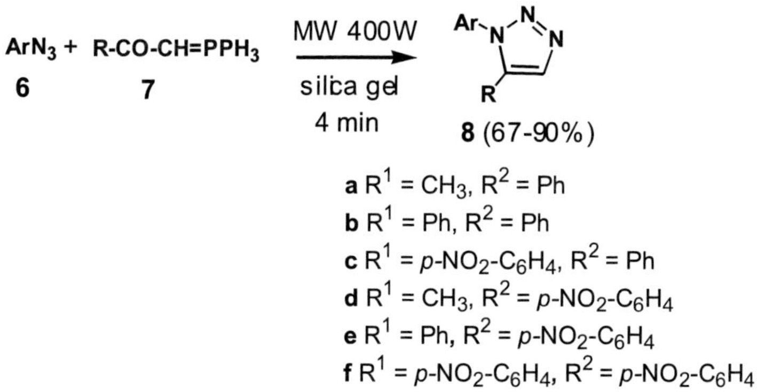

Scheme 2.

Another early example of microwave enhanced dipolar cycloaddition of azides is given by Tao. [20] Reaction is solvent-free and regioselective, wherein microwave promoted rapid synthesis of 1-aryl-1,2,3-triazoles *8* from aryl azides *6* and α-keto phosphorus ylides *7* on the silica gel support (Scheme 2).

Katritzky [21] showed that acetylenic amides *11* or *12* act as good dipolarophiles in 1,3-dipolar cycloaddition with organic azides *9* or *10* under solvent-free microwave irradiation. The corresponding *N*-substituted *C*-carbamoyl-1,2,3-triazoles *13-16* were produced in good to excellent yields (Scheme 3). Reactions of to *N*-benzyl-2-propynamide *11* required mild heating, while 1-piperidino-2-propyn-1-one *12* is less dipolarophilic, and reactions were carried out at higher temperature. Triazoles were formed as a mixture of two regioisomers *13a-16a* and *13b-16b*, where regioisomers *13a-16a* predominate in 3:1 ratio. In these conditions, classical heating experiments failed to produce any product.

Katritzky [22] extended the use of 1,3-dipolar cycloadditions of organic diazides to ester or benzotriazolylcarbonyl activated di-acetylenic amides. Employing microwave method substituted triazoles were synthetised by the 1,3-dipolar cycloaddition of azides with acetylenes and bis-triazoles from mono-azides and di-acetylenes or di-azides and mono-acetylenes. For instance, microwave reaction of benzyl azide *9* with bis-carbamoylpropiolate *17a* gave a regioisomeric mixture of bis-triazoles *18a'* and *18a''* (Scheme 4). Analogous reactions of azide *9* with bis-carbamoylpropiolates *17b-d* gave the corresponding bis-triazoles *17b-d* as the major regioisomers. Accordingly, microwave reactions of 3-(azidomethyl)-3-methyloxetane *19* with bis-carbamoylpropiolates *17a* or *17d* furnished the bis-triazoles *20a* and *20b* as the major isomers in moderate yields.

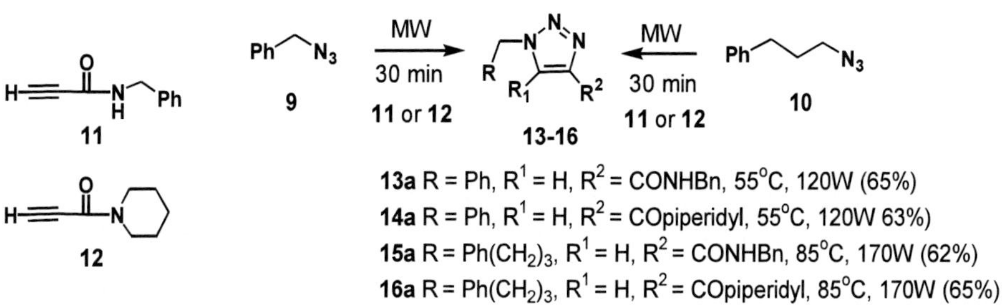

Scheme 3.

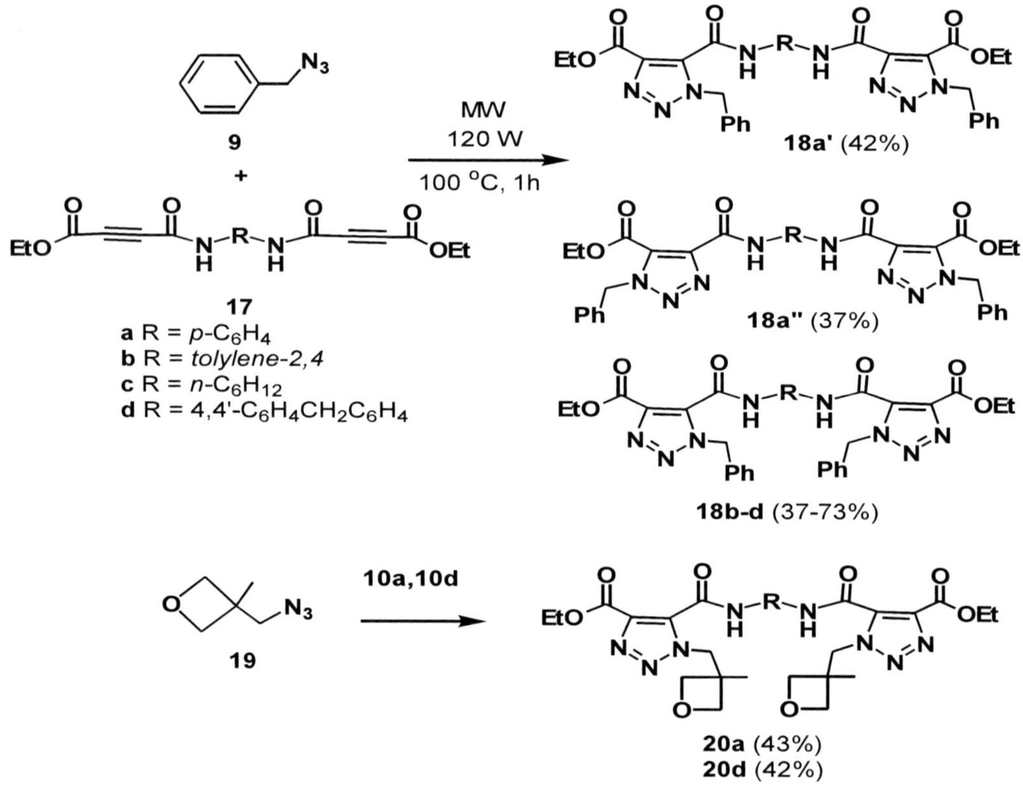

Scheme 4.

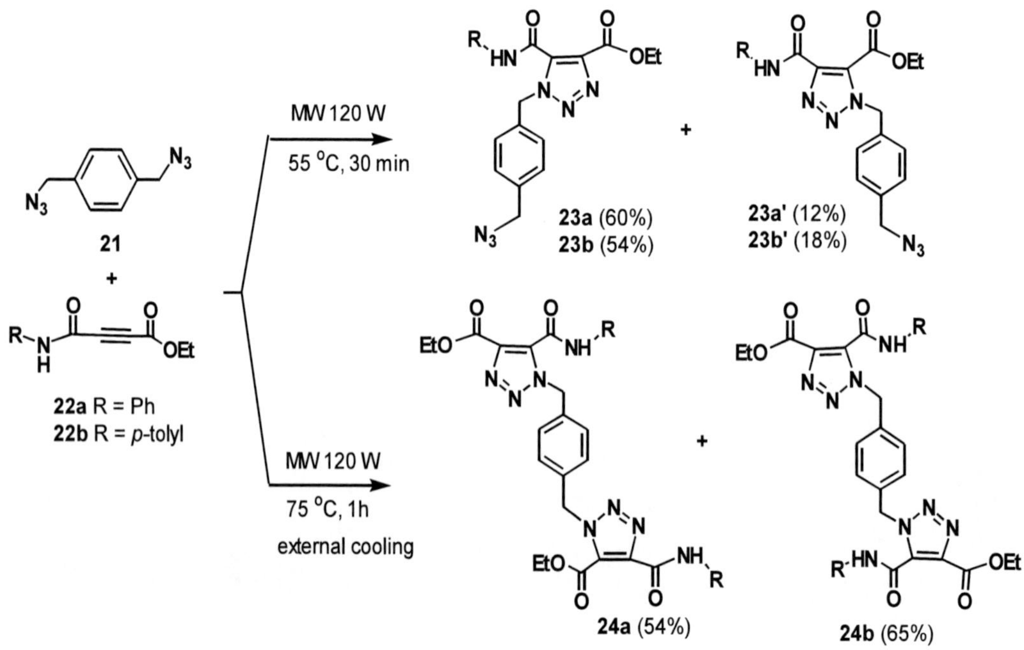

Scheme 5.

The other synthetic route to produce bis-triazoles is the 1,3-dipolar cycloaddition of di-azides and mono-acetylenes. Under mild reaction conditions, cycloaddition could be stopped at the 1:1 stage. Microwave reactions of bis-azide *21* with carbamoylpropiolates *22a* or *22b* at 55°C resulted in 1,3-dipolar cycloaddition of only one of the azido functionalities to give the regioisomeric mixtures of azido-triazoles *23a* and *23a'* or *23b* and *23b'* (Scheme 5). The use of a microwave synthesizer equipped with simultaneous irradiation and external air-cooling system afforded different results. The reaction of 1,4-bis(azidomethyl)benzene *21* with ethyl 4-anilino-4-oxo-2-butynoate *22a* in toluene under continuous microwave irradiation with simultaneous cooling at 75°C furnished a mixture of regioisomeric bis-triazoles; the major regioisomers were *24a* and *24b*.

Interestingly, the presence of a benzotriazolylcarbonyl group on the acetylene provides the required activation for 1,3-dipolar cycloaddition with an organic azide. When benzyl azide *9* was refluxed with *N*-propioloylbenzotriazole *25* in acetone, the benzotriazolylcarbonyl substituted 1,2,3-triazole *26* was formed. Alternatively, the microwave reaction of benzyl azide with *25* provided *26* in an improved yield of 75% (Scheme 6).

Microwave azide click chemistry was also employed by Katritzky [23] for formation of triazoles under mild conditions at low temperature. The advantage of this protocol is the fact that traditional synthesis of triazoles by heating has a problem with thermally labile molecules. Under microwave irradiation, terminal acetylene carboxamides *27* (electron-withdrawing acetylenes) react with azides *28* at 55°C without solvent for 30 min to give mixture of regioisomers *29a,b* in high yields (Scheme 7). Identical conditions were used for the reaction of 4-amido-4-oxo-2-butynoate and benzyl azide to regioselectively produce triazole *29a* (R^1=H, R^2=p-$CH_3C_6H_4CH_2$, R^3=Bn, R^4=CO_2Et, in 73% yield).

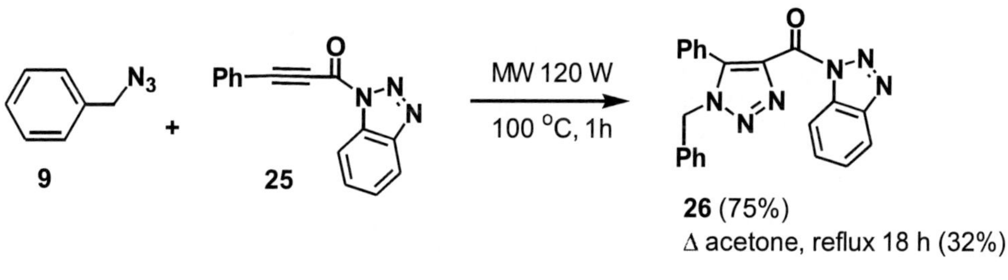

Scheme 6.

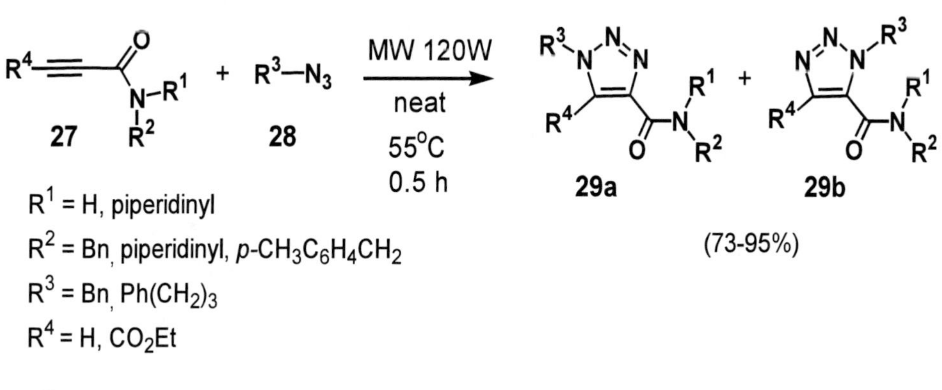

Scheme 7.

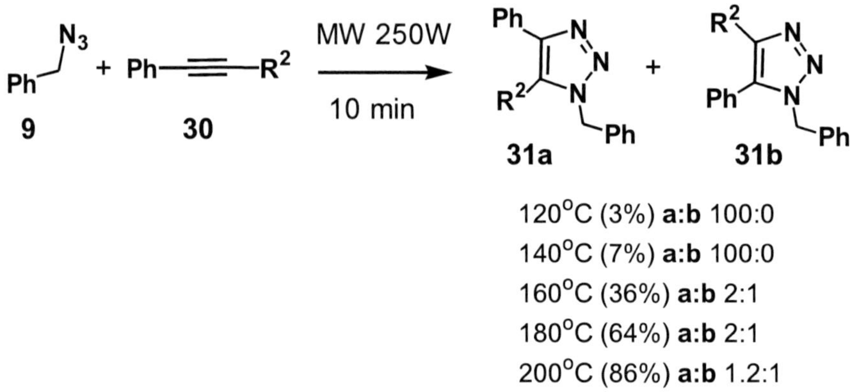

Scheme 8.

In detailed study of the microwave synthesis of triazoles *31*, Savin [24] has found that the rate accelerated, but also the reaction of benzyl azide and acetylene *30* generated fewer side products (Scheme 8). Using microwave energy at elevated temperatures in minutes were obtained yields that took conventional heating at least 24 hours. As the temperature was increased, a slight deterioration in the regioisomeric ratio of product triazoles was observed.

Concentration has a profound effect on the outcome of the reaction: on going from 0.33 M solution in toluene to 1.0 M, and neat reaction, yields are improving (to 99% for neat), while the regioselectivity ratio *a:b* also increases in favor of *a*. It was also found that solvent changes can drastically alter the reaction outcome. The use of more polar solvents like ethanol did improve the reaction performance in some cases, however, the use of ethanol in combination with acetylenic dipolarophiles possessing methyl ester substituents led to complications as a result of transesterification, and the use of methanol as solvent did not lead to any significant improvements. Dichloroethane improved the reaction rate (74%), while THF performs poorly (53%). When reaction of *9* with DMAD was exposed to solvent-free microwave irradiation (200W, 120°C) by suspending the material in Montmorillonite (K-10) clay, high yields of 89-93% were obtained.

One of the examples where reactants were used as solvent, i.e. solvent free reaction, with rarely used silyl substituted azides is published by Kheradmand. [25] Fast one-pot microwave-assisted (60 W, 10 min) solvent free synthesis of simple alkyl 1,2,3-triazole-4-carboxylate derivatives *35-37* was achieved by 1,3-dipolar cycloaddition reactions with trimethylsilyl azide on the alkylpropyolates *32,33* and DMAD *34* in high yields (Scheme 9).

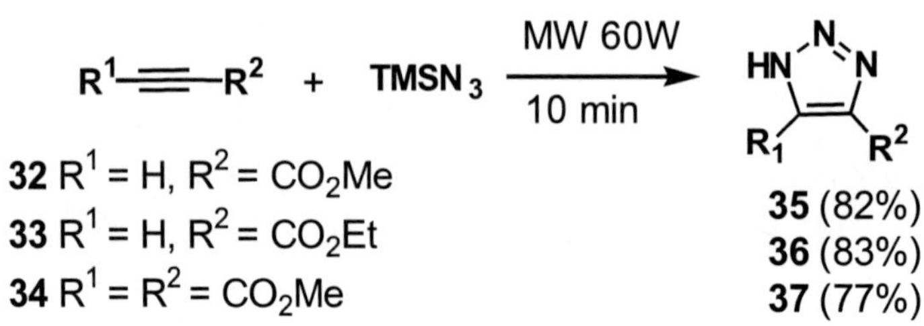

Scheme 9.

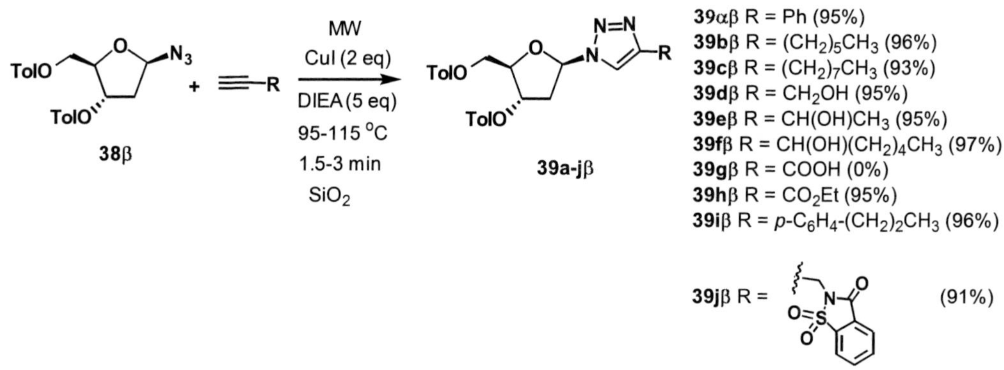

Scheme 10.

Immediately after the discovery, the reaction conditions of the copper catalyzed 1,3-dipolar reactions of azides were transferred to microwave settings. Today copper catalyzed 1,3-dipolar reactions represent the majority of publications in the field, giving products from simple triazoles to heavily functionalized complex structures. It was found that the presence of copper catalyst is essential for success of reaction, and the best conditions and catalysts were sought after, for instance by Benhida [26] in microwave-assisted solvent free synthesis of α- and β-2'-deoxy-1,2,3-triazolyl-nucleosides *39β* (Table 1). Microwave irradiation significantly reduced reaction times required for conventional heating, while the replacement of solvent by SiO_2 as a solid support further increased the yield of single regioisomer without epimerization of the C1' stereocenter. A series of α- and β-2'-deoxy-1,2,3-triazolyl-nucleosides *39β* was subsequently prepared in optimized reaction conditions (Scheme 10). Reaction between propargylic acid and azido-sugar *38β* did not proceed probably due to the strong chelation of Cu(I) with acid functionality, but proceeds well with other functionalities, including ester, hydroxy and sulfonamide groups.

Table 1. Variation of reaction conditions used for synthesis of 39α

Conven-tional	Heating				MW	Irradi-ation			
Catalyst /Ligand	Solvent	Time	T (°C)	Yield (%)	Catalyst /Ligand	Solvent	Time	T (°C)	Yield (%)
CuI /DIEA	Toluene	24 h	110	87	-	Toluene	5 min	90	0
$CuSO_4$ /NaAsc	H_2O /tBuOH	24 h	rt	86	-	Toluene /DMF	5 min	140	<5
$CuSO_4$ /NaAsc	H_2O /tBuOH	24 h	rt	86	CuI /DIEA	SiO_2	3 min	115	95

Following these results, Benhida [27] carried out efficient synthesis of cytostatic 4-substituted triazolyl-nucleosides *41* starting from protected azido-ribose *40* and ethyl propiolate (Scheme 11). Fast 1,3-dipolar cycloaddition takes place under the cooperative effect of microwave activation and copper(I) catalyst. When performed under solvent free microwave activation, Cu(I) catalysis and SiO_2 support, provided the desired cycloadduct *41a* (R=CO_2Et) stereo- and regioselectively and almost quantitatively. In this cycloaddition, only the β-anomer-4-substituted-triazole *41a* was obtained within a very short reaction time (1 min). The SiO_2 support or Brønsted acid have an important role in the activation of the key 1,3-dipolar cycloaddition step. When the reaction was performed directly on Al_2O_3 or sand (a poorly hydroxylated silicate) *41a* was obtained with yields comparable to those obtained with SiO_2 as support. Classical thermal reactions in solution required longer times obtaining product in lower yield.

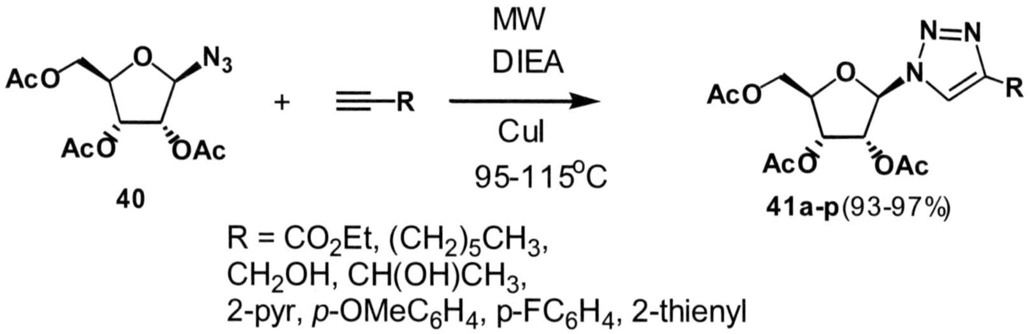

Scheme 11.

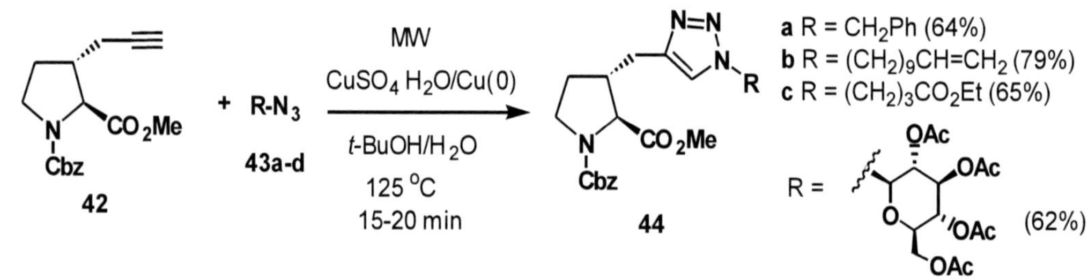

Scheme 12.

Standard reaction conditions for copper-catalyzed 'click' reaction of azides and alkynes are: *t*-BuOH/H_2O solution, $CuSO_4$ (0.7 equiv), copper turnings (0.2 equiv). The synthesis of α-substituted prolines *44* by Cordero [28] has been accomplished by microwave-assisted Huisgen 1,3-dipolar cycloaddition between azides and orthogonally protected α-propynyl proline *42* in the presence of Cu(I). The click chemistry reaction was performed in a microwave reactor with 1:1 mixtures of propargyl proline *42* and the appropriate azide *43a-d* by irradiation at 100 W, at 125 °C for 15-20 min. (Scheme 12).

For preparation of the enantiomerically pure *N*-benzyl triazole amide derivatives *46a-h*, Botta [29] used a three component copper(I)-catalyzed approach: racemic propargylamines (*R* or *S*)-*45a-h*, benzyl azide, in a 1:1 mixture of *t*-BuOH and water, Cu(I) catalyst were irradiated under microwave conditions for 10 min at 125°C. Triazole products (*R* or *S*)-*46a-h*, were formed in a completely regioselective manner, as a 1,4-regioisomer (Scheme 13). The unprotected (*S*)-amines did not gave the triazole derivatives, since the presence of the free amino group proved detrimental to the outcome of the 1,3-dipolar cycloaddition reaction most likely as a consequence of the chelation of the copper ions.

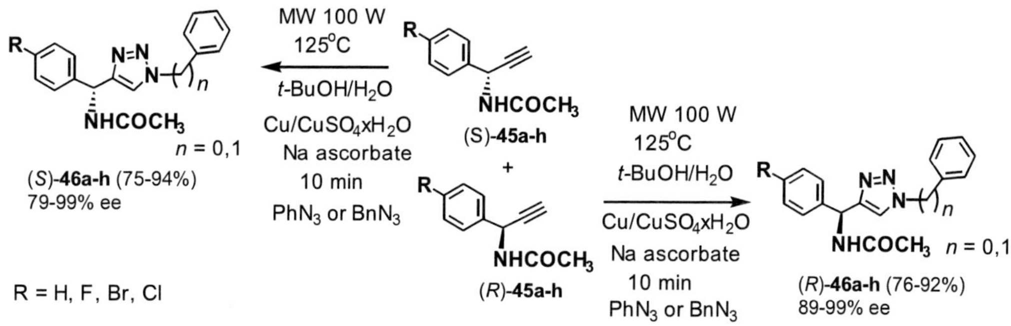

Scheme 13.

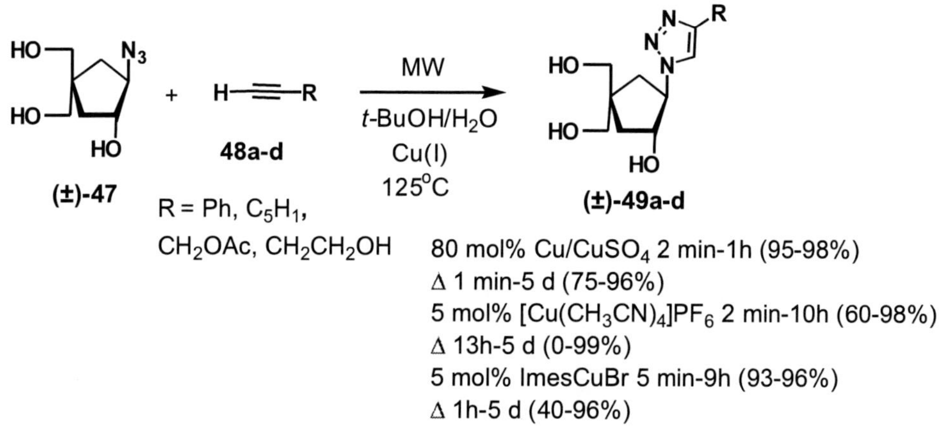

Scheme 14.

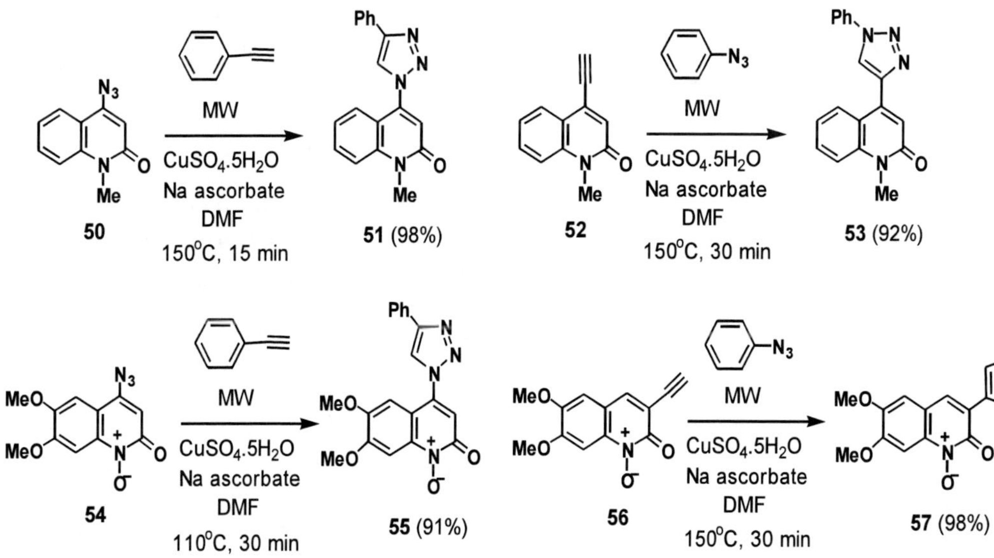

Scheme 15.

Agrofoglio [30] has compared the efficiency three copper (I) catalysts used for the microwave synthesis of 1,4-disubstitued-1,2,3-triazolo-carbanucleosides *49* through 1,3-dipolar cycloaddition of azide *47*: the tetrakis(acetonitrile)copper hexafluorophosphate ($[Cu(CH_3CN)_4]PF_6$), the imidazoline(mesythyl)copper bromide (Imes)CuBr, and the copper/copper sulfate Cu(0)/$CuSO_4$(II) mixture (Scheme 14). The microwave irradiation considerably reduced the reaction time and increased the conversion compared to the thermal results. Polar substituents (acetate and ethanol) reacted faster than triazoles with non-polar substituents (phenyl and pentyne). The catalytic efficiency of Cu(0)/$CuSO_4$(II) and $[Cu(CH_3CN)_4]PF_6$ is comparable in term of percentage of conversion. Reaction times are generally longer with the $[Cu(CH_3CN)_4]PF_6$ catalyst. For a high percentage of conversion, longer times were needed for ImesCuBr catalyzed reactions.

Preparation of 3- and 4-triazolyl-2(1*H*)-quinolones by the 1,3-dipolar cycloaddition reaction substituted 3- and 4-azido-, and 3- and 4-ethynylquinolin-2(1*H*)-ones utilizes the standard $CuSO_4$/sodium ascorbate coupling conditions. Kappe [31] reported that the best conversions and product yields were achieved by using DMF as a solvent at 110°C for 15 min (Scheme 15). Installation of the triazolyl moiety onto the quinolin-2(1*H*)-one core could be achieved by reversing the functionalities of the coupling partners. Hence, the 4-ethynylquinoline-2(1*H*)-one *52* reacted with azide to give the desired isomeric triazolo product *53* in 92% yield. Application of the same reaction conditions to azido quinoline-*N*-oxide *54* and phenylacetylene led to cycloaddition, providing the 6,7-dimethoxy-4-(4-phenyl-1*H*-1,2,3-triazol-1-yl)quinoline 1-oxide *55* in excellent yield. This is the first example of polar *N*-oxide microwave cycloaddition, which reacts nicely since polar species are enhanced by microwave. The preparation of the isomeric triazole-1-oxide *57* was carried out by applying the click cycloaddition of the isomeric quinoline-2(1*H*)-one-1-oxide *56*.

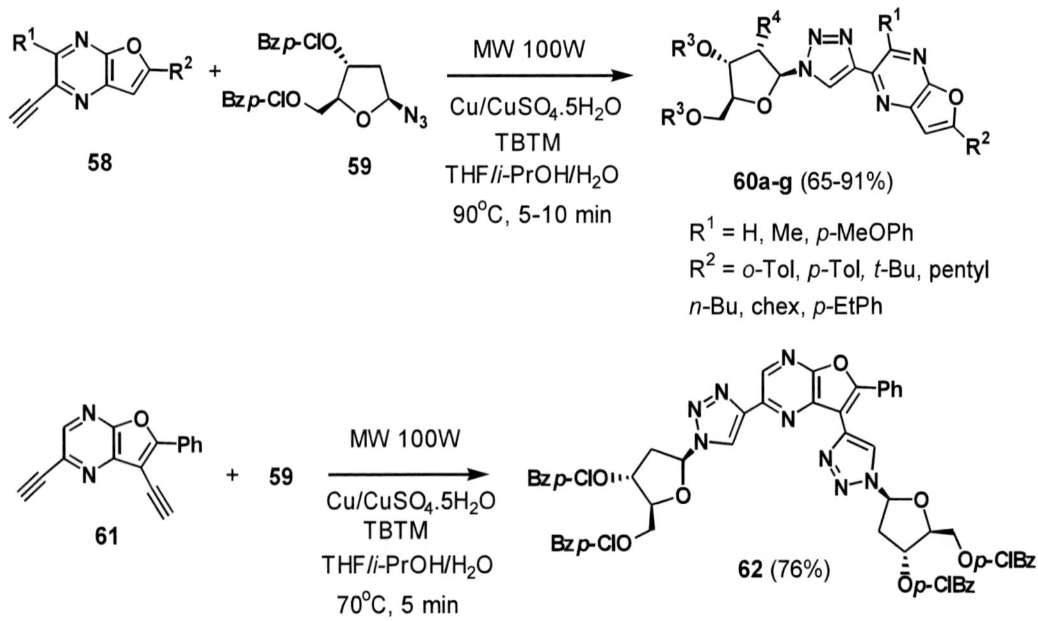

Scheme 16.

Synthesis of furo[2,3-b]pyrazine nucleoside analogues with 1,2,3-triazole linkage was achieved by coupling of 2-ethynylfuro[2,3-b]pyrazines *58* with D-ribose and 2-deoxy-D-ribose *59*. For this purpose, Van der Eycken [32] used a microwave-assisted Cu(I) catalyzed [2+3] dipolar cycloadditions in the presence of *N,N,N*-tris[(1-benzyl-1*H*-1,2,3-triazol-4-yl)-methyl]amine ligand (TBTM) generating a library of new nucleoside analogues (Scheme 16). In the process, full regioselectivity was obtained. Yields refer to overall yields over three steps (TMS protection, Sonogashira coupling, 1,3-dipolar cycloaddition). When 2,7-bisethynyl-6-phenylfuro[2,3-b]pyrazine *61* was coupled with 2-deoxy-D-ribosyl-β-azide *59* upon microwave irradiation, the bis-2-deoxy-*D*-ribose-1,2,3-triazole conjugate *62* was obtained in 76% yield over three steps.

Microwave solvent free regioselective 1,3-dipolar cycloaddition was employed by Santagada [33] in the synthesis of 1,4-substituted 1,2,3-triazoles as amide bond isosters. Fmoc amino-azides *63a-d*, acetylenic amides *64a-c*, and copper (I) iodide were mixed with the basic alumina in a 4:1 ratio and irradiated by microwave. The main advantage of procedure, in comparison to literature, is that a short time of irradiation provided the final compounds *65a-l* in good yields without the presence of the 1,5-isomer. The copper (I) iodide catalyzed microwave assisted thermal synthesis is regioselective and not dependent from steric effects that play a central role in the uncatalyzed thermal reactions where the hindrance of substituents in the amino azide side chain can allow the formation of both the 1,4- and 1,5-substituted triazoles or completely inhibit the cycloaddition leading to 1,5-substituted triazole. If the application of microwave energy was performed for longer time (>20 min) the reagents were not completely transformed in 1,2,3-trizoles due to the decomposition of the starting reagents. In the same way, no further yield increase was evidenced when the irradiation power was increased from 800W to 1600W, which is related to decomposition of the starting materials (Scheme 17).

Van der Eycken [34] applied 'click chemistry' for the functionalization of 2(1*H*)-pyrazinone scaffold. The conditions were optimized for the reaction of the substrate *66*, applying the electron-rich 4-methoxyphenyl azide *67a*. The Cu(I) catalyst was generated *in situ* using copper wire and $CuSO_4{\cdot}5H_2O$ (2 mol%). In standard thermal reaction solvent mixture of *t*-BuOH/H_2O/THF was used, at room temperature, the reaction proceeded slowly, requiring 1.5 days, affording the triazole as the sole product, completely regioselective with only one regioisomer formed. Upon microwave irradiation at 100°C, the reaction time was reduced to 15 min, without the use THF as a cosolvent. The reaction could be carried out in water without influencing the yield but resulting in a slightly longer reaction time. The catalysis by Cu(II) seemed to be critical to stabilize the air-sensitive Cu(I), while the addition of one of the ligands *71* (0.8 mol %) or *72* [35] (1-2 mol %) resulted in an increase of the rate without influence on the yield. Using the optimized procedure, aromatic azides *67b-f* possessing electron-donating and withdrawing substituents were compared, the ones bearing an electron-donating group *67a* and *67c* reacted faster. In the case of diester *67e* lower yields are associated with hydrolysis of the ester groups. Reaction of the pentaerythritol-derived bisazide *69* resulted in the formation of a dimer *70*. All the conversions performed in a *t*-BuOH/H_2O mixture can be carried out in water as the sole solvent, although the reaction time was longer.

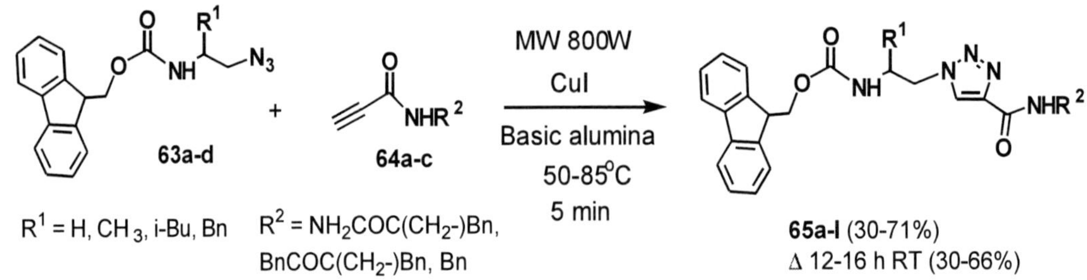

Scheme 17.

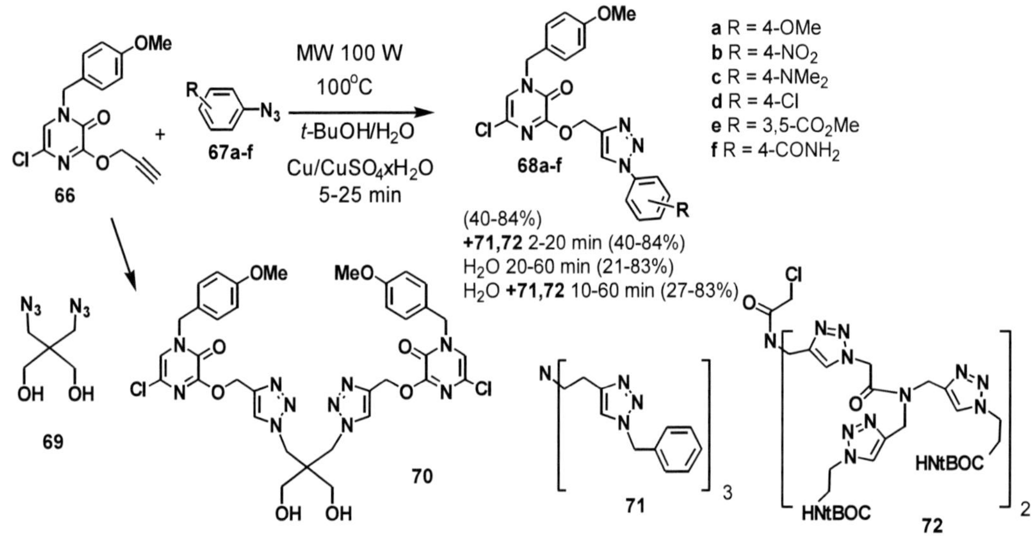

Scheme 18.

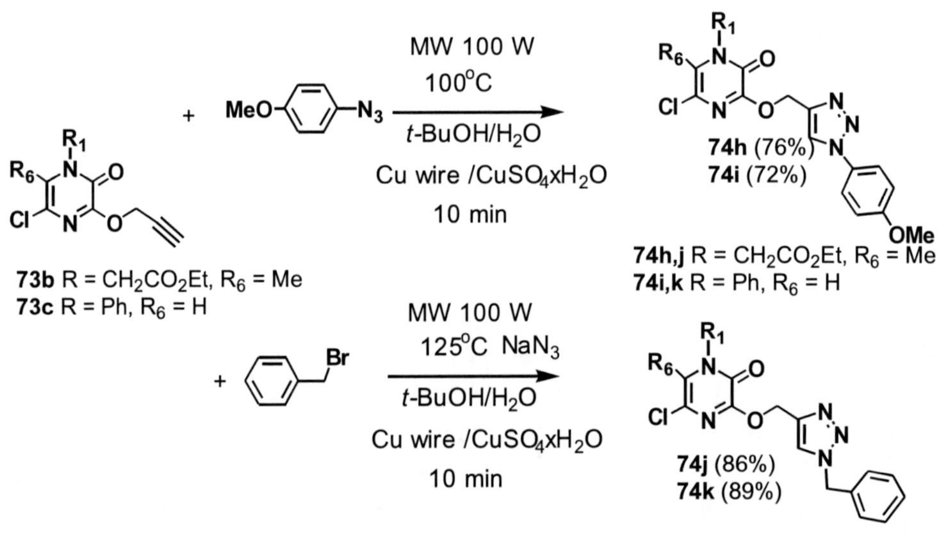

Scheme 19.

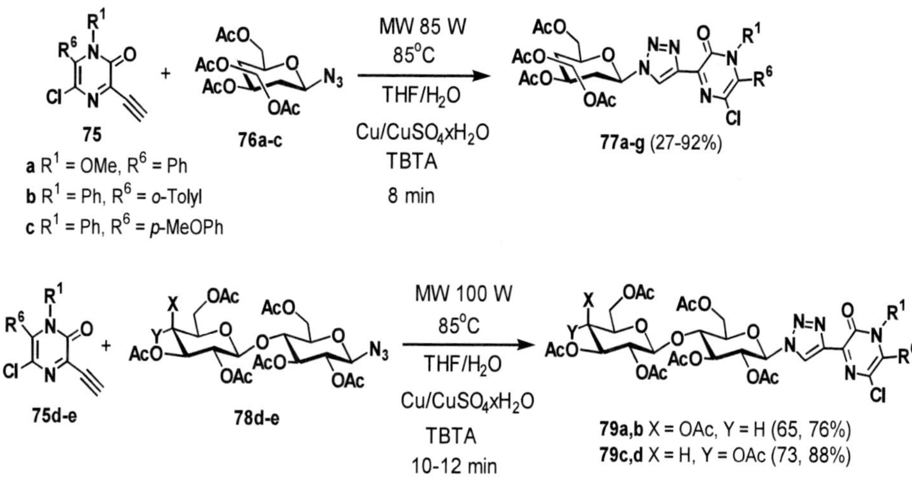

Scheme 20.

Accordingly, the propargyl derivatives of 2-(1*H*)-pyrazinones *73b,c* in the 1,3-dipolar cycloadditions generated the corresponding triazoles *74h-k*. The reactions proceed well in 10 min at 100°C using an irradiation power of 100 W, yielding the products in good yields (Scheme 19). For the preparation of the triazoles *74j,k* microwave-enhanced three-component coupling protocol was employed, with *in situ* generation of benzyl azide from benzyl bromide and sodium azide. Obtained yields were superior to the synthetic method used for preparation of triazoles *74h,i*. For cycloadditions where reactants experience solubility problems, to standard solvent mixture of *t*-BuOH/H_2O, THF was added as a cosolvent.

Microwave enhanced 'click chemistry' has been extensively used in glycopeptidomimetics, Van der Eycken [36] coupled 3-ethynyl-2(1*H*)-pyrazinones with various (oligo)saccharides (glycosyl β-azides *76a-c*) with the Cu/$CuSO_4$ catalyzed reactions, supported by *N,N,N*-tris[1-benzyl-1*H*-(1,2,3-triazol-4-yl)methyl]amine (TBTA) ligand. Cyccloadditions occur with full regioselectivity resulting in the corresponding 1,4-disubstituted 1,2,3-triazoles *77a-g* (Scheme 20). When microwave irradiation at 100°C was applied, reaction time significantly shortens, but obtained yields were lower, due to polymerization. Employing optimized reaction conditions, the Cu(I) catalyzed 1,3-dipolar cycloaddition reactions of 2,3,6,2',3',4',6'-hepta-*O*-acetyl-β-azido-lactose *78d* and 2,3,6,2',3',4',6'-hepta-*O*-acetyl-β-azido-cellobiose *78e* produced 1,4-disubstituted 1,2,3-triazoles *79a-d*.

Brimble [37] has effectively employed microwave-enhanced 'click' glycoconjugation of a propargylated GalNAc$(OAc)_3$ *80* sugar moiety with an azido-functionalized amino acid, Fmoc-L-azidoalanine *81* (or multiazido functionalized peptides) using a catalytic quantity of Cu(I) enabling a high-yielding and rapid synthesis of biologically relevant neoglycopeptides *82* (Scheme 21). Catalyst loadings of 1-5 mol % in presence of sodium ascorbate in H_2O/tBuOH are standard conditions for click reactions at room temperature. However, higher catalyst loadings of 10 mol % are used for microwave reactions. [38,67] A considerably lower catalyst loading than previously reported was found to be sufficient for transformation (even 2 mol % Cu catalyst and 5 mol % sodium ascorbate). Full conversion took place within 30 min. $CuSO_4{\cdot}5H_2O$ in H_2O/*t*BuOH gave 42% conversion, whereas an excellent yield of 88% was obtained using DCM as cosolvent. Other catalyst systems such as $(EtO)_3PCuI$ or CuI did not result in high conversion. Next, azidoalanine incorporated into the peptides *83a* and *83b* was submitted to click reaction with *80* under optimized conditions, with 1.5 equiv of *80* per azide group. Full conversion was obtained in less than 10 min. Similarly, *83b* was converted to *84b* with 20 mol% $CuSO_4 \cdot 5H_2O$ and 50 mol% sodium ascorbate. Such conventional 'click' reactions on peptide [39] and peptoid [40] backbones were usually performed with large excesses of copper.

Microwave-assisted click polymerization was used by Liskamp [41] for the synthesis of Aβ(16–22) cyclic oligomers. The synthesis of the cyclic oligomers was performed *via* a microwave-assisted Cu(I)-catalyzed 1,3-dipolar cycloaddition reaction of azido-Lys-Leu-Val-Phe-Phe-Ala-Glu-propargyl amide *85* as the monomer. The formation of cyclic oligomers, up to pentamers (35 amino acid residues), was verified by MS analysis and the individual cyclic monomer *86* (an intramolecular cycloadduct) and dimer *87* were isolated as a major products (Scheme 22).

Pedersen [42] showed that modification of unprotected oligonucleotides could be achieved via 1,2,3-triazole linkers by microwave-accelerated click chemistry of pyren-1-yl and benzyl azide with oligonucleotides incorporating internal insertions of (*R*)-3-O-{4-[1-(pyren-1-yl)-1*H*-1,2,3-triazol-4-yl]benzyl}glycerol, and (*R*)-3-O-{4-[1-(benzyl)-1*H*-1,2,3-triazol-4-yl]benzyl}glycerol (Scheme 23). Thus, the 1,3-dipolar cycloaddition between *89* and pyren-1-yl azide *88* in DMSO/H_2O in the presence of triethylammonium acetate buffer (TEAA) was achieved by microwave irradiation at 70°C for two minutes to obtain oligonucleotide *90a* in 45% yield. Identical reaction conditions were employed for a series of oligonucleotides, and for reactions with benzyl azide, producing cycloadducts *90b-i* in 20-36% yields.

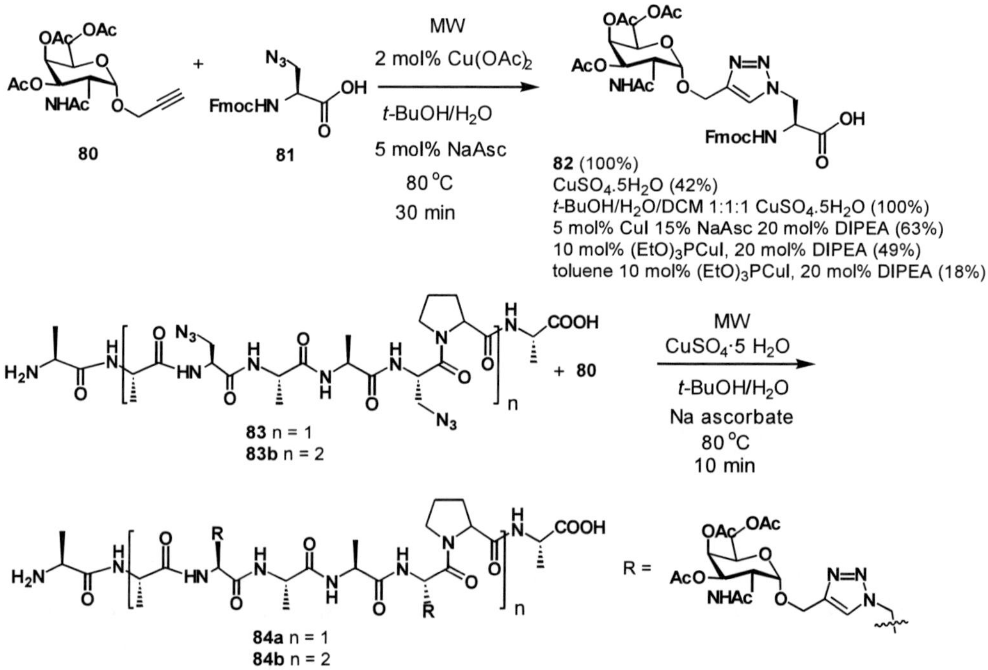

Scheme 21.

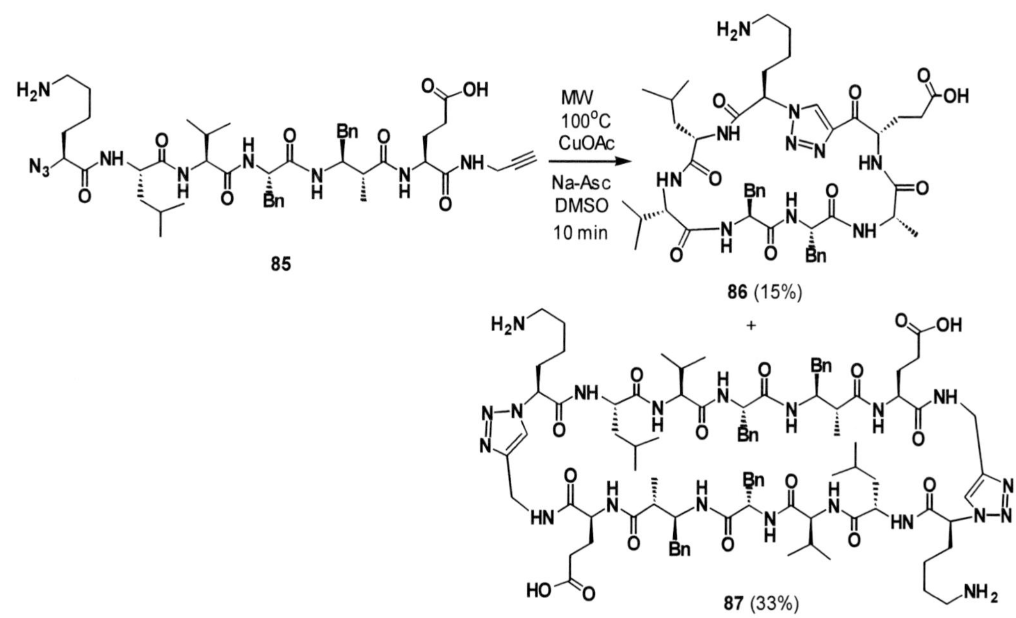

Scheme 22.

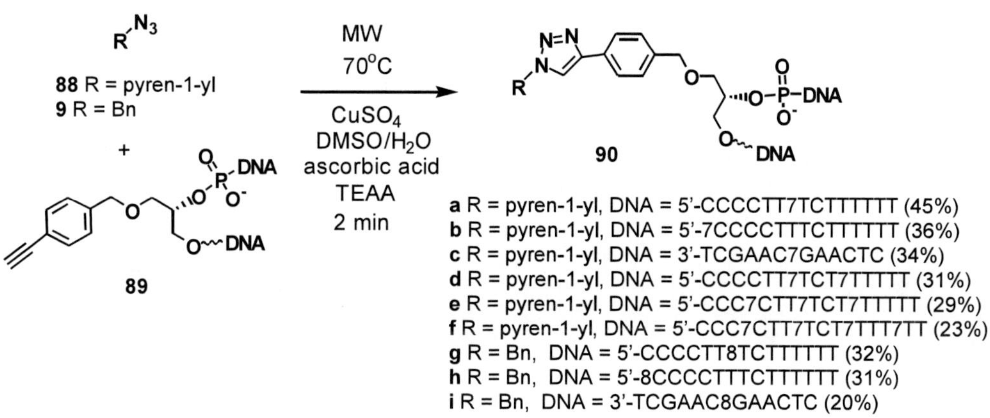

Scheme 23.

Microwave assisted click chemistry on solid support could be used for the synthesis of multiple labeled-carbohydrate oligonucleotides, as demonstrated by Morvan. [43] Solid-supported synthesis of a T12 trispropargyl phosphoramidate oligonucleotide *92* and its conjugation to galactosyl azide *91* by microwave-assisted click chemistry afforded trigalactosylated glycoconjugate *93* with labelled dodecathymidine (T12*) (Scheme 24). Microwave activation significantly improved the reaction kinetic compared to standard conditions (classical, room temp. 420 min, 73%). Subsequent treatment of cycloadduct *93* with aqueous ammonia afforded the fully deprotected trigalactosylated product. Solid-supported reactions provided better results in terms of purity when compared to similar solution phase conditions where some phosphoramidate hydrolysis was observed.

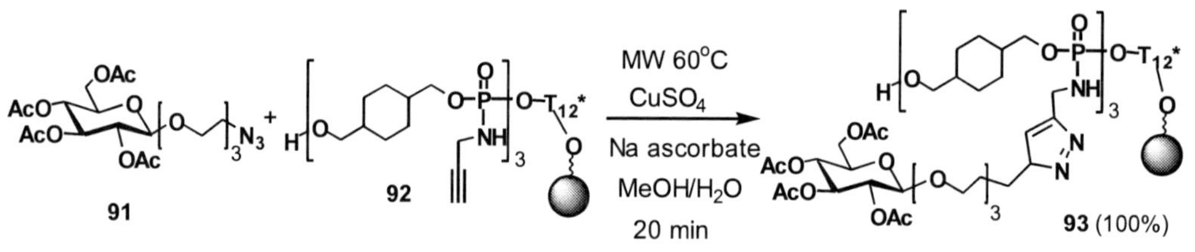

Scheme 24.

The Cu(I)-catalyzed alkyne-azide cycloaddition could be iteratively applied to synthetise oligotriazole products, such as triazolomannoses reported by Dondoni. [44] For the synthesis of 1,2,3-triazolo-linked (1,6)-α-D-oligomannoses, the *O*-perbenzylated ethynyl α-*C*-mannoside acetylene *94* was reacted with azide *95*. The coupling reaction between *1* and *2* was carried out in the presence of CuI and *N,N*-diisopropylethylamine, in toluene at room temperature for 10 h affording triazolo-dimannose *96* as a single cycloadduct in good yield (Scheme 25). Then the free hydroxy group of *96* was transformed into the azido group by treatment with diphenyl phosphoryl azide (DPPA) and DBU in DMF. The azide *97* served as substrate in the subsequent cycloaddition with *94*. The coupling/azidation sequence was repeated over three more cycles with similar efficiency to afford product *102* (triazolo-pentamannose). The disadvantage with this methodology was long reaction time of 30 h per reaction cycle, which was significantly shortened by the use of microwave irradiation at 100 ^{o}C for the cycloaddition and 120 ^{o}C for the azidation. At the same time, the yields of isolated products were very close or identical to those in the reactions carried out under conventional thermal conditions.

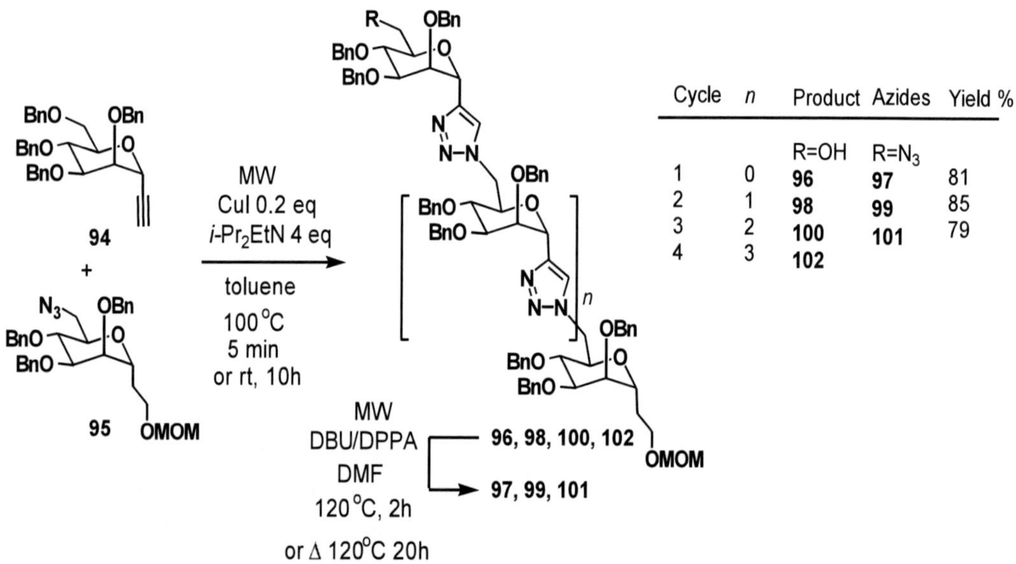

Cycle	*n*	Product R=OH	Azides R=N_3	Yield %
1	0	**96**	**97**	81
2	1	**98**	**99**	85
3	2	**100**	**101**	79
4	3	**102**		

Scheme 25.

Identical synthetic strategy, the iterative Cu(I)-catalyzed alkyne-azide cycloaddition, was applied by Zerrouki [45] for the synthesis of 3',5'-pentathymidine. Optimized reaction conditions for [3+2] addition [46] starting from 5'-*O*-tosylthymidine, sequential azidation and

Cu-catalyzed azide-alkyne 1,3-dipolar cycloaddition led to the formation of a 3',5'-pentathymidine in high yield. The whole process needed only work-up/precipitation steps and was completed within just 18 minutes, due to microwave activation (Scheme 26). Hence, derivatives *103* and *104* were coupled using Cu(I)-catalyzed cycloaddition in water and ethanol to obtain 1,4 dithymidine-substituted-1,2,3-triazole *105* after 3 min of irradiation in excellent yield. Tosyldithymidine *105* was then converted into 'azido' dimer *106* and the next click reaction between compounds *106* and *104* carried out using the same reagents and activation gave *107*. Since tetramer *109* is insoluble in H_2O/EtOH, CuI/DIPEA/DMF system was used for [3+2] cycloaddition and gave pentamer *111* after microwave activation.

Multiple "click" chemistry was applied by Bew [47] for the synthesis of hybrid calix[4]arenes appended on the upper rim with carbohydrate and N,C-protected α-amino acids. When the reaction of tetraazido calix[4]arene *112* and (*S*)-tyrosine derivative *113a* was conducted using combination of copper, copper-(II) sulfate, and sodium ascorbate, the tetrazole *114* was afforded as a single regioisomer (Scheme 27). The [3+2]-dipolar cycloaddition using *112* was also carried out with other α-amino acid derivatives: *N*-trityl-*O*-propargyl-(*S*)-serine ethyl ester, *N*-Boc-(*S*)-alanine propargyl ester, *N*-Boc-(*S*)-propargylglycine ethyl ester and (*N*-Boc-(*S*)-alanine)-*O*-propargyl-(*S*)-tyrosine methyl ester *113b*-e affording tetrazoles *115-118* in good to excellent yields. In the process, the *N*-trityl and the thermally labile *N*-Boc protecting group remained intact. Employing identical reaction conditions, the corresponding 1,4-disubstituted-[1,2,3]-triazole disaccharide and 1,3-bis(disaccharide) *119* were afforded in excellent 74% and 88% yields, respectively. Substituting *112* for 1,3-diazidocalix-[4]arene and employing the same reaction conditions and catalysts, a series of 1,3-di-*N,C*-protected-α-amino acids and saccharide was attached to calix[4]arenes in 47-60% yields.

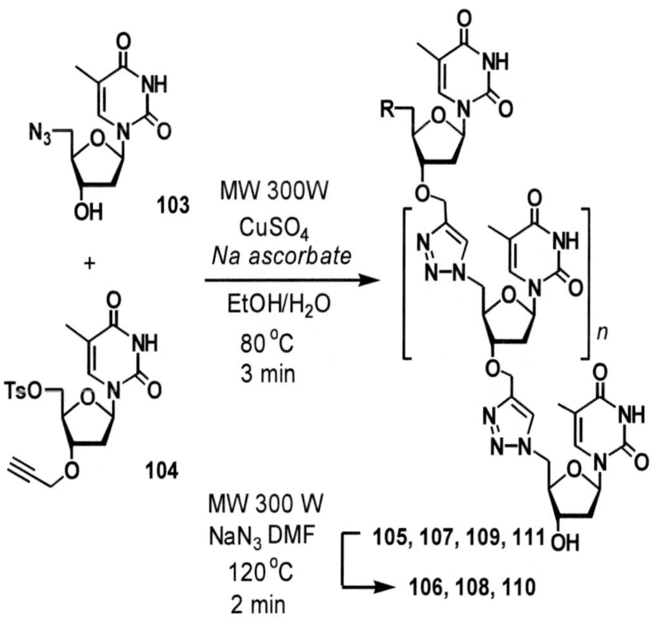

Cycle	*n*	Product	Azides	Yield %
		R=OTs	R=N_3	
1	0	**105**	**106**	92
2	1	**107**	**108**	80
3	2	**109**	**110**	72
4	3	**111**		99

Scheme 26.

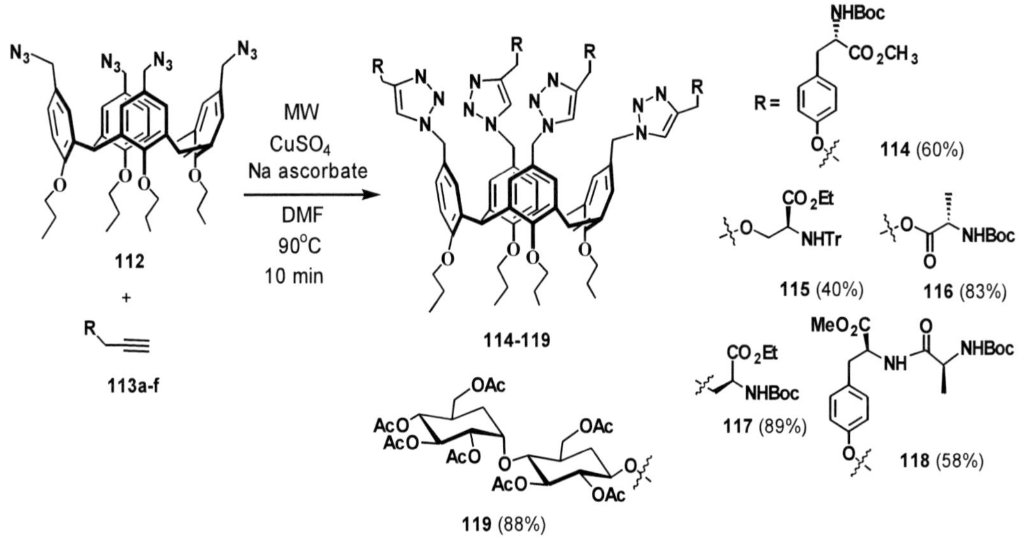

Scheme 27.

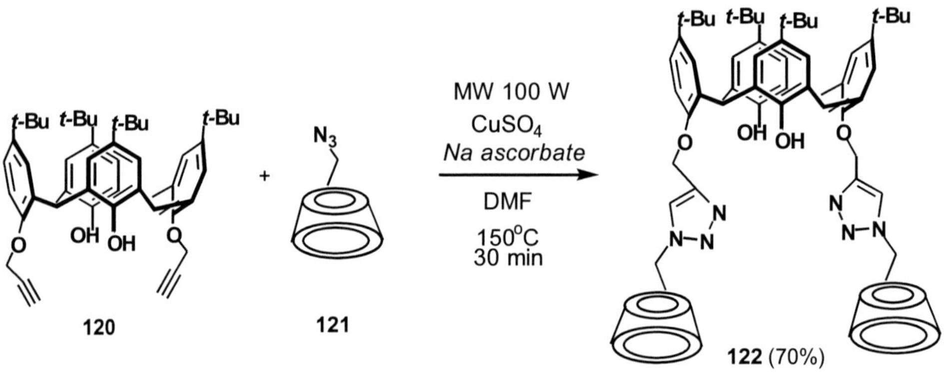

Scheme 28.

The lower rim of calix[4]arenes was similarly modified employing 1,3-dipolar cycloadditions by Ritter. [48] Synthesis of calix[4]arene-cyclodextrin adduct *122* from dipropargyl calix[4]arene *120* and 6I-azido-6I-deoxycyclo maltoheptaose *121* was carried out in DMF in the presence of Cu(I) catalyst, under microwave conditions (Scheme 28).

Liskamp [49] used 'multi-click' 1,3-dipolar cycloaddition approach to efficiently attach azido peptides to dendrimeric alkynes, which was assisted by microwave irradiation to ensure a complete modification of the alkyne endgroups. Conventional reactions carried out in the presence of $CuSO_4$/Na-ascorbate/Cu-wire require 16 h at room temperature to obtain fair yields (43-56%). A significant improvement was achieved by running this reaction under microwave irradiation, within 10 min at 100°C in the presence of $CuSO_4$/Na-ascorbate, *132* was isolated in 96% yield (Scheme 29). Using these conditions, divalent peptide cycloadducts *136-139* were obtained in fair to good yields (48-72%). When amino acid azide *123* was reacted with second generation dendrimer *2G* the tetravalent amino acid cycloadduct *133* was obtained in excellent yield (91%). Octavalent peptide *141* (48%) as well as octavalent and hexadecavalent systems *134* and *135* were prepared in 69 and 94% yield, respectively.

Azidopeptide *128* when coupled under microwave assistance obtained divalent cycloadduct *142* in nearly quantitative yield. Azido-Leu-enkephalin *129*, azide *130* and the cyclic azido peptide *131* were coupled to alkynes *1G* and *2G*, to afford the divalent and tetravalent peptides *143*, *145*, *147* and *144*, *146*, *148*, in yields ranging from 14 to 84%. It could be noted that variety of functional groups tolerate these reaction conditions, even unprotected guanidine functionality could be present.

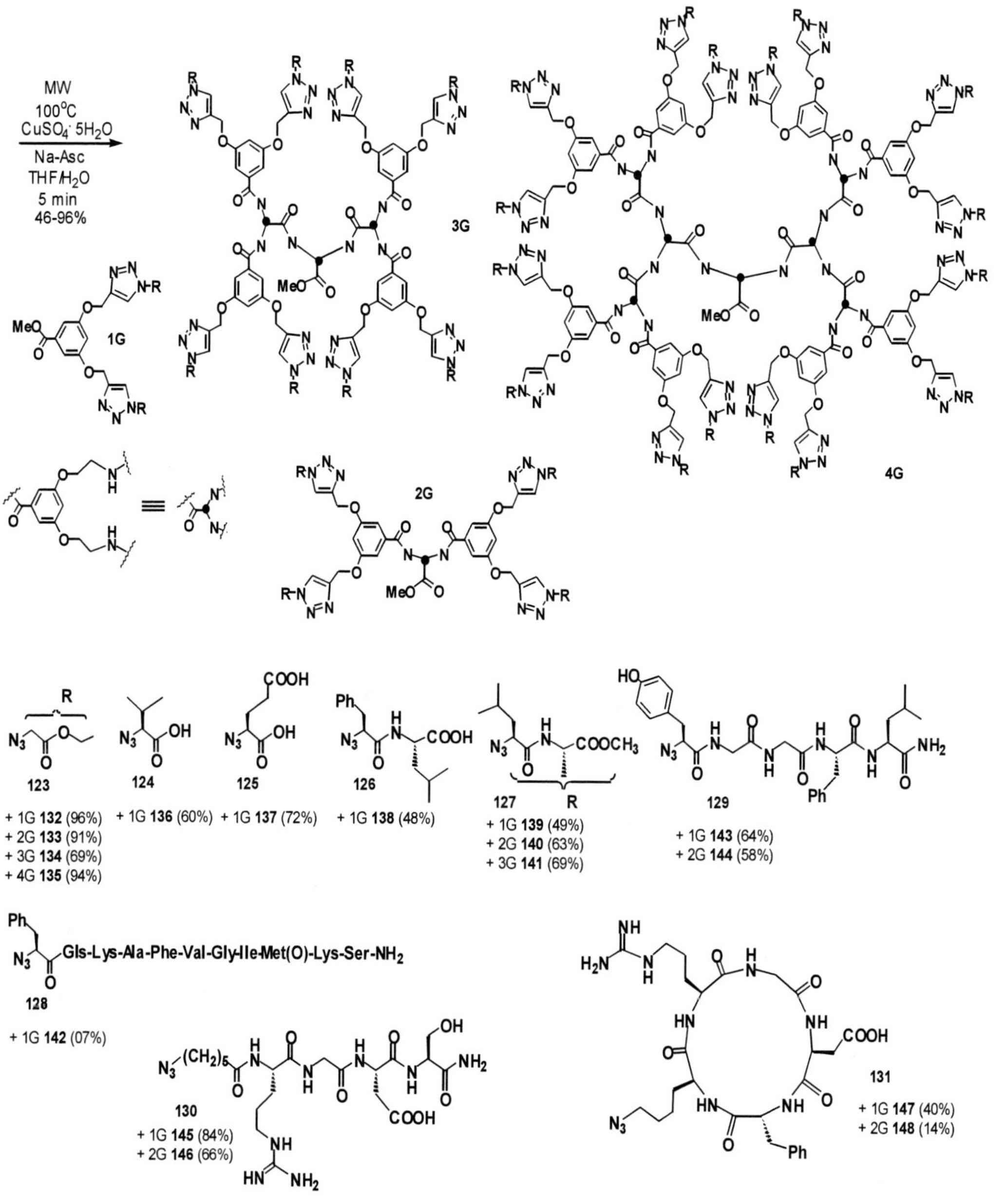

Scheme 29.

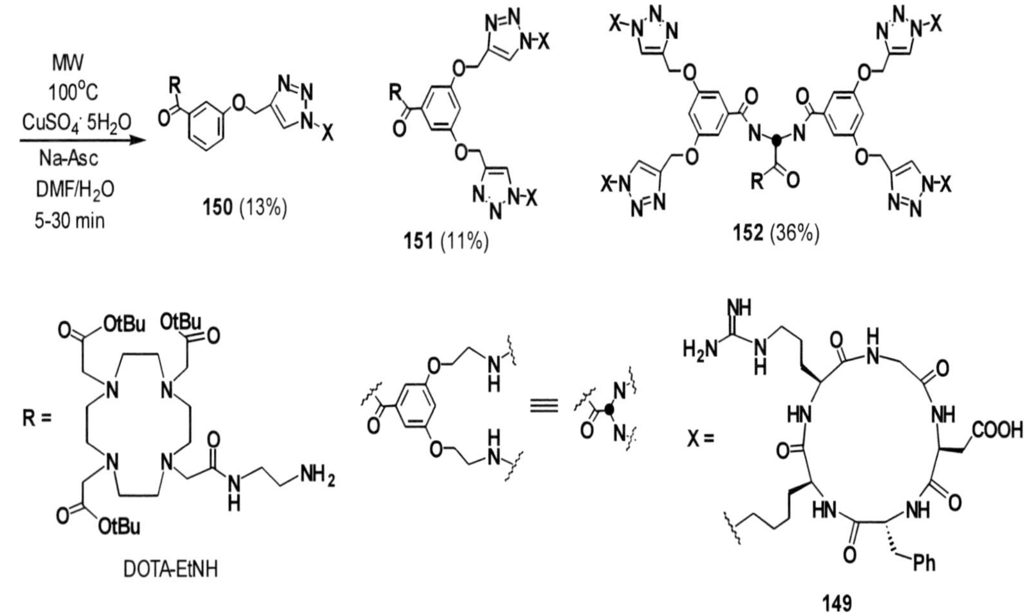

Scheme 30.

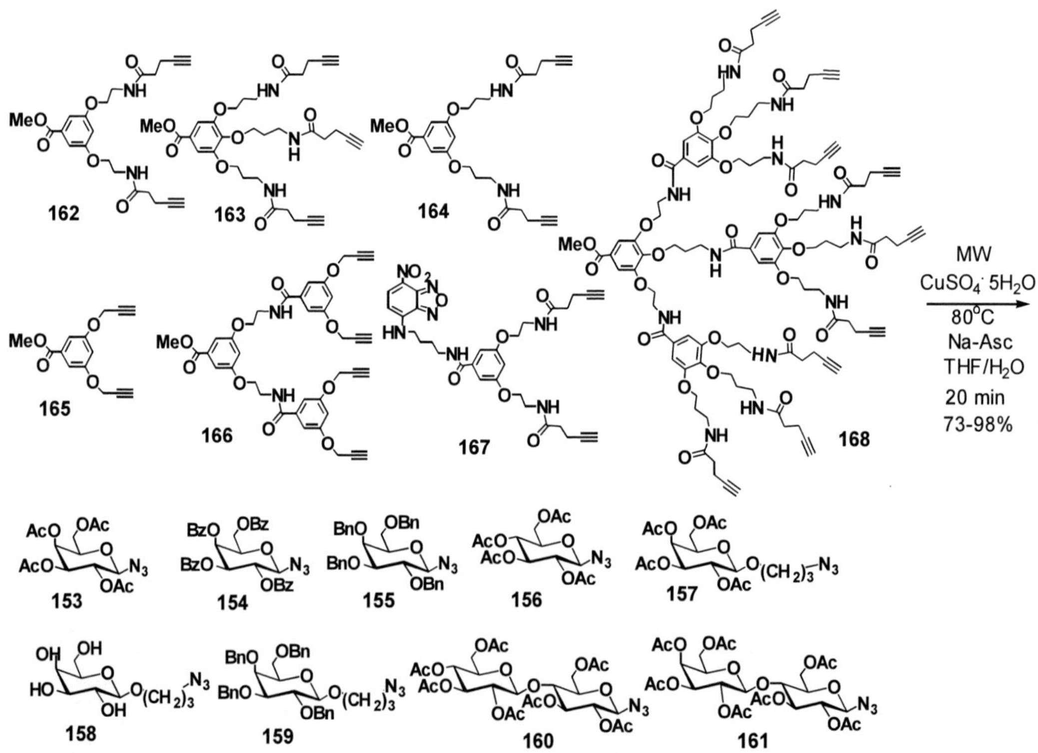

Scheme 31.

In the follow-up work Liskamp [50] has shown that DMF is equally acceptable solvent for this type of substrates. The *N*-ε-azido *cyclo*(Arg-Gly-Asp-D-Phe-Lys) peptide *149* was the entry point to the chemoselective coupling of the different dendrimeric alkynes to the cyclic RGD azido peptide *149* to furnish the DOTA-conjugated dendrimeric *cyclo*-RGD peptides acting as $\alpha_V\beta_3$ integrin antagonists: the mono-, di- and tetravalent DOTA-conjugated *cyclo*[RGDfK]peptide dendrimers *150-152* (Scheme 30).

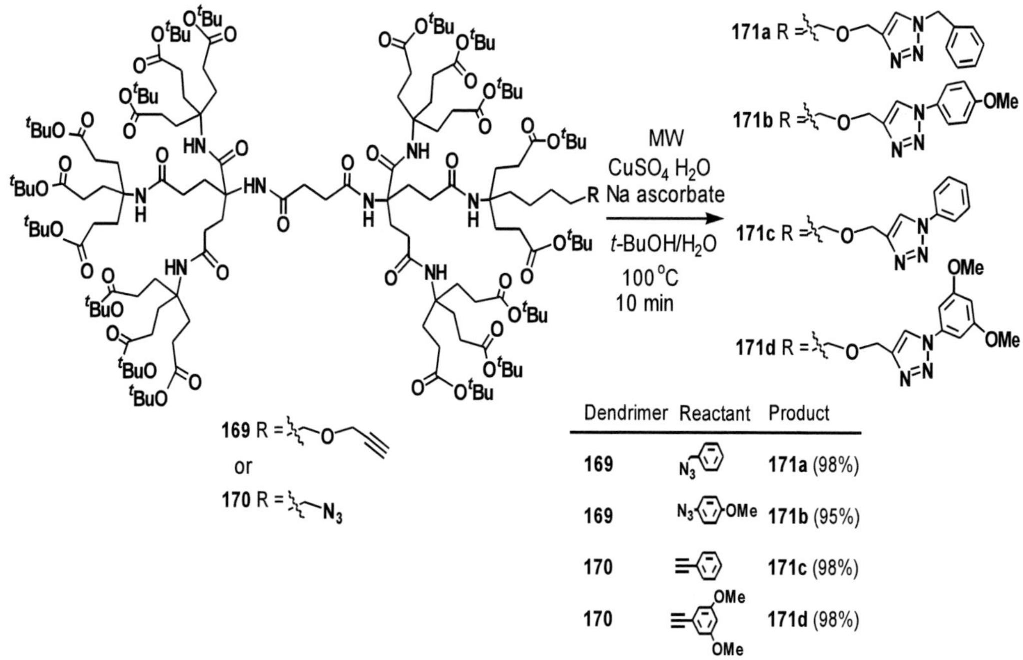

Scheme 32.

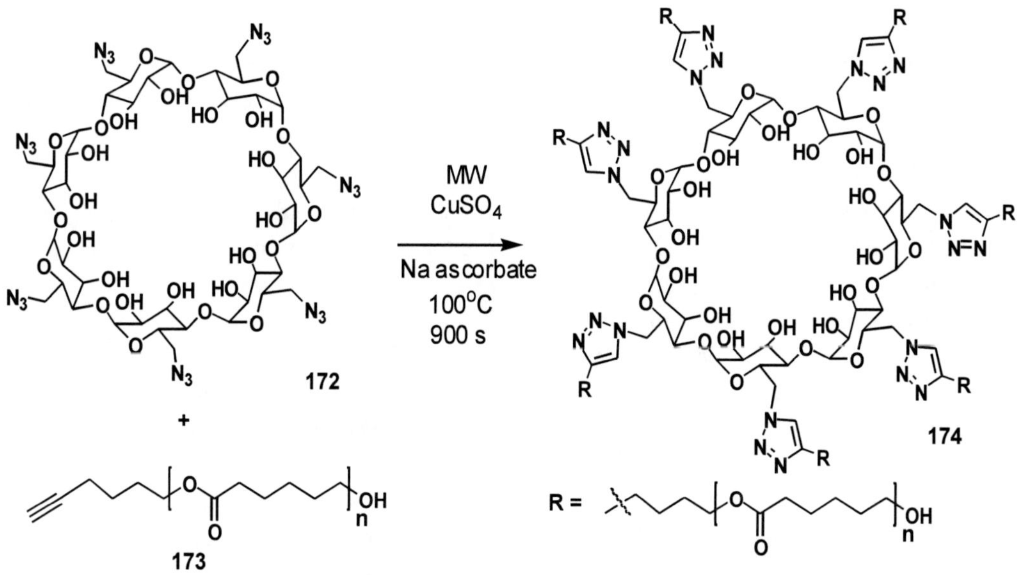

Scheme 33.

Synthetic utility of these reaction conditions was further demonstrated in Liskamp's [51] mulitclick synthesis of multivalent dendrimeric 1,4-disubstituted 1,2,3-triazole-linked glycodendrimers (Scheme 31). The first and the second generation alkyne dendrimers *166* (tetravalent) and *168* (nonavalent) were employed. The carbohydrate derivatives included the monosaccharide galactose that was used without *153-155* or with *157-159* a spacer to the azido function, with their hydroxy protecting groups varied. Besides galactose *153-155*, the monosaccharide glucose *156*, the disaccharides cellobiose *160* and lactose *161* were included. Treatment of dendrimeric alkynes with sacaccharides in microwave conditions yielded corresponding cycloadducts in high yields (86-98%). Even the unprotected saccharide glucose *158* reacted well (73-94%), therefore protection/deprotection steps could be reduced and the efficiency of the whole synthesis significantly increased.

The microwave-assisted 1,3-dipolar cycloaddition reactions that were carried out by Weck [52] on dendrimers *169* and *170* proceeded in excellent yields to obtain monofunctionalized triazole dendrimers *171a-d* (Scheme 32). For the microwave experiments *169* was treated with benzyl azide, using Na ascorbate and $CuSO_4.5H_2O$ (5 mol. %) at 100 W irradiation power at temperature of 100°C for 10 min. While the yields are similar to the ones obtained with conventional heating methods, the microwave irradiation significantly cuts down the reaction time from hours to minutes. The 1,3-dipolar cycloaddition using acetylene dendrimer *169* and benzyl azide under the standard reaction conditions and temperatures of 80°C needs 20 h to obtain quantitative yields.

Rare example of click chemistry concerning β- cyclodextrins has appeared in the literature, where in a microwave -assisted procedure Schubert [53] afforded a mixture of heptakis-4-[poly(ε-caprolactone)]-1*H*-1,2,3-triazol-1-yl-β-CD and heptakis-5-[poly(ε-caprolactone)]-1*H*-1,2,3-triazol-1-yl-β-CD. The acetylene functionalized pεCL *173* was coupled to heptakis-azido-β-cyclodextrin *172* resulting in the formation of the seven-armed star-shaped heptakis-pεCL-β-cyclodextrin *174* (Scheme 33).

Cravotto [54] optimized synthesis of homo- and heterodimers of α-, β- and γ-cyclodextrin by microwave -promoted Huisgen cycloaddition. The three asymmetric dimers *179-181* ('head to tail') were obtained by reaction 6^I-azido-6^I-deoxy-β-CD *175* with 2^I-*O*-monopropargyl-6-*O*-(*tert*-butyldimethylsilyl)-α-, β- and γ-CD *176-178* (Scheme 14). Cycloaddition was carried out in *tert*-butanol/water (1:1) with $CuSO_4$ as precatalyst and L-ascorbic acid as reducing agent and products deprotected under acidic conditions. A series of symmetric 'tail-to-tail' dimers *185-187* was obtained accordingly by coupling of azido-α, β- and γ-CDs *182-184* with *177* (Scheme 15).

The same synthetic protocol featuring the coupling of acetylenic and azido CD derivatives Cravotto [55] used to prepare cyclodextrin dimers and trimers. Two alternative strategies were employed: the first was to directly react two CD derivatives bearing an azido and an alkyne group (as in Scheme 34); the second one used a bifunctional cross-linkers, either 1,3-bis(azidomethyl)benzene *188* or 1,3-bis(propargyloxy)benzene *189* to join together two appropriate CD derivatives by generating two triazole bridges of symmetric homodimers *190* and *191* (Scheme 35).

Synthesis of a CD-trimers from dipropargylated derivatives *192* and 6-monoazido-βCD *183*, gave product *193* in 68% yield. When 6A,6^D-dideoxy-6^A,6^D-diazido-βCD *194*, and mono-*O*-propargyl-6-TBDMS-βCD *177*, were reacted, the product was identifiable as *195* (Scheme 36).

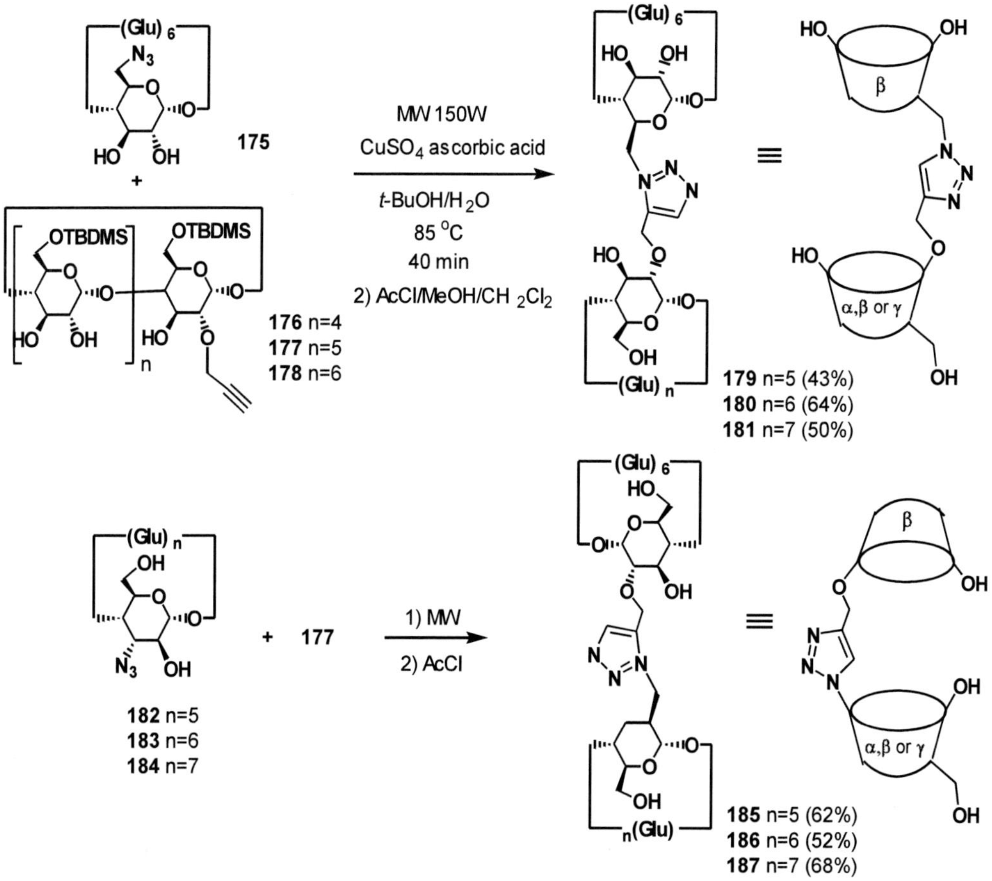

Scheme 34.

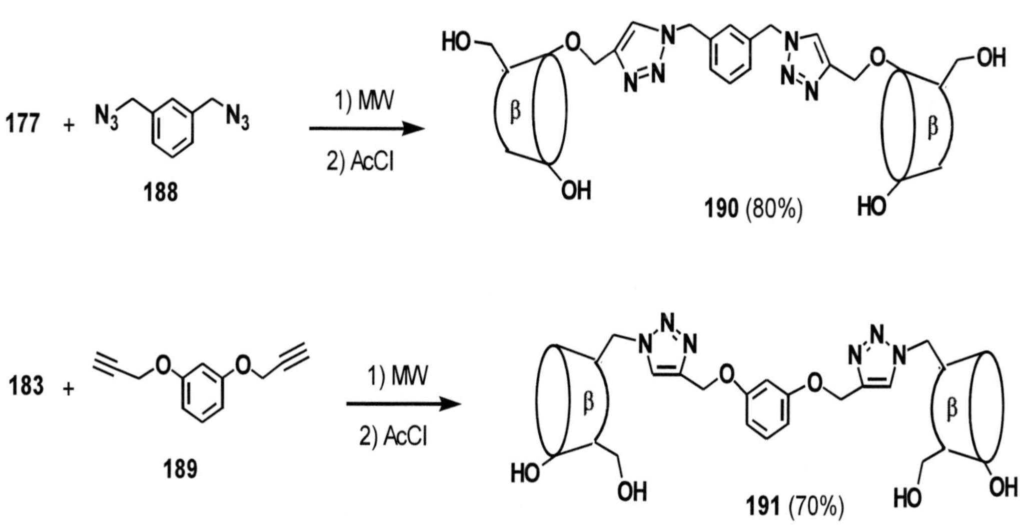

Scheme 35.

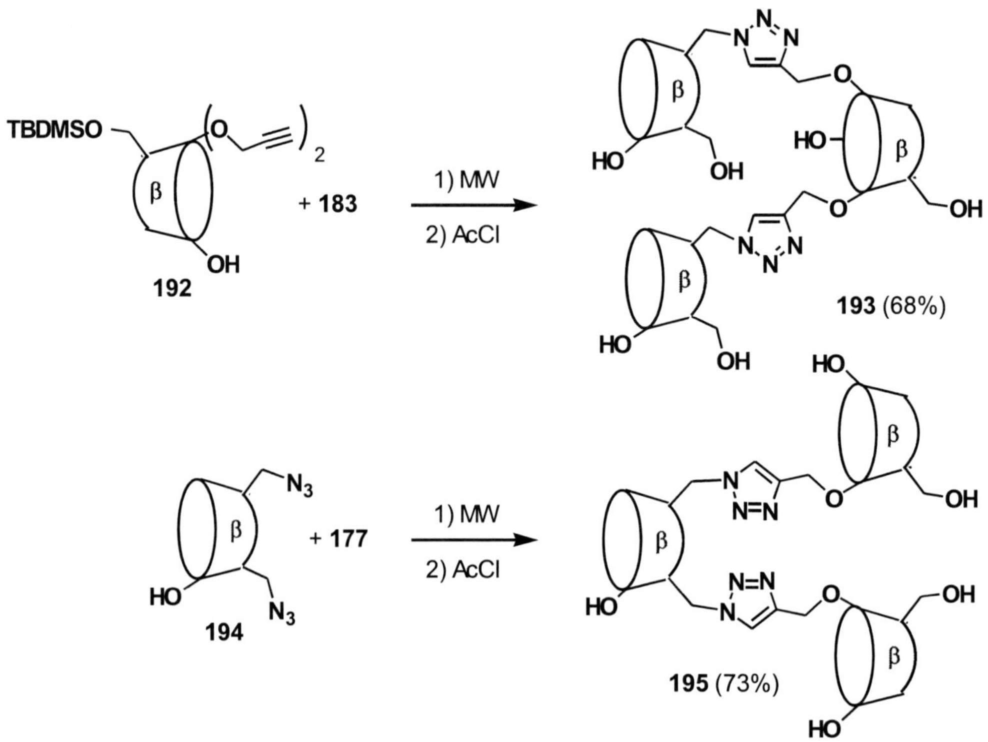

Scheme 36.

Identical synthetic protocol was enabled Cravotto [56] to prepare a novel reticulated CD polymer *193* (crosslinked through the secondary OH groups) was prepared by 1,3-dipolar cycloaddition under microwave conditions by the reaction of a randomly propargylated β-CD *192* with 1,3-bis(azidomethyl)benzene *188* (Scheme 37). In this reaction, unprotected propargylated β-CD was employed, avoiding protection/deprotection steps.

Efficient synthesis of a fluorescent tripod detection system for pesticides by microwave-assisted click chemistry was obtained by Delattre. [57] This is an interesting example of the combination of cyclodextrin synthesis with microwave-assisted triple-click chemistry. The new triazine tripod system *196* was obtained by the Cu(I)-catalyzed azide-alkyne cycloaddition reaction between an excess of 6'-azido-6'-deoxy-cyclomaltoheptaose peracetate (6-azido-6-deoxy-peracetylated-β-CD) *194* (5 equiv) and 1,3,5-tris(2-propynyloxy) benzene *195*, CuI (3 equiv), *N*-diisoproplyethylamine (9 equiv) in 15% isolated yield (Scheme 22).

Besides the classical reagents $CuSO_4$/Na-ascorbate, other catalysts were ocassionally employed for 1,3-dipolar click reactions. The construction of multivalent neoglycoconjugates is efficiently achieved by the catalytic cycloaddition of alkynes and azides using the organic-soluble copper complexes $(Ph_3P)_3$.CuBr and $(EtO)_3$P.CuI (Scheme 39). Santoyo-González [58] simultaneously used microwave irradiation, which notably shortened the reaction times. The regiospecific synthesis of 1,4-disubstituted 1,2,3-triazoles possessing divalent mannopyranosylated ligands followed two different approaches. The direct assembly of functionalized sugars was achieved by reaction of alkynyl glycosides *197a-c* and azide glycoside *198a*. Alkyne *197a*, which supports the more electronegative substituent, reacts well with $(Ph_3P)_3$.CuBr/DIPEA, but no reaction was detected with *197b* and *197c*. Cycloaddition of *197c* was achieved by the use of a stronger base DBU, or employing the $(EtO)_3$P.CuI/DIPEA catalytic system, to obtain the 1,2,3-triazoles *199b* and *199c* in almost quantitative yields. Without microwave irradiation, $(EtO)_3$P.CuI catalyst proved to be equally efficient and regiospecific but longer reaction times were required (24-72 h). As a second approach for the synthesis of divalent mannosylated ligands, the diazides were used in the cycloaddition reaction with the propargyl mannosides obtaining bis-triazoles. Furthermore, the synthesis of polyavalent mannopyranosylated ligands was achieved by using the polyazides *200a-e*. Propargyl mannoside *197a* was scaffolded on those aromatic cores, and the multivalent ligands *201a-c* were obtained, as well as monovalent and heptavalent glycocyclodextrins *201d,e*.

For calix[4]arene functionalization by mulitiple click reaction Chiappe and Dondoni [59] reacted a tetra-azido calix[4]arene derivative *202* with ethynyl tetra-*O*-benzyl-*C*-galactoside *203a* in the presence of CuI and *i*-Pr_2EtN in three different ionic liquids, $[C_8dabco][N(CN)_2]$, $[C_8dabco][Br]$, and Ammoeng110 (Scheme 40). Reactions were performed at 80°C by microwave heating to obtain tetra glycoclusters on a calix[4]arene molecular platform *204* in 90, 73, 55% yields, respectively. Here multiple cycloadditions gave a triazole-linked tetra-*C*-galactosyl-calix[4]arene *204* in high yield. Obtained yields are better than thermal heating, especially in the case of *C*-galactosyl acetylene *203a*. The $[C_8dabco][N(CN)_2]$ ionic liquid was also used to perform the multiclick reactions with propargyl *O*-lactoside *209* and *S*-sialoside *210*.

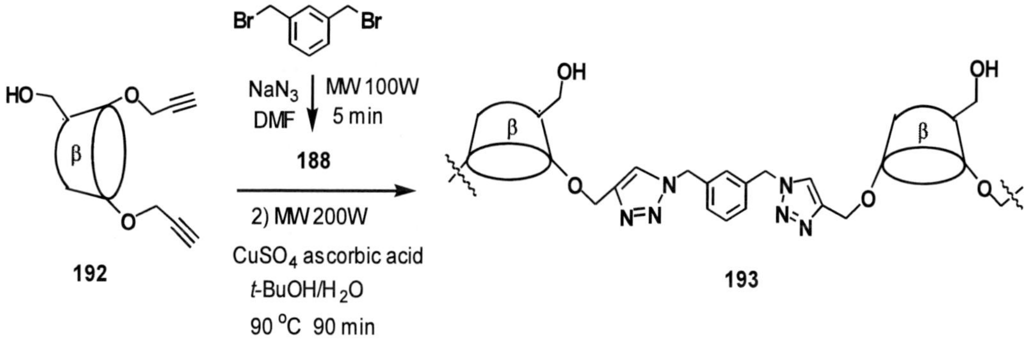

Scheme 37.

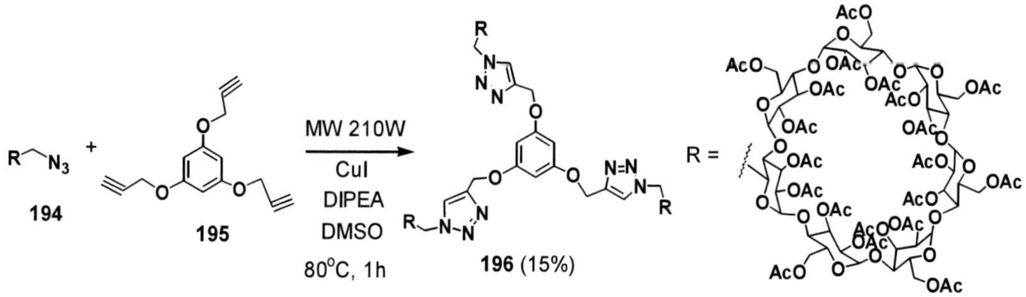

Scheme 38.

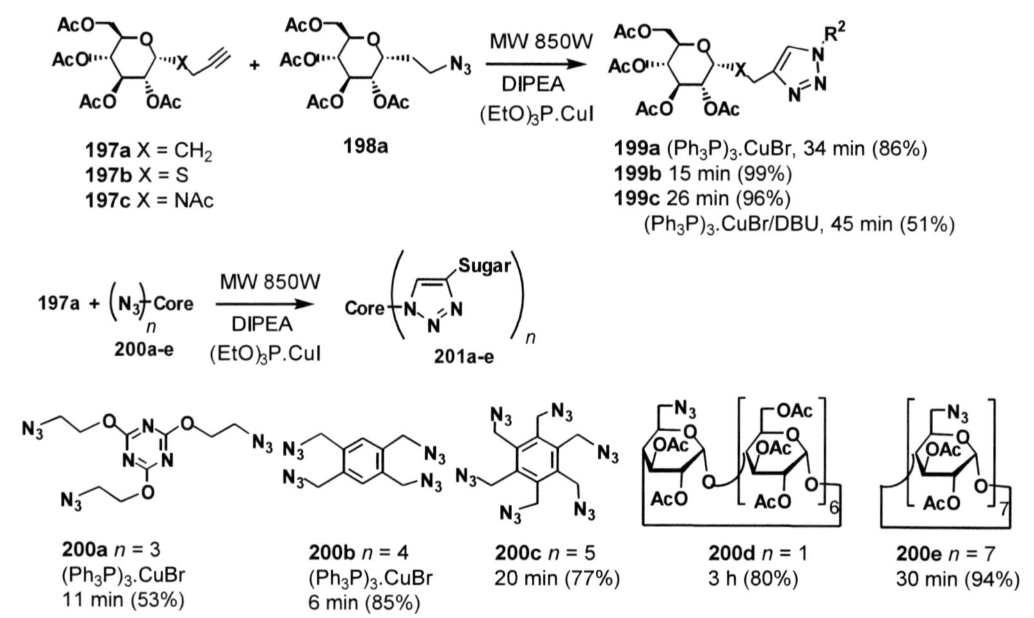

Scheme 39.

An accelerating mechanism was postulated, in which the firstly formed triazole served as accelerating element for the subsequent cycloaddition. The first click reaction on *203* leads to the formation of a copper(I) triazolide complex intermediate on the the sugar alkyne, with subsequent rapid intramolecular capture by an adjacent azido group. The suitable cone conformation of the calix[4]arene platform and the flexibility of the azidopropyl chain are suitable geometrical factors for the intramolecular click process. The same mechanism can be applied to the subsequent third and fourth click reactions.

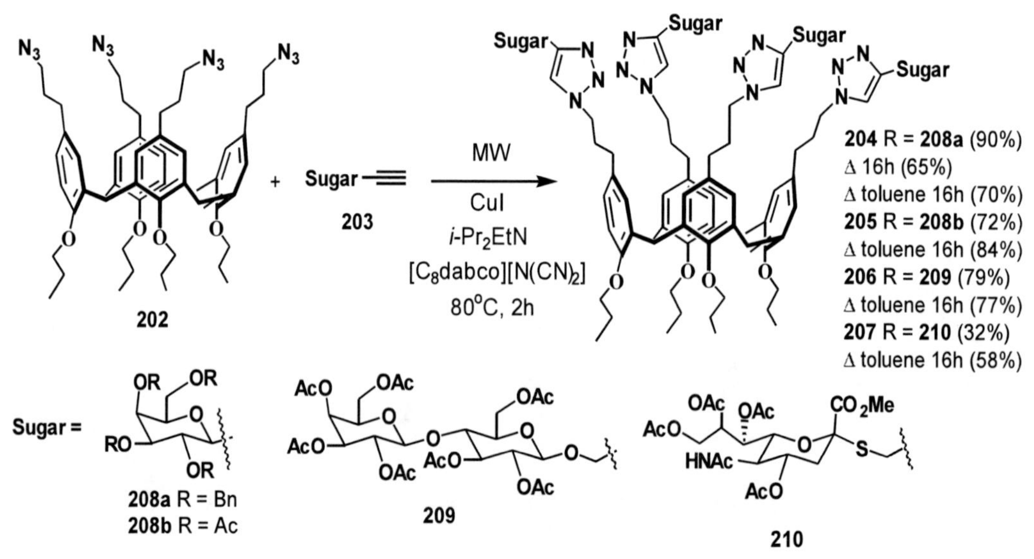

Scheme 40.

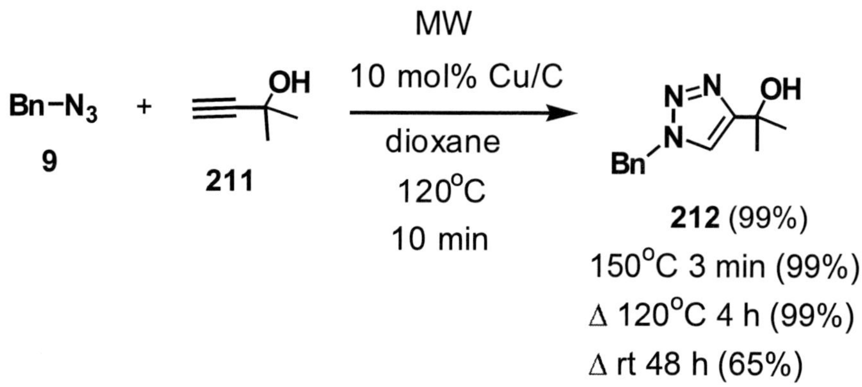

Scheme 41.

Lipshutz [60] pointed out the advantages of copper in charcoal (Cu/C) as a simple, inexpensive, general and efficient heterogeneous catalyst for use in click cycloadditions. Catalyst preparation is simple, the impregnation of charcoal with $Cu(NO_3)$ leads to the nanoparticle-sized Cu/C. No bases, ligands, or other additives are required for Cu/C to effectively catalyze these cycloadditions. When reactions of benzyl azide *9* to acetylene *211* were exposed to microwave irradiation with Cu/C catalysis in the absence of any additive, the time needed for cycloaddition resulting in triazole *212* was reduced from 4 h to less than 10 min, with the reaction temperature being increased from 60°C to 120°C (Scheme 2).

Fokin [61] has found that instead of the commonly used CuAAC catalytic system, the formation of 1,5-disubstituted 1,2,3-triazoles from aryl azides and alkynes can be readily accomplished using pentamethylcyclopentadienyl ruthenium(II) chloride tetramer $[Cp^*RuCl]_4$ catalyst in dimethylformamide, while the use of other solvents resulted in significantly lower efficiency (Scheme 42). While CuAAC process works only with terminal alkynes and produces 1,4-disubstituted 1,2,3-triazoles, it was known from literature that the ruthenium-catalyzed reaction provides ready access to the complementary 1,5-regioisomers of 1,2,3-triazole *215a-f*, *218a-f*. [62] It was demonstrated that the reaction provided higher yields, cleaner product, and shorter reaction times when carried out under microwave irradiation. Both electron-rich (R^1=Me, OMe) and moderately electron-poor (R^1=CO_2Et) aryl azides *214a-f*,*217* readily participated in the reaction. However, those containing sterically demanding substituents or strongly electron withdrawing groups, such as (4-nitro)azidobenzene, failed to participate in the reaction, likely due to catalyst deactivation and/or decomposition of the azide at the elevated temperature. A broad range of functional groups could be employed without significantly affecting the yield of the triazole product. Thus, alkynes containing an amine, alcohol, nitrile, and heterocyclic amine functionality readily participated in the reaction.

Multistep synthetic procedure to 6-(1,2,3-triazol-1-yl)-dihydropyrimidones in which the every synthetic step was carried with the assistance of microwave irradiation was developed by Kappe. [63] Initially, the 6-azidomethyldihydropyrimidone *219* was treated with terminal acetylenes under CuAAC azide-acetylene ligation, to provide 6-(1,2,3-triazol-1-yl)-dihydropyrimidones *220a-g* in a regiospecific fashion (1,4-triazoles) in moderate overall yield utilizing controlled microwave irradiation (10 min) (Scheme 43). Azide acetylene cycloaddition *219* under strictly thermal conditions (DMF, microwave, 150°C) in the absence of the Cu(I) catalyst afforded mixtures of both isomeric cycloadducts *220a* (*anti* and *syn* triazoles), and several hours of irradiation were required to achieve full conversion. Other compounds *224a-m* were prepared by microwave assisted Biginelli multicomponent condensations, bromination, azidation and 1,3-dipolar addition. In this synthetic protocol, the dihydropyrimidone analogues *221* were subjected to the flowthrough bromination/azidation sequence to provide the corresponding azides *223*, followed by isolation of the crude azides and their cycloaddition to triazoles. These reactions led to an attractive linkage of two important *N*-heterocyclic pharmacophores with four points of diversity (R^1-R^4) that can easily be addressed by selecting the appropriate building blocks. Employing these two synthetic procedures, 10 aryl, heteroaryl, and alkylacetylenes were reacted to prepare a library of 27 compounds with 4 points of diversity.

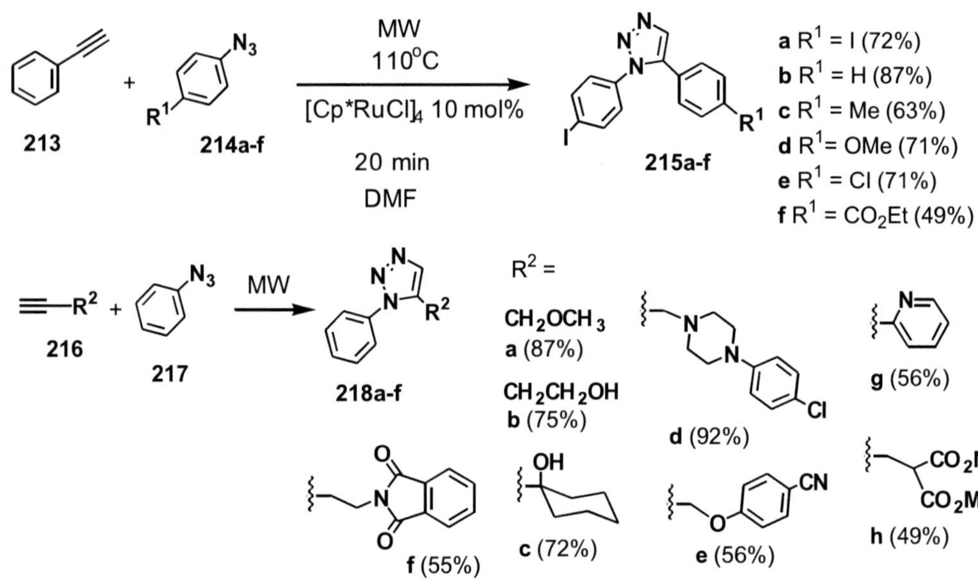

Scheme 42.

Different synthetic approach was used by Wittmann [64] involving a one-pot reaction for Cu(II)-catalyzed diazo transfer and Cu(I)-catalyzed azide-alkyne 1,3-dipolar cycloaddition. 1,4-Disubstituted 1,2,3-triazoles are obtained in excellent yields from a variety of amines using trifluoromethanesulfonyl azide for diazo transfer, without the need for isolation of the azide intermediates. The reaction featuring *in situ* formation of azides is especially advantageous since procedure does not require isolation of unstable and potentially hazardous azides. In the first step, the diazo transfer was performed with sodium bicarbonate as base at room temperature. After conversion to benzylazide, phenylacetylene *226,* the sodium ascorbate as reducing agent were added and heated to 80°C by microwave irradiation. The short reaction times could be achieved only by addition of the Cu(I)-stabilizing ligand tris(benzyltriazolylmethyl)amine (TBTA). When the Cu/$CuSO_4$ system was used, comparable reaction rates and yields of *227* without TBTA additive were obtained (89%) (Scheme 44). The fact that Fmoc-protected lysine could be converted to *229* and propargyl glucoside to the divalent glucoconjugate *233* illustrates the potential of the procedure for the functionalization.

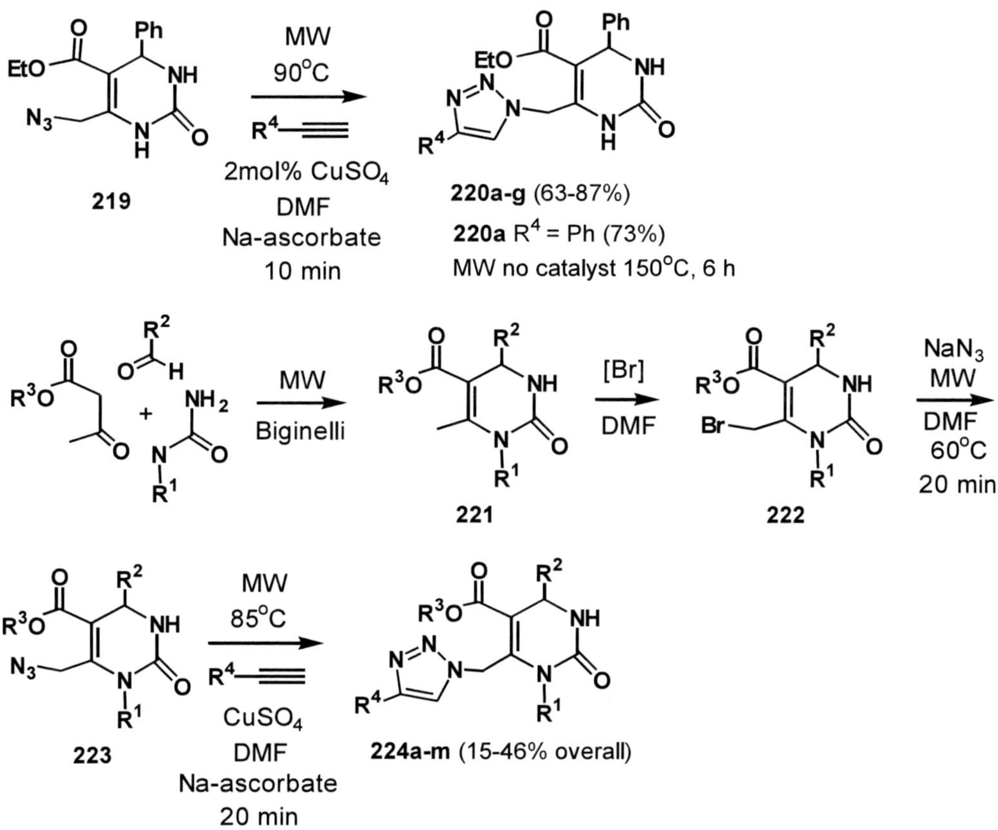

Scheme 43.

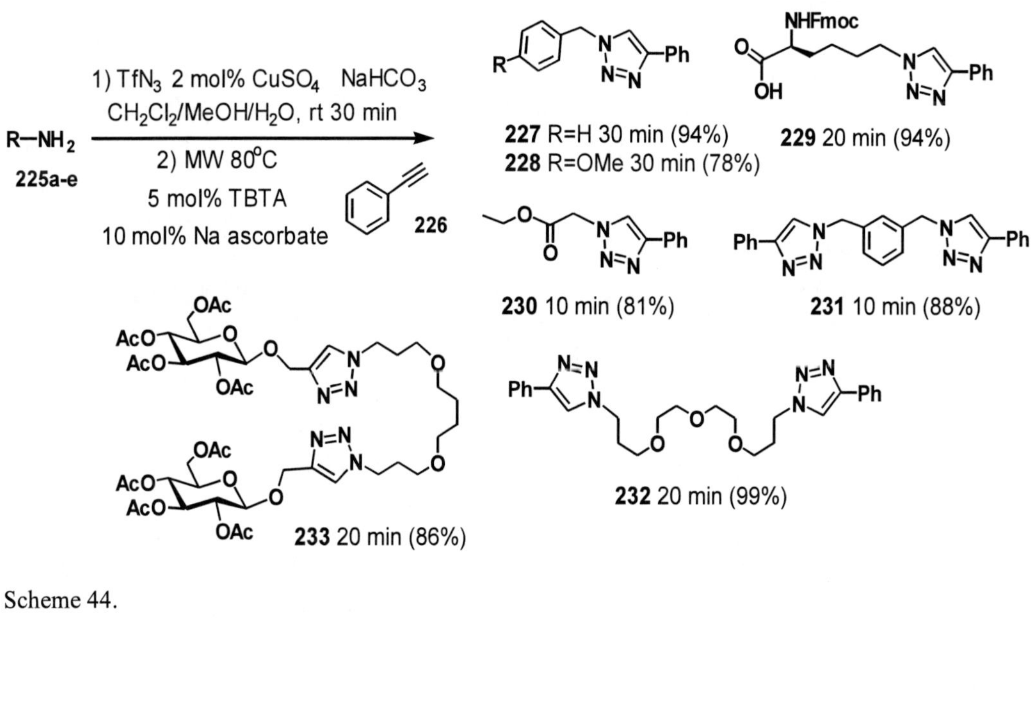

Scheme 44.

Diazo transfer and cycloaddition could be performed as a one-step, one-pot reaction, mixing all components together (at 80°C, 40 min), where product *227* was formed in 93% yield, slightly lower than of the sequential one-pot procedure. The possibility that azides react with dichloromethane to form explosive diazidomethane was circumvented by replacement of dichloromethane with toluene. With the system toluene/2-propanol/water product *227* was obtained in 77% yield (80°C, 40 min). Although this yield is lower, the procedure represents a safe alternative for large scale synthesis.

A series of modified nucleosides, adenosine derivatives have been synthesised using click chemistry, as reported by Grøtli [65] (Scheme 45). The hydroxy groups in the 3'- and 5'-position of compound *234* were protected with a tetraisopropyldisilyl group making it much more soluble in organic solvents, which could be easily removed in post-coupling functionalizations. Compounds *235a-c* were prepared by a copper-catalyzed [3+2]-cycloaddition as a two-step one-pot reaction in which compound *234* was treated with triflic azide (diazo transfer reagent) in the presence of tris[(1-benzyl-1*H*-1,2,3-triazol-4-yl)methyl]amine (TBTA) and sodium ascorbate by microwave-assisted heating (method A). Compounds *235d,e* were prepared according to the method B, in which substrate *234* was treated with the crude azide using $CuSO_4$/sodium ascorbate and TBTA in DMSO. Finally, for preparation of *235f-i* a proline-promoted CuI-catalyzed coupling reaction in DMSO was used (method C). Here, compound *234*, CuI and TBTA were added directly to the reaction mixture containing the crude azide and irradiated by microwave. This two-step one pot procedure generated *235f-i* in moderate yields.

For the synthesis of oligomers of triazoles Arora[66] used two approaches: (a) *in situ* diazo generation procedure and (b) a sequential single-pot procedure. In the *in situ* method, all reagents for the diazotransfer and the cycloaddition steps were added simultaneously. Hence, amine, alkyne, TfN_3, catalyst ($CuSO_4{\cdot}5H_2O$, $ZnCl_2$), TEA, Na ascorbate, tris-(benzyltriazolylmethyl)amine (TBTA) were heated by microwave (80°C) in CH_2Cl_2/MeOH/H_2O) to obtain unsatisfactory yields (14-30%). The sequential procedure was much better (yielding up to 96%), involving addition of the alkyne and cycloaddition catalyst to the flask containing freshly generated azide but without the isolation or purification of the azide (Scheme 46). When 4-phenyl-1-butyne *238* was replaced with Boc-Phe-alkyne *240*, $CuSO_4{\cdot}5H_2O$ catalyst gave much better yield of *241* than $ZnCl_2$. Temperature was higher (100°C), but more importantly, *N*Boc protection withstands the reaction conditions, while $ZnCl_2$ also removed Boc group.

Sequential one-pot methodology was then used in the synthesis of triazole oligomers employing amino acid derivatives and iterative syntheses of triazole rings. Trimer *245* was afforded in 62% overall yield (Scheme 47).

The conversion of amines to azides via Cu(II)-catalyzed yields for peptido-triazoles were higher in the presence of $ZnCl_2$ as the diazotransfer catalyst on solid phase. On PAM resin, the solid-phase methodology affords excellent yields of triazolamer *246* and tetramer *247* with average yield 85-97% per diazotransfer/cycloaddition step (Scheme 48).

Moses [67] used the reagent combination *t*-BuONO and $TMSN_3$ for diazo transfer, in microwave enhanced one-pot tandem azidation/'click' cycloaddition of anilines. These reaction conditions offer practical and efficient one-pot azidation as useful addition to the click-chemistry toolbox. Use of microwave radiation significantly enhances the rate of formation of 1,4-triazoles from *in situ* generated azides, and minimises the potential explosive hazard (Scheme 49). The other notable feature of these reaction conditions is the regioselectivity and site-selectivity, since azides usually react with nitriles to form tetrazoles. This approach offers a quick and reliable procedure for the synthesis of a variety of 1,4-triazoles (80-99% yields), which is particularly amenable to electron-deficient anilines, but also works well with a variety of alkynes including aromatic, conjugated, aliphatic, electron-rich and electron-deficient varieties. The following molar ration was used: *t*-BuONO (1-1.5 equiv), $TMSN_3$ (1-1.2 equiv), $CuSO_4 \cdot 5H_2O$ (0.1 equiv), sodium ascorbate (0.5 equiv), and alkyne (1.5 equiv).

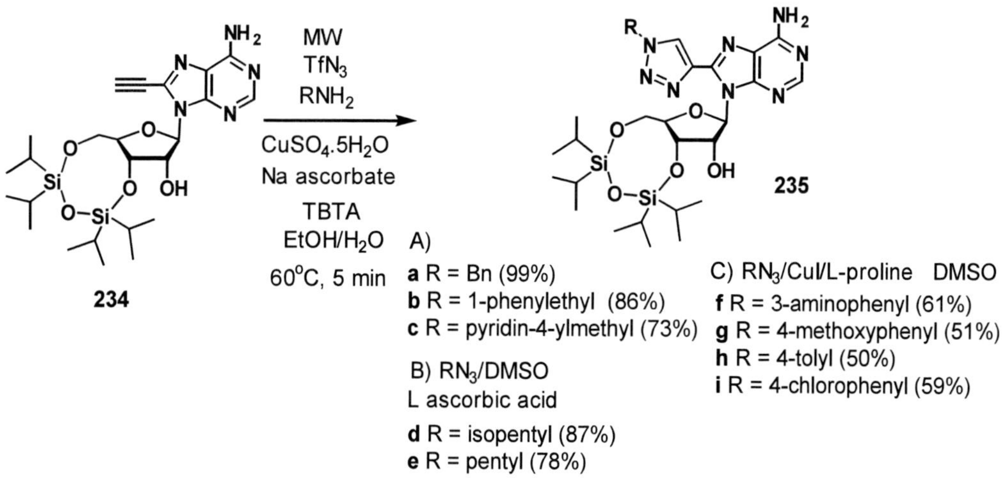

Scheme 45.

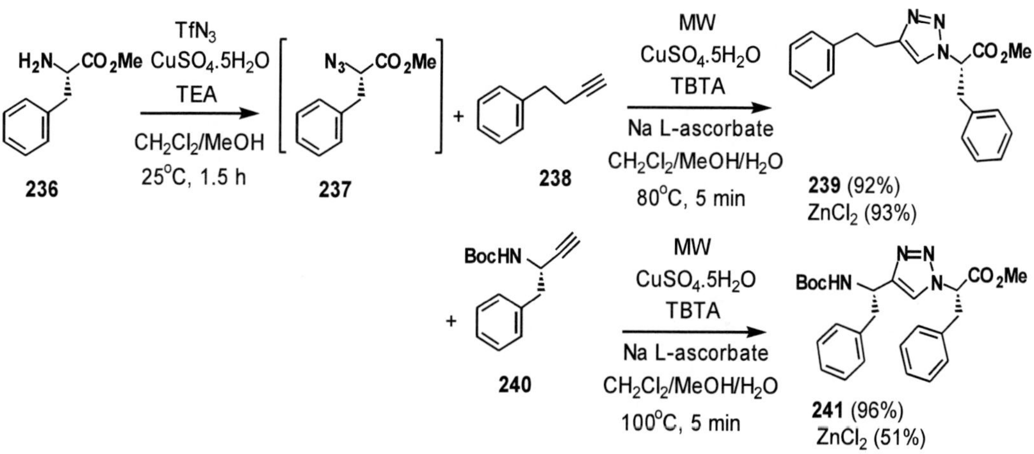

Scheme 46.

Sodium azide and alkyl bromide were used by Cravotto [68] for diazo transfer and one-pot generation of azides and 1,3-dipolar cycloaddition in microwave conditions. Best results were obtained by heterogeneous reaction with charcoal-supported Cu(I) catalyst, carried out in DMF. The Cu(I)/C was simply prepared by impregnation of activated charcoal with CuI. Sodium azide, alkyl bromide and acetylene were mixed with catalyst and irradiated. Cu(I)/C catalysis gave after 10 min the final product *255* in 89% yield (Scheme 50). Novel synthetic protocol was also used with simultaneous microwave/ultrasound irradiation, and small improvement was found (93%). Obtained yields were better than with soluble $CuSO_4$ or Cu(II)//L-ascorbic acid protocols in *t*-BuOH/H_2O. In addition, two-step reaction protocol which involves in the first step preparation of azide from sodium azide and alkyl bromide (microwave, DMF), followed by Cu catalyzed 1,3-dipolar addition was also tested. Similar results were obtained in the case of dibromo-*m*-xylene and phenylacetylene, giving corresponding bistriazole derivatives. Two step protocol NaN_3/DMF//$CuSO_4$//L-ascorbic acid/*t*-BuOH/H_2O in microwave (85°C, 30 min) gave 71%, while novel conditions microwave/US (85°C, 20 min) increased yield to 76%. Even better results were obtained by one-step protocol Cu(I)/C/DMF/microwave (110°C, 30 min, 78%).

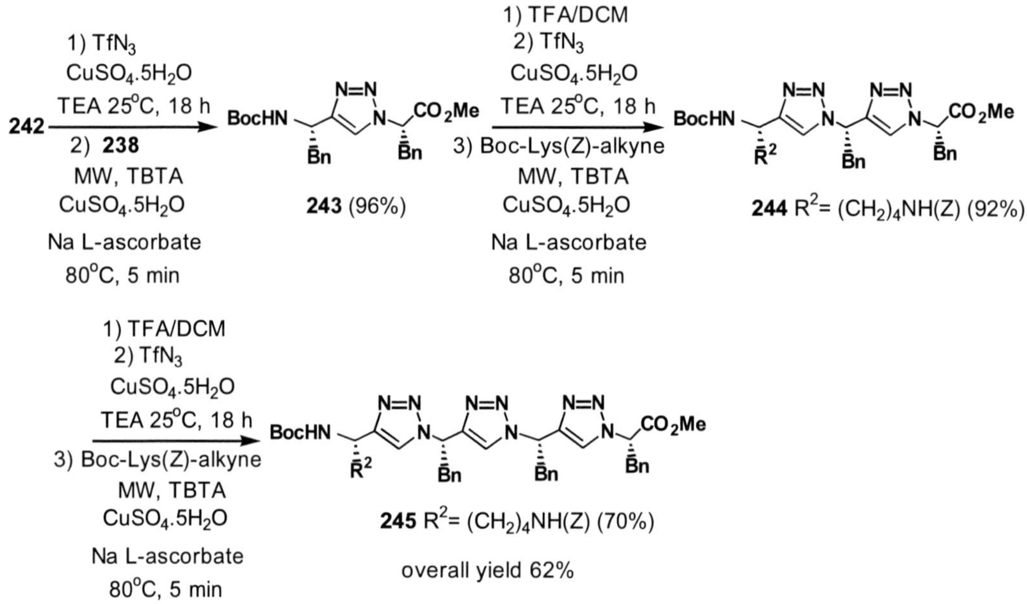

Scheme 47.

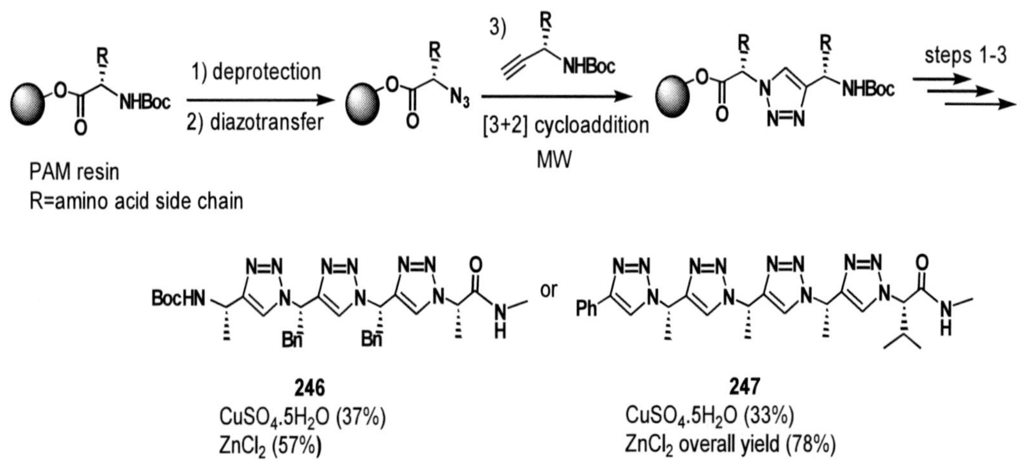

Scheme 48.

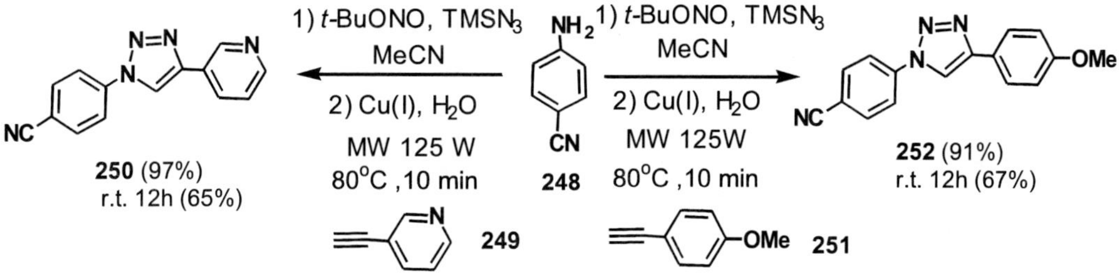

Scheme 49.

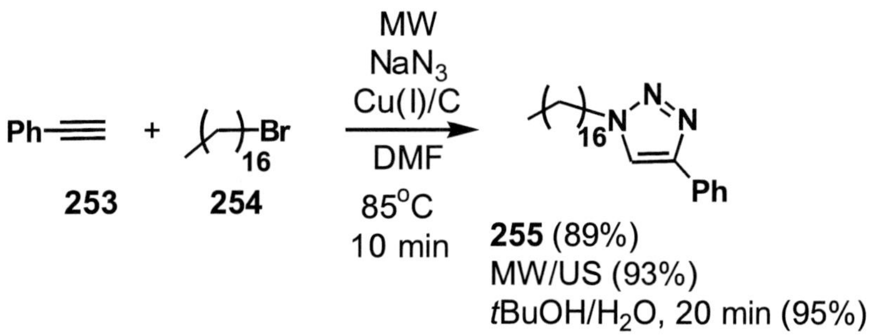

Scheme 50.

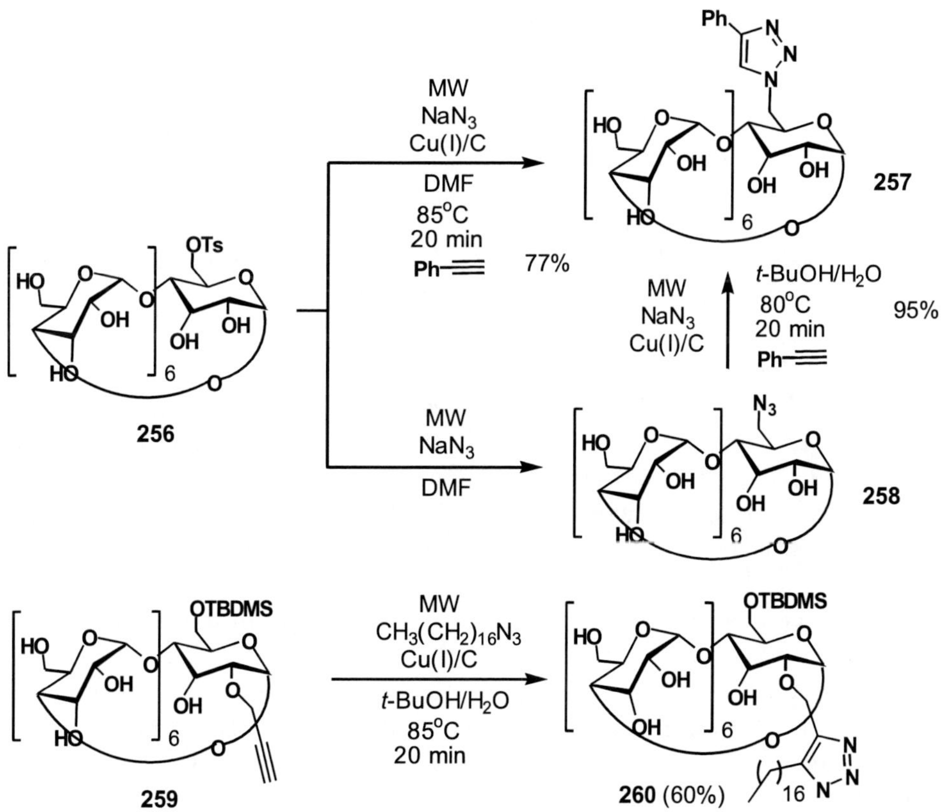

Scheme 51.

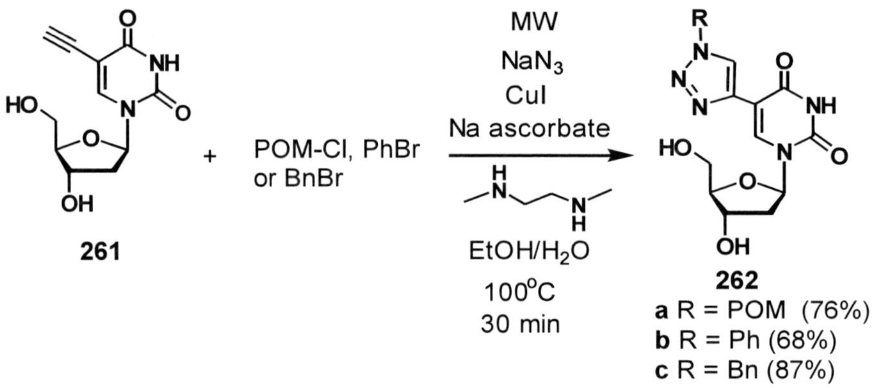

Scheme 52.

On the other hand, the one-pot reaction using 6^1-*O*-tosyl-β-CD *256* under microwave irradiation gave *257* in 77% yield (85°C, 20 min), while two step protocol involving 6^1-azido-61-deoxy-β-CD *258* provided more product (95%) (Scheme 51). Synthetic alternative to triazole substituted CD is the 1,3-dipolar cycloaddition of propargyl CD with an azide, where the *O*-propargyl-β-CD *259* reacted with 1-azidoheptadecane using Cu(I)/C as catalyst. After 20 min of microwave irradiation, conversion was complete and compound *260* was isolated in 60% yield.

In situ azidations and [3+2] cycloadditions as one-pot reaction were carried out by Nielsen [69] starting with 5-ethynyl-2'-deoxyuridine *261*, pivaloyloxymethyl (POM) chloride, bromobenzene or benzyl bromide. Treatment with sodium azide in microwave-heated ethanol/water mixtures in the presence of catalytic amounts of copper iodide, sodium ascorbate and *N,N'*-dimethylethylenediamine furnished products *262a-c* in high yields (Scheme 52).

2.1.2. Synthesis of Triazolines

Although 1,3-dipolar cycloaddition reactions of azides and alkenes usually give triazoline products, Barros [70] reported that microwave-assisted 1,3-dipolar cycloaddition reactions of vinylic glycosides with aryl azides unexpectedly produce triazolines and acetyl group migration. Three carbohydrate derivatives *263-266* were reacted with *p*-methoxybenzylazide *266* and benzylazide *9*, under microwave irradiation in the absence of solvent (Scheme 53). The triazolines obtained are 1,4-disubstituted, without evidence of its regioisomer. This regioselectivity is induced by the presence of an electrondonating and leaving group (the sugar) and an electron withdrawing group (the ester) in each side of the olefin, favoring the attack of the terminal nitrogen atom of the azide to the more nucleophilic carbon atom. For glucose derivatives *264* and *265*, different powers and times were screened. The use of 200W power promotes occurrence of reaction and after 15 min two products were obtained (sugar with free OH group *272*, and triazole *270*). By extending the reaction time up to 30 min, the conversion was total and the yields obtained were good. However, when increasing the power to 300 W, after only 15 min acetyl group migration from carbon position 4 to C-6 of the glucose derivatives was observed.

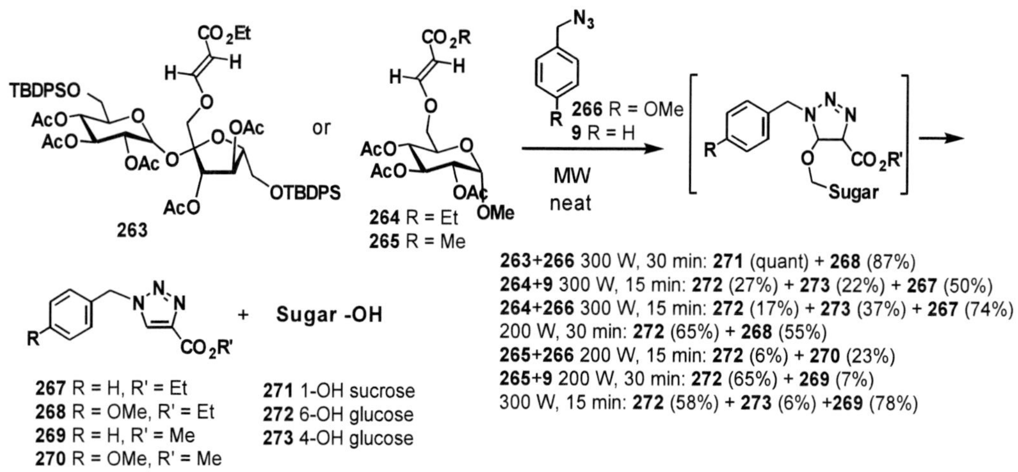

Scheme 53.

Thermal instability of the triazoline *249* formed by solvent-free microwave assisted cycloaddition of benzyl amine to alkene *248* led to its facile rearrangement cascade to polycyclic products (Scheme 22). [71] Starting with 1,3-dipolar cycloreversion, diazoalkane *250* was formed, which undergoes facile [1,3]hydrogen migration and formation of 1,5-dipole *251*, followed by intramolecular ring closure which leads to benzylimine dihydro pyrazole *252*.

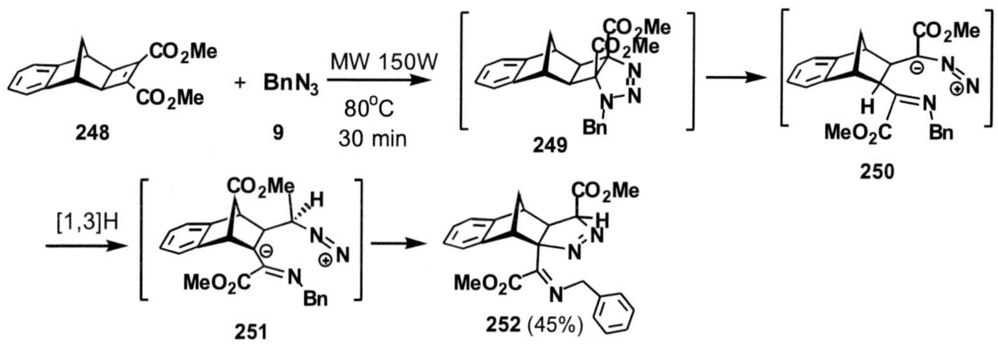

Scheme 54.

2.1.3. Synthesis of Tetrazoles

Synthesis of tetrazoles carried out by Hallberg [72] features the treatment of nitriles *253a-h* with sodium azide and ammonium chloride in DMF. A microwave power of 20 W was applied to avoid formation of the side products and a temperature of 220°C was reached after 10 min. The yields were generally good and comparable to those previously reported (Scheme 55). In addition, the microwave promoted heating technique is also suitable for conversions of iodides to tetrazoles on solid support, as demonstrated for *255* where a Rink linker on Tentagel was used as solid support. A high yield of the aryl tetrazole *256* was thus obtained. Finally, a potent C_2-symmetric HIV-1 protease inhibitor *258*, with two carboxyl group bioisosteres was prepared in one pot from the corresponding aryl bromo precursor *257*. It is notable that the bisfunctionalization could be accomplished smoothly in good yield and that no side products were detected.

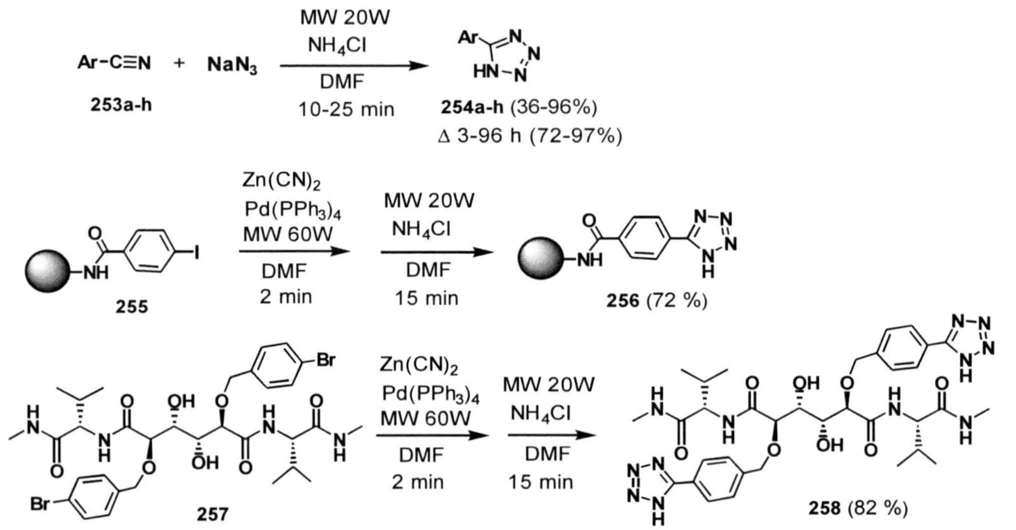

Scheme 55.

Vilarrasa [73] has found that $Cu_2(OTf)_2$-C_6H_6 catalyst is efficient for the preparation of tetrazoles from electron withdrawing nitriles and organic azides (Scheme 56). To shorten the reaction times to 1-2 h, reactions were also conducted at 80°C in microwave reactor. Organic azide/nitrile/catalyst was used in a 1.0:1.1:0.1 molar ratio in CH_2Cl_2 (1M solutions). Reactions of benzyl and phenylpropyl azide gave a single regioisomeric triazole product *261a*, while sterically more hindered PMB azide gave a mixture of *264a* with minor 1,4-regioisomer *264b*. No change in regioselectivity was observed by switching from classical to microwave heating.

Zhao syntethised novel series of dual Akt1/Akt2 kinase inhibitors, based on a 2,3,5-trisubstituted pyridine scaffold (Scheme 57). [74] Utilizing a microwave enhanced [3+2] cycloaddition of nitrile *265a-h* with NaN_3 under Sharpless conditions in water as solvent, the cyanopyridine library was rapidly converted to another tetrazolylpyridine library *266a-h* (32 compounds total).

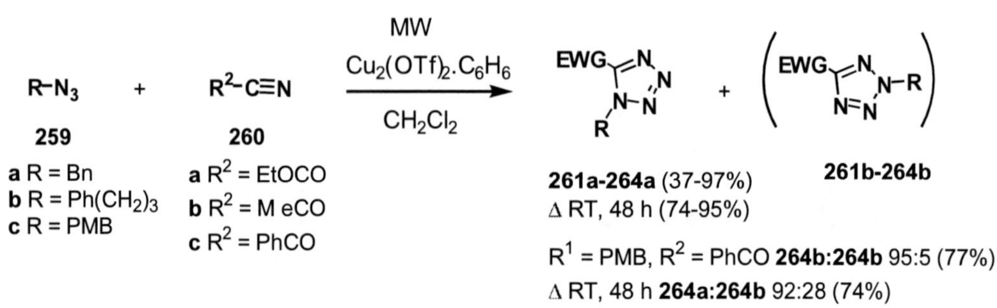

Scheme 56.

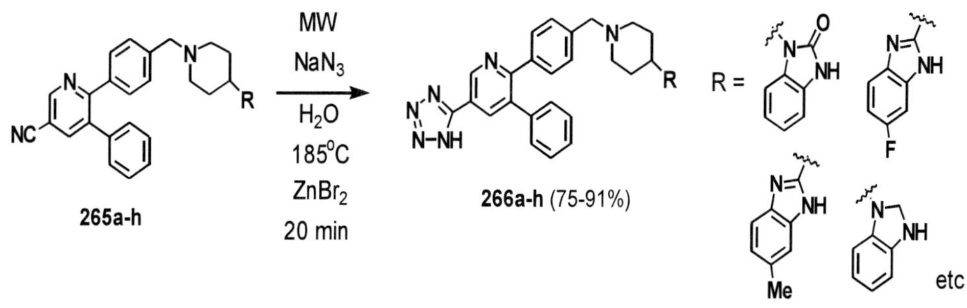

Scheme 57.

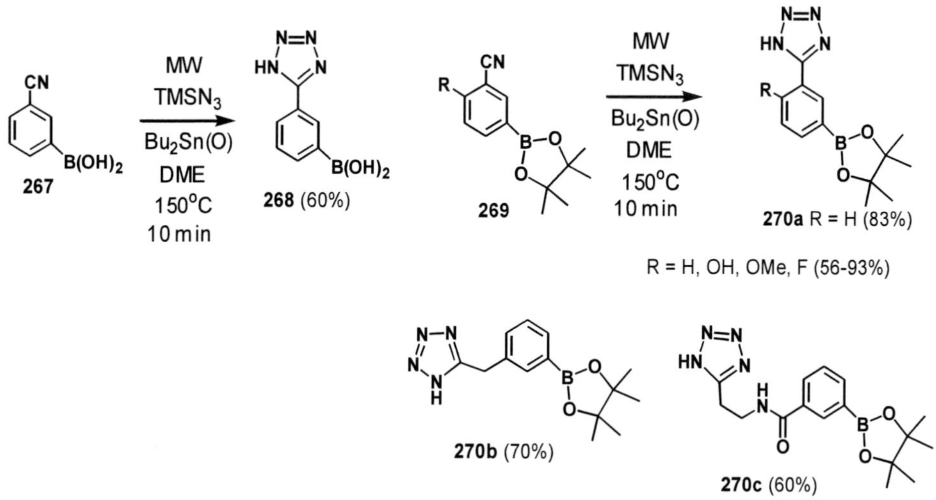

Scheme 58.

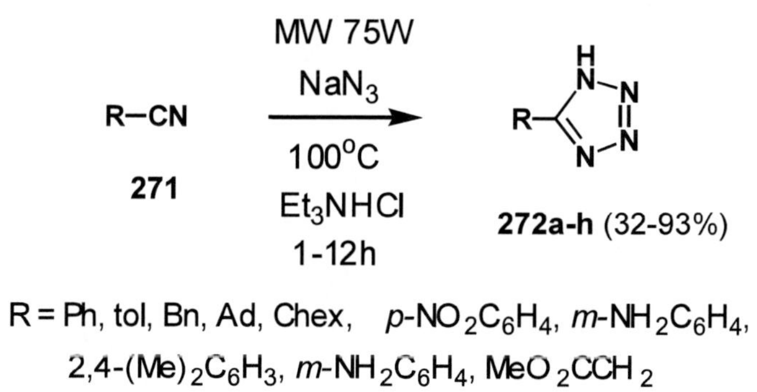

Scheme 59.

Schultz [75] has found that tetrazoles boronate ester functionality survives the reaction conditions applied for cycloaddition of azides to nitriles. The addition of azido trimethylsilane to arylnitrileboronate esters by microwave-assisted reaction proved to be useful methodology for the preparation of aryltetrazoleboronate esters (Scheme 58). Reaction was carried out in dimethoxyethane, in the presence of 10 mol% dibutyltin oxide. The protection of boronate ester as pinacol ester in *269* helped to achieve near quantitative conversion of tetrazole *270c* after two irradiation cycles. This synthetic methodology was used to prepare a series of functionalized tetrazoles *270a-h*, with *o*-, *m*- or *p*- positioned cyano functionality.

Roh [76] has developed practical synthesis of 5-substituted tetrazoles under microwave irradiation. To take advantage of microwave irradiation, hydrophobic solvents with large dipole moments were used. Toluene which has low absorption of microwave radiation (μ=1.0x10^{30} Cm) was found unsuitable, chlorobenzene and anisole gave better results, while nitrobenzene was found to be the best solvent due to its high dipole moment and minimal water solubility, which facilitates the extraction of tetrazoles in the form of their salts, into aqueous solution (Scheme 59). The advantage of this method is the simple isolation of products by aqueous extraction and recycling of the solvent. Variety of nitriles was converted to tetrazoles, including electron rich nitriles (alkyl or cycloalkyl), sterically hindered (adamantyl) or combined electron-poor/sterically hindered 5-(2,4-dimethoxybenzonitrile), and benzonitriles containing chemically sensitive ester bonds.

Koldobskii [77] has transferred Finnegan's [78] classical thermal method for synthesis of 5-substituted tetrazoles to microwave conditions in water. It was found that nitriles *273* cleanly reacted with sodium azide in water in the presence of Lewis acid ($ZnCl_2$) and microwave irradiation. Under these conditions the 5-substituted tetrazoles *274* were formed in shorter time and in high yield, regioselectively, and the reaction in water is practically secure (Scheme 51).

Sterically hindered 2,4-disubstituted 3-(5-tetrazolyl)pyridines were synthesized from corresponding nicotinonitriles using microwave technology. Lukyanov [79] converted the 2-azidomethyl-3-cyanopyridines *275a-d* to 3-(5-tetrazolyl)pyridines *276a-d* via 1,3-dipolar cycloadditions with the $TMSN_3/Bu_2SnO$ reagent system. 1,4-dioxane was used instead of toluene as the solvent more suitable for microwave conditions (Scheme 61). These reaction conditions were applied in the preparation of the 2,4-disubstituted 3-(5-tetrazolyl)pyridines *278a-f* from the corresponding sterically hindered nicotinonitriles *277a-f*. [80] Straightforward conversion of the cyanoalcohol *279* into 2-hydroxymethyl-3-(5-tetrazolyl)pyridine *280* is notable since no protection of hydroxyl group was required. The reaction of nitrile 2-vinyl derivative *281* under microwave irradiation afforded a mixture of two products, tetrazole *282a* and 5,6-dihydrotetrazolo[5,1-f]-1,6-naphthyridine *282b*, which was formed by an intramolecular *N*-alkylation of tetrazoles with olefins under neutral conditions.

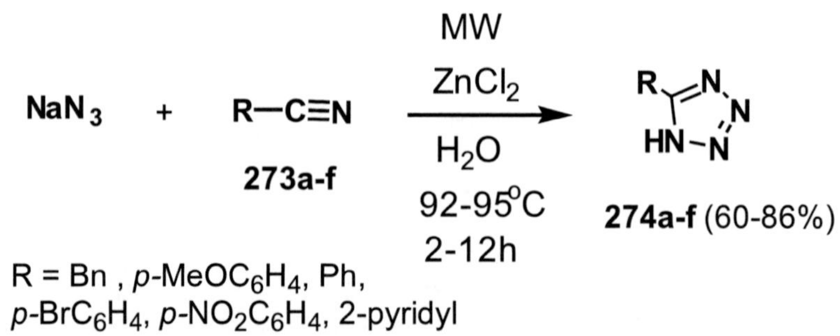

Scheme 60.

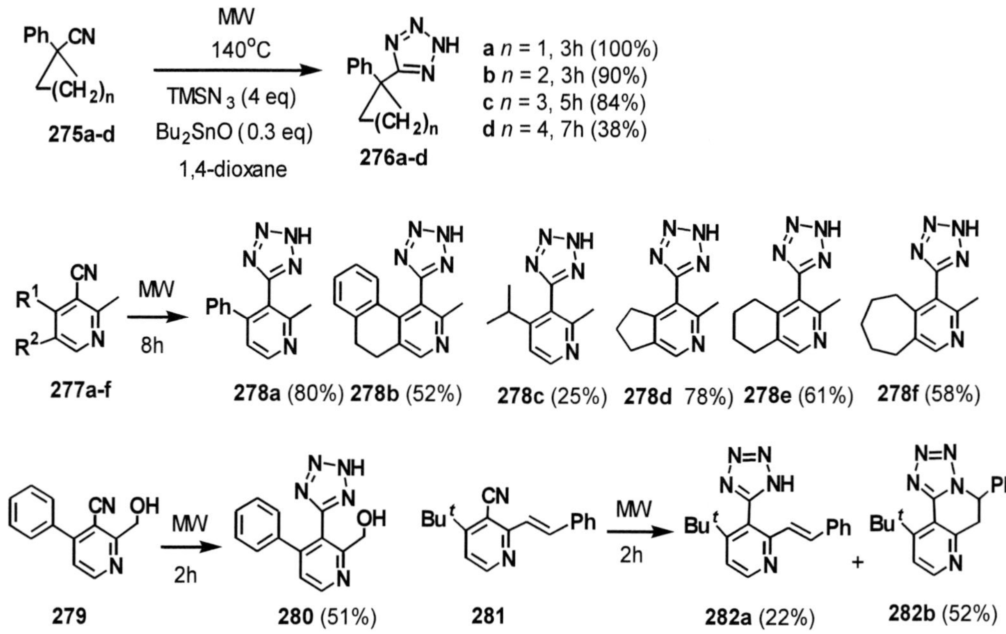

Scheme 61.

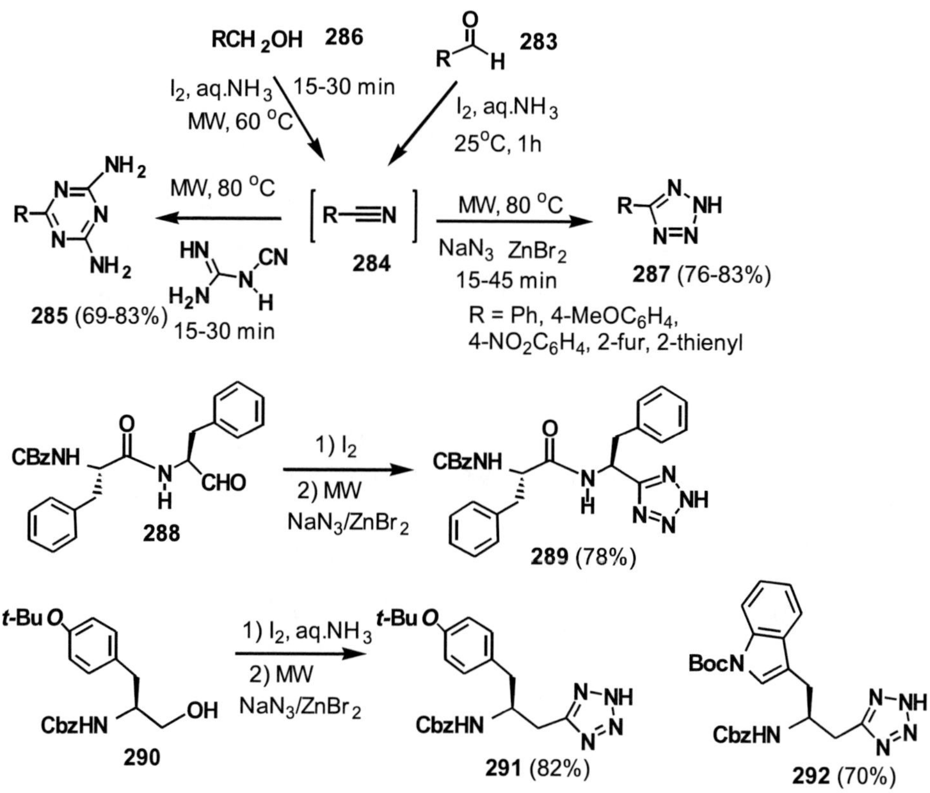

Scheme 62.

Fang [81] used $NaN_3/ZnBr_2$ reagents in one pot synthesis of 5-aryl-1,2,3,4-tetrazoles from azide and nitrile (generated *in situ* from aldehydes or alcohols). Treatment of an aldehyde (alcohol) with iodine in aqueous ammonia generated an intermediate nitrile *284*, which was subsequently reacted with azide to obtain tetrazoles *287*, while heating of nitriles with dicyandiamide furnished the [2+3] cycloaddition product 2,6-diamino-1,3,5-triazine *285* (Scheme 62). The 1,3-dipolar cycloaddition of nitrile compounds with dicyandiamide proceeded smoothly at 80°C under microwave irradiation, and the conventional heating reaction time was shortened to 15-30 min. The drastic reaction conditions were reduced by the microwave-assisted reaction in aqueous media, where the high heat capacity of water is able to mitigate the explosion hazards on sodium azide use. In comparison with the conventional heating method for tetrazole synthesis using prolonged reflux (17-48h) at a high temperature (>100°C), the microwave accelerated reaction in aqueous media is safer and more efficient. In similar manner dipeptidyl aldehyde *288* was treated with I_2-$NaN_3/ZnBr_2$ to furnish tetrazole *289*. The scope of this method was further broadened by combination with the direct oxidative conversion of primary alcohols to nitriles in iodine-ammonia water. For instance, the microwave promoted reactions of the tyrosine-derived primary alcohol *290* with iodine (4 equiv) gave the corresponding nitriles, which underwent cycloadditions with sodium azide in one pot to furnish tetrazole *291*. The direct conversion of the tryptophan-derived primary alcohol was similarly performed, albeit by using a larger quantity of NaN_3 (18 equiv) and $ZnBr_2$ (9 equiv) for a longer period (45 min), to obtain tryptophan tetrazole *292*.

Another safe and high yielding conditions for the formation of tetrazole in ionic liquids was devised by Schmidt. [82] The microwave assisted [2+3] cycloaddition of azides to nitriles in 1-butyl-3-methylimidazolium (BMIM) ionic liquid obtained tetrazoles *293-298* in significantly shorter raction times than used for thermal reactions (Scheme 63).

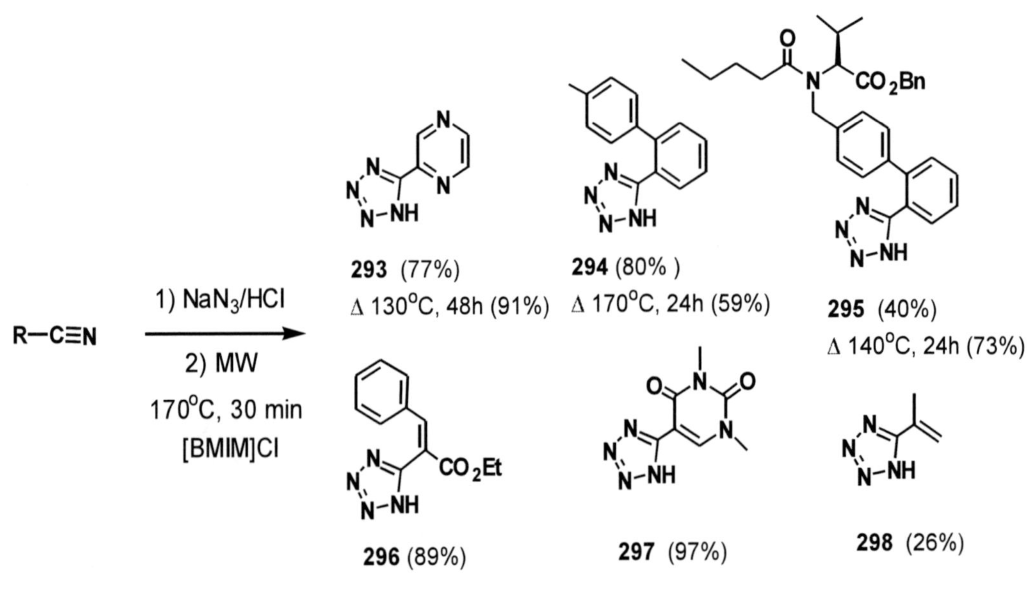

Scheme 63.

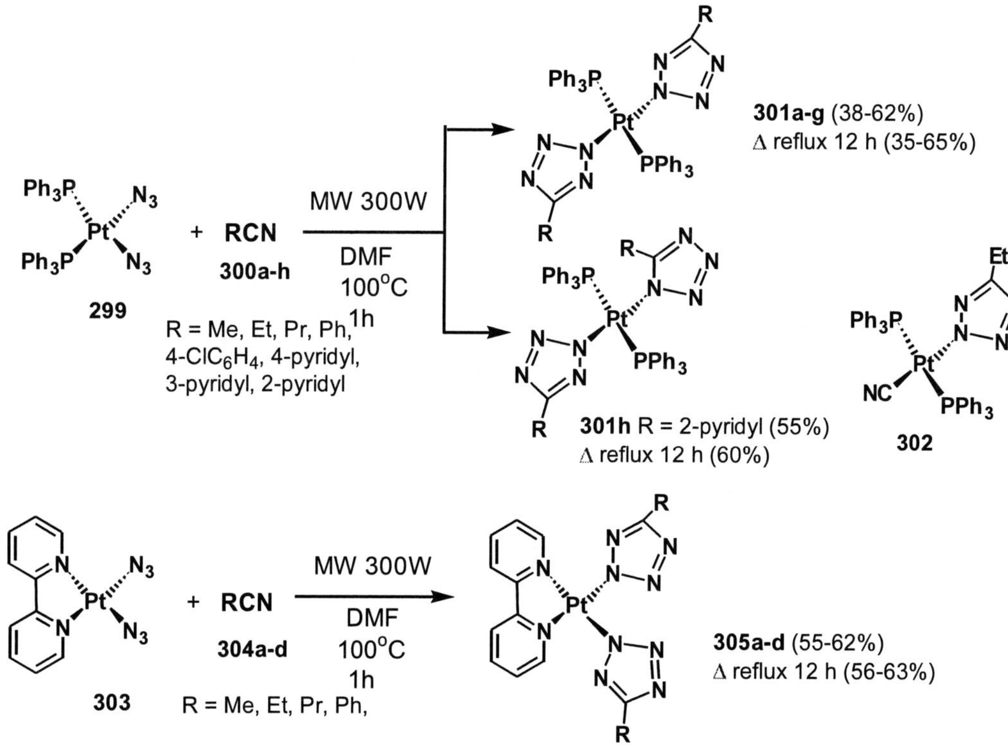

Scheme 64.

Conventional heating of the diazidoplatinum(II) complex *299*, *cis*-[Pt(N_3)$_2$(PPh_3)$_2$] with nitriles *300a-h* in refluxing DMF for 12 h furnishes the *trans-* bis(tetrazolato) complexes *301a-g* in moderate yields (Scheme 64). As shown by Pombeiro, [83, 84] these reactions are greatly accelerated under microwave irradiation (100°C, 300 W), affording yields that are comparable to those obtained under conventional heating. While cycloaddition reactions of nitriles *300a-g* produce N^2N^2-bonded isomer, nitrile *300h* furnishes N^1N^2-bonded isomer *302h.* Unexpected is the fact that thermal reaction of propionitrile *300b* with *299*, gives not only the expected *trans*-[Pt(5-ethyltetratetrazolato)$_2$(PPh_3)$_2$] *301b*, but also the cyano-complex *trans*-[Pt(CN)(5-ethyltetratetrazolato)(PPh_3)$_2$] *302.* On the other hand, microwave irradiation is, besides reaction acceleration, highly advantageous in selective formation of complex *301b.* Treatment of diazidoplatinum(II) complex *cis*-[Pt(N_3)$_2$(2,2'-bipy)] *303* with nitriles, using both methods of heating (traditional and microwave). The cycloaddition occurs to yield the bis(tetrazolato) complexes *cis*-[Pt(N_4CR)$_2$(2,2'-bipy)] *305.* Yields of microwave reactions are comparable to these obtained by classical thermal reactions.

2.1.4. Intramolecular Reactions

For the synthesis of aza-analogues of the naturally occurring bisbenzocyclooctadiene lignan lactone (-)-steganacin, Van der Eycken [85] employed one-pot three-step procedure, involving intramolecular 1,3-dipolar cycloaddition of azide with alkyne. This one-pot three-step cyclization starts with bromination of alcohols *306* (Scheme 65). The corresponding benzyl bromides *307* were not isolated, but the nucleophilic displacement of the bromo group with azide functionality (using NaN_3 under microwave irradiation) resulted in the formation of the corresponding azides *308*. Reaction sequence is finished by the intramolecular 1,3-dipolar cycloaddition reaction upon microwave irradiation furnishing the target molecules *309a-l* in good overall yields. The three steps were performed consecutively, without isolation of the intermediates and diastereoselectively.

New (-)-steganacin and (-)-steganone aza analogues were also prepared by Van der Eycken [86] *via* microwave-enhanced synthesis employing isolated azides (Scheme 66). Intramolecular heterocyclization of intermediate *310* to triazole *311* was achieved by microwave-assisted 1,3-dipolar cycloaddition at 210 °C in *o*-dichlorobenzene. This cyclization, when carried out under conventional heating in refluxing *o*-dichlorobenzene for 48 h, failed to furnish the product *311*. The cyclization of the azide *312* into dibenzotriazolo[1,5-*a*]azocine *313* was possible at a much lower temperature and lower irradiation power than for compound *311*, due to the lower rigidity of the eight-member ring, resulting in lower activation energy. The cyclization of *312* was found to be highly diastereoselective, generating the isomers *313* in a ratio of 9:1, the major isomer was assigned to be diastereomer *313*-(*S*).

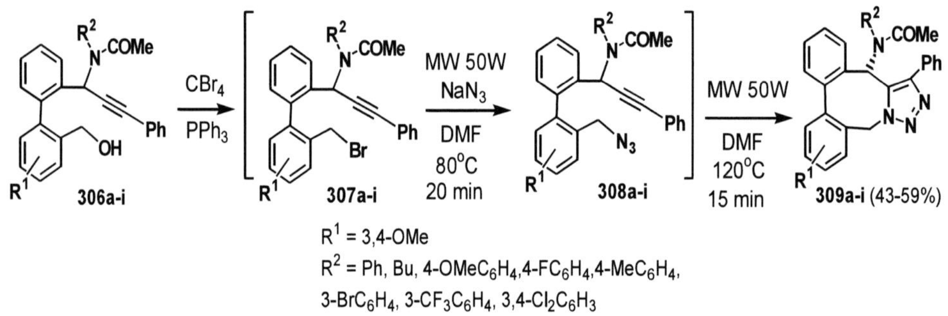

Scheme 65.

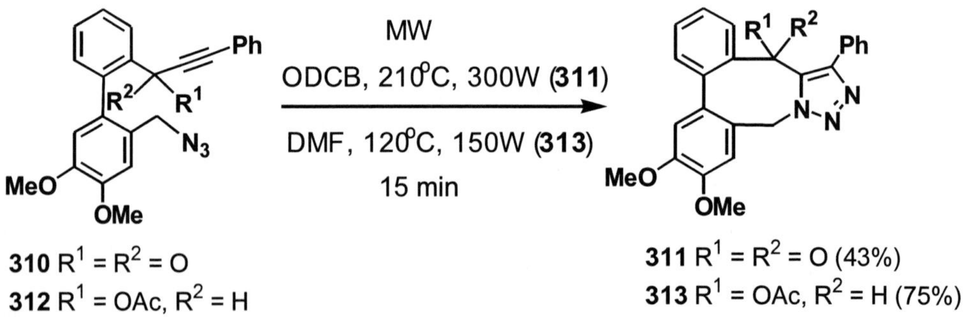

Scheme 66.

New strategies for cyclization of oligonucleotides by click chemistry assisted by microwaves were devised by Morvan. [87] Linear oligonucleotide without *311b* and with the disulfide linker *310b* were subjected to microwave copper-catalyzed azide alkyne cycloaddition in solution (Scheme 67). Two main products *313b* and *312b* correspond to the cyclic form without and with the disulfide linker. A treatment with TCEP reduced the disulfide bound to the cyclic T_9 (cT_9) *313b*. Similarly, cyclization of the heavily functionalized oligonucleotide *314* by intramolecular CuAAC reaction assisted by microwave at 60°C was complete within 1 h, to afford cyclic oligonucleotide *315*.

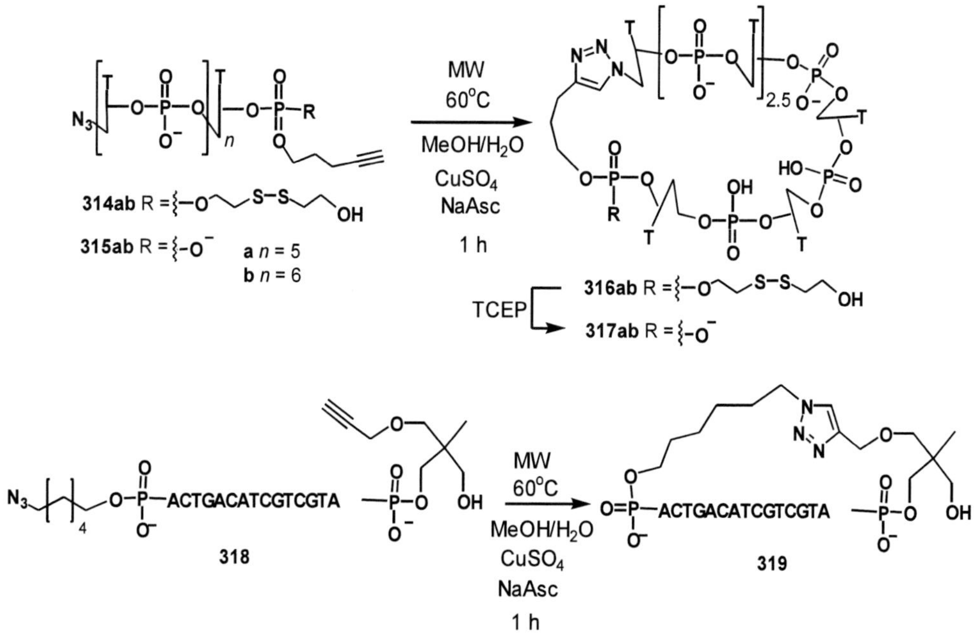

Scheme 67.

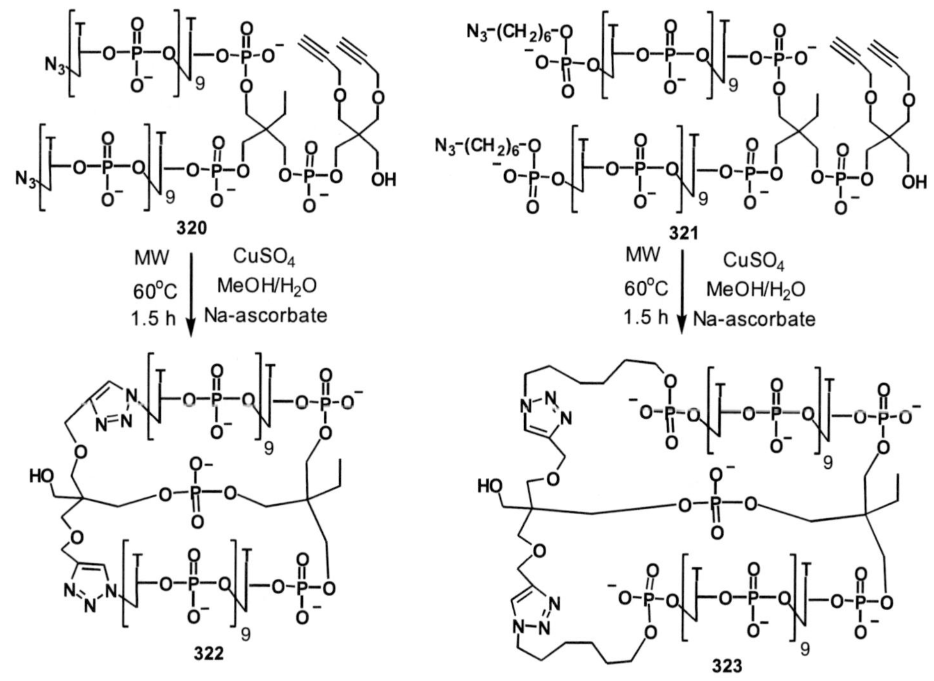

Scheme 68.

The most remarkable achievement by this synthetic methodology is the double-intramolecular 1,3-dipolar cycloaddition reaction performed on substrates *320* and *321*, possessing two azide and one alkyne functions, obtaining branched bicyclic oligonucleotides *322* and *323* (Scheme 68).

Taddei [88] has found that propargyl α-azido amides *324* undergo intramolecular 1,3-dipolar cycloaddition under microwave heating to provide access to new condensed triazoles *325*. On the other hand, α-azido propargyl esters *326* derived from α-amino acids undergo intermolecular [3+2] cycloaddition, forming the oligomeric cyclic polyester *327* (Scheme 69). It was found that 7-substituted-4,5-dihydro[1,2,3]triazolo[1,5-*a*]pyrazin-6(7*H*)-ones *325b-f* were obtained in good yields and ee >95%. Compound *325a* was instead obtained with ee=70%, due to partial racemization of the phenylalanine azide during the heating. In some case of allylic α-azido amide cyclisation was repeated three times to achieve acceptable yields of triazoles *325g-j*. Cyclization at the same temperature under conventional heating, after 24 h at 160°C led to complete consumption of the starting material without formation of *325g*. Ester compound *326*, when subjected to similar reaction conditions for cyclization gave oligomer *327*. Finally, aminolysis of ester bond of oligomer *327* resulted in the isolation of triazole *328*.

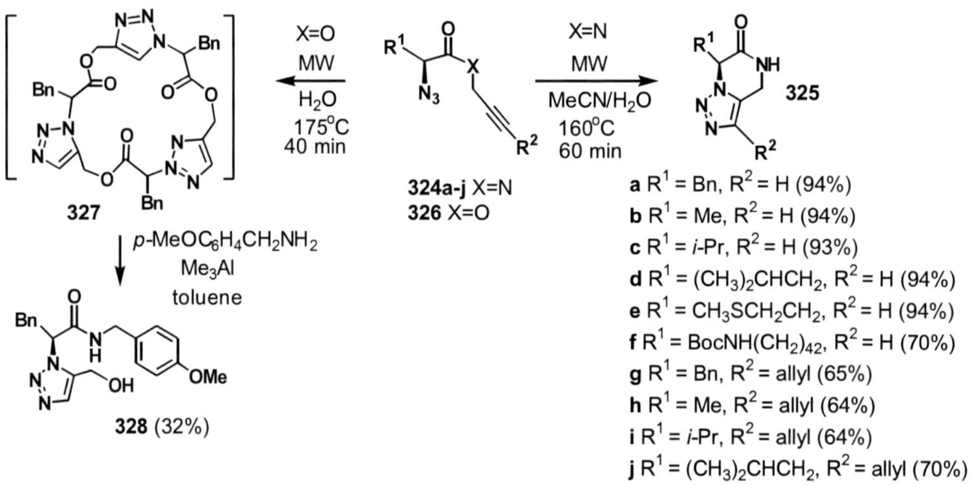

Scheme 69.

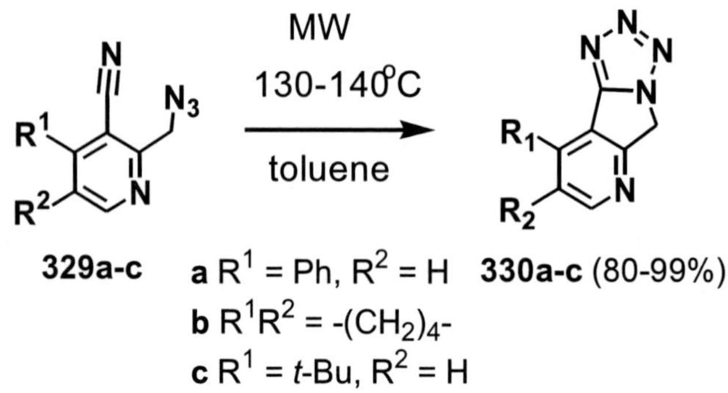

Scheme 70.

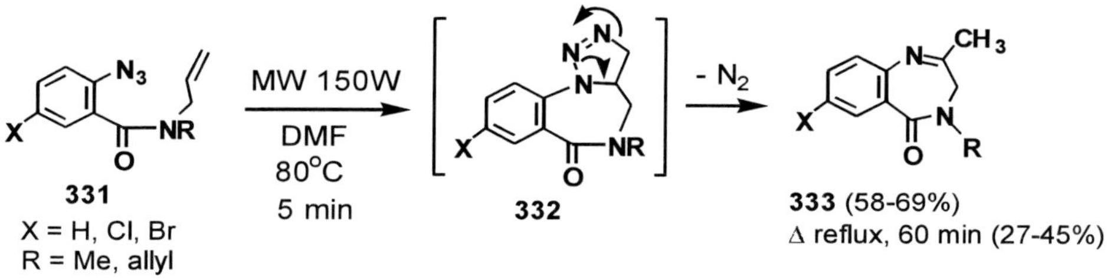

Scheme 71.

The 2-azidomethyl-3-cyanopyridines *329a-c* were effectively cyclized intramolecularly by Lukyanov [79] *via* microwave heating in the toluene solution at 130-140°C to obtain (tetrazolo)azaisoindols *330a-c* in high yields (Scheme 70).

Santagada [89] used microwave enhanced solution synthesis of 1,4-benzodiazepin-5-ones *331* by intramolecular 1,3-dipolar cycloaddition of an azide with alkene functionalities. The intermediate triazolines *332* which were obtained *in situ* either by conventional refluxing in DMF or by exposure to microwave irradiation, rearranged spontaneously to *333* with yields better than these obtained by conventional thermal reaction in refluxing DMF (Scheme 71).

References

[1] Huisgen, R. *Angew. Chem. Int. Ed. 1963*, *2*, 565-632.

[2] Sustmann, R. *Heterocycles 1995*, *40*, 1-68.

[3] Padwa, A.; Pearson, W. H. (Eds.), *Synthetic Applications of 1,3-dipolar Cycloaddition Chemistry toward Heterocycles and Natural Products*, Wiley, New York, *2002*.

[4] Scriven, E. F. V. (ed.), *Azides and Nitrenes, Reactivity and Utility*, Academic Press, New York, *1984*.

[5] Tufariello, J. J., Nitrones, in *1,3-Dipolar Cycloaddition Chemistry*, Vol. 2, Padwa, A. (ed.), Wiley, New York, 1*984*, pp. 83-168.

[6] Torssell, K. B. G. Nitrile Oxides, Nitrones, and Nitronates in Organic Synthesis, in *Organic Nitro Chemistry*, Feuer, H. (ed.), VCH, Weinheim, *1988*.

[7] Grigg, R.; Sridharan, V. in *Advances in Cycloaddition,* Vol. 3, Curran D.P. (ed.), JAI Press, Greenwich, *1993*, pp. 161-204.

[8] Doyle, M. P.; McKervey, M. A.; Ye, T. Modern Catalytic Methods for Organic Synthesis with Diazo Compounds-From Cyclopropanes to Ylides, Wiley, New York, *1998*.

[9] Gribble, G. W. Mesoionic Ring Systems, Chapter 10. in *Synthetic Applications of 1,3-Dipolar Cycloaddition Chemistry Toward Heterocycles and Natural Products*, Vol. 59, Padwa, A., Pearson, W. H. (eds.), Wiley, New York, *2003*, pp 681-754.

[10] Pineiro, M.; Pinho e Melo, T. M. V. D. *Eur. J. Org. Chem. 2009*, *31*, 5287-5307.

[11] de la Hoz, A.; Díaz-Ortíz, A.; Moreno, A.; Langa, F. *Eur. J. Org. Chem. 2000*, 3659-3673.

[12] Appukkuttan, P.; Van der Eycken, E. *Eur. J. Org. Chem. 2008*, 1133-1155.

[13] Bräse, S.; Gil, C.; Knepper, K.; Zimmermann, V. *Angew. Chem. Int. Ed. 2005*, *44*, 5188-5240.

[14] Margetić, D.; Jin, C.-M.; Warrener, R. N.; Butler, D. N., 8th *Electronic Conference on Synthetic Organic Chemistry* (ECSOC-8), Seijas Vázquez, J. A.; Lin, S.-K.; Ji, D-C. (Eds.), http://www.lugo.usc.es/~qoseijas/ECSOC-8/, 1-30 November *2004*.
[15] Anderson, G. T.; Henry, J. R.; Weinreb, S. M. *J. Org. Chem. 1991*, *56*, 6946-6948; Butler, D. N.; Malpass, J. R.; Margetić, D.; Russell, R. A.; Sun, G.; Warrener, R. N. *Synlett 1998*, 588-589.
[16] Tornøe, C. W.; Christensen, C.; Meldal, M. *J. Org. Chem. 2002*, *67*, 3057-3064; Rostovtsev, V. V.; Green, L. G.; Fokin, V. V.; Sharpless, K. B. *Angew. Chem., Int. Ed. 2002*, *41*, 2596-2599.
[17] Lahann, J. (Ed.), *Click Chemistry for Biotechnology and Materials Science*, Wiley, Chichester, *2009*.
[18] Kappe, C. O.; Van der Eycken, E. V. *Chem. Soc. Rev. 2010*, *39*, 1280-1290.
[19] Louërat, F.; Bougrin, K.; Loupy, A.; Ochoa de Retana, A. M.; Pagalday, J.; Palacios, F. *Heterocycles 1998*, *48*, 161-170.
[20] Tao, L.; Zhang, L. L.; Shen, S. J.; Han, X. P. *Chin. Chem. Lett. 2001*, *12*, 763-764.
[21] Katritzky, A. R.; Singh, S. K. *J. Org. Chem. 2002*, *67*, 9077-9079.
[22] Katritzky, A. R.; Yuming Zhang, Y.; Singh, S. K.; Steel, P. J. *ARKIVOC 2003*, *xv*, 47-64.
[23] Katritzky, A. R.; Zhang, Y.; Singh, S. K. *Heterocycles 2003*, *61*, 1225-1240.
[24] Savin, K. A.; Robertson, M.; Gernert, D.; Green, S.; Hembre, E. J.; Bishop, J. *Mol. Diversity 2003*, *7*, 171-174.
[25] Taherpour, A. A.; Kheradmand, K. *J. Het. Chem. 2009*, *46*, 131-133.
[26] Guezguez, R.; Bougrin, K.; El Akria, K.; Benhida, R. *Tetrahedron Lett. 2006*, *47*, 4807-4811.
[27] El Akri, K.; Bougrin, K.; Balzarini J.; Faraj, A.; Benhida, R. *Bioorg. Med. Chem. Lett. 2007*, *17*, 6656-6659.
[28] Pisaneschi, F.; Cordero, F. M.; Lumini, M.; Brandi, A. Synlett *2007*, 2882-2884..
[29] Castagnolo, D.; Dessì, F.; Radi, M.; Botta, M. *Tetrahedron: Asymmetry 2007*, *18*, 1345-1350.
[30] Broggi, J.; Joubert, N.; Aucagne, V.; Zevaco, T.; Berteina-Raboin, S.; Nolan, S. P.; Agrofoglio, L. A. *Nucleos. Nucleot. Nucl. Acids 2007*, *26*, 779-783.
[31] Glasnov, T. N.; Kappe, C. O. *QSAR Comb. Sci. 2007*, *26*, 1261-1265.
[32] Ermolat'ev, D. S.; Pravinchandra Mehta, V.; Van der Eycken, E. V. *QSAR Comb. Sci. 2007*, *26*, 1266-1273.
[33] Perissutti, P.; Frecentese, F.; Fiorino, F.; Severino, B.; Cirillo, D.; Santagada, V.; Caliendo, G. *J. Heterocyclic Chem. 2006*, *43*, 1-7.
[34] Kaval, N; Ermolat'ev, D.; Appukkuttan, P.; Dehaen, W.; Kappe, C. O.; Van der Eycken, E. *J. Comb. Chem. 2005, 7,* 490-502.
[35] Chan, T. R.; Hilgraf, R.; Sharpless, K. B.; Fokin, V. V. *Org. Lett. 2004*, *6*, 2853-2856.
[36] Ermolat'ev, D.; Dehaen, W.; Van der Eycken, E. *QSAR Comb. Sci. 2004*, *23*, 915-918.
[37] Miller, N.; Williams, G. M.; Brimble, M. A. *Org. Lett. 2009*, *11*, 2409-2412.
[38] Pietrzik, N.; Schips, C.; Ziegler, T. *Synthesis 2008*, 519-526.
[39] van Kasteren, S. I.; Kramer, H. B.; Jensen, H. H.; Campbell, S. J.; Kirkpatrick, J.; Oldham, N. J.; Anthony, D. C.; Davis, B. G. *Nature 2007*, *446*, 1105–1109.
[40] Jang, H. J.; Fafarman, A.; Holub, J. M.; Kirshenbaum, K. *Org. Lett. 2005*, *7*, 1951-1954.

[41] Elgersma, R. C.; van Dijk, M.; Dechesne, A. C.; van Nostrum, C. F.; Hennink, W. E.; Rijkers, D. T. S.; Liskamp, R. M. J. *Org. Biomol. Chem. 2009*, *7*, 4517-4525.

[42] Géci, I.; Filichev, V. V.; Pedersen, E. B. *Chem. Eur. J. 2007*, *13*, 6379-6386.

[43] Bouillon, C.; Meyer, A.; Vidal, S.; Jochum, A.; Chevolot, Y.; Cloarec, J.-P.; Praly, J.-P.; Vasseur, J.-J.; Morvan F. J. Org. Chem. 2006, 71, 4700-4702.

[44] Cheshev, P.; Marra, A; Dondoni, A. *Org. Biomol. Chem. 2006*, *4*, 3225-3227.

[45] Lucas, R.; Zerrouki, Z.; Granet, R.; Krausz, P.; Champavier, Y. *Tetrahedron 2008*, *64*, 5467-5471.

[46] Lucas, R.; Neto, V.; Bouazza, A. H.; Zerrouki, R.; Granet, R.; Krausz, P.; Champavier, Y. *Tetrahedron Lett. 2008*, *49*, 1004-1007.

[47] Bew, S. P.; Brimage, R. A.; L'Hermite, N.; Sharma, S. V. *Org. Lett. 2007*, 9, 3713-3716.

[48] Garska, B.; Tabatabai, M.; Ritter, H. *Beilstein J. Org. Chem. 2010, 6*. doi:10.3762/bjoc.6.83

[49] Rijkers, D. T. S.; van Esse, G. W.; Merkx, R.; Brouwer, A. J.; Jacobs, H. J. F.; Pieters, R.; Liskamp, R. M. J. *Chem. Commun. 2005*, 4581-4583.

[50] Dijkgraaf, I.; Rijnders, A. Y.; Soede, A.; Dechesne, A. C.; van Esse, G. W.; Brouwer, A. J.; Corstens, F. H. M.; Boerman, O. C.; Rijkers, D. T. S.; Liskamp, R. M. J. *Org. Biomol. Chem. 2007*, *5*, 935-944.

[51] Joosten, J. A. F.; Tholen, N. T. H.; El Maate, F. A.; Brouwer, A. J.; van Esse, G. W.; Rijkers, D. T. S.; Liskamp, R. M.; Pieters, R. J. *Eur. J. Org. Chem. 2005*, 3182-3185.

[52] Yoon, K.; Goyal, P.; Weck, M. *Org. Lett. 2007*, *9*, 2051-2054.

[53] Hoogenboom R., Moore B., Schubert U. *Chem. Commun.* 2006, 4010-4012.

[54] Cravotto, G.; Mendicuti, F.; Martina, K.; Tagliapietra, S.; Robaldo, B.; Barge, A. *Synlett 2008*, 2642-2646.

[55] Aime, S.; Gianolio, E.; Arena, F.; Barge, A.; Martina, K.; Heropoulos, G.; Cravotto, G. *Org. Biomol. Chem. 2009*, *7*, 370-379.

[56] Binello, A.; Robaldo, B.; Barge, A.; Cavalli, R.; Cravotto, G. *J. Appl. Polym. Sci. 2008*, *107*, 2549-2557.

[57] Mallard-Favier, I.; Blach, P.; Cazier, F.; Delattre, F. *Carbohydr. Res. 2009*, *344*, 161-166.

[58] Pérez-Balderas, F.; Ortega-Muñoz, M.; Morales-Sanfrutos, J.; Hernández-Mateo, F.; Calvo-Flores, F. G.; Calvo-Asín, J. A.; Isac-García, J.; Santoyo-González, F. *Org. Lett. 2003*, *5*, 1951-1954.

[59] Vecchi, A.; Melai, B.; Marra, A.: Chiappe, C.; Dondoni, A. *J. Org. Chem. 2008*, *73*, 6437-6440.

[60] Lipshutz, B. H.; Taft, B. R. *Angew. Chem. Int. Ed. 2006*, *45*, 8235-8238.

[61] Rasmussen, L. K.; Boren, B. C.; V. Fokin, V. V. *Org. Lett. 2007*, *9*, 5337-5339.

[62] Maireck, M, M.; Weinreb, S. M. *J. Org. Chem. 2006*, *71*, 8680-8683.

[63] Khanetskyy, B.; Dallinger, D.; Kappe, C. O. *J. Comb. Chem. 2004, 6,* 884-892.

[64] Beckmann, H. S. G.; Wittmann, V. *Org. Lett. 2007*, *9*, 1-4.

[65] Dyrager, C.; Börjesson, K.; Dinér, P.; Elf, A.; Albinsson, B.; Wilhelmsson, L. M.; Grøtli, M. *Eur. J. Org. Chem. 2009*, 1515-1521.

[66] Angelo, N. G.; Arora, P. S. *J. Org. Chem. 2007*, *72*, 7963-7967.

[67] Moorhouse, A. D.; Moses, J. E. *Synlett 2008*, 2089-2092.

[68] Cintas, P.; Marina, K.; Robaldo, B.; Garella, D.; Boffa, L.; Cravotto, G. *Collect. Czech Chem. Commun. 2007*, *72*, 1014-1024.

[69] Kočalka, P.; Andersen, N. K.; Jensen, F.; Nielsen, P. *ChemBioChem 2007, 8*, 2106-2116.

[70] Andrade, M. M.; Barros, M. T. *ARKIVOC 2009*, *xi*, 299-306.

[71] Margetić, D. unpublished results.

[72] Alterman, M.; Hallberg, A. *J. Org. Chem. 2000*, *65*, 7984-7989.

[73] Bosch, L.; Vilarrasa, J. *Angew. Chem. Int. Ed. 2007*, *46*, 3926-3930.

[74] Zhao, Z.; Leister, W. H.; Robinson, R. G.; Barnett, S. F.; Defeo-Jones, D.; Jones, R. E.; Hartman, G. D.; Huff, J. R.; Huber, H. E.; Duggana, M. E.; Lindsley, C. W. *Bioorg. Med. Chem. Lett. 2005*, *15*, 905-909.

[75] Schultz, M. J.; Coats, S. J.; Hlasta, D. J. *Org. Lett. 2004*, *6*, 3265-3268.

[76] Roh, J.; Artamonova, T. V.; Vávrová, K.; Koldobskii, G. I.; Hrabálek, A. *Synthesis 2009*. 2175-2178.

[77] Myznikova, L. V.; Roh, J.; Artamonova, T. V.; Hrabalek, A.; Koldobskii, G. I. *Russ. J. Org. Chem. 2007*, *43*, 765-767.

[78] Finnegan, W. G.; Henry, R. A.; Lofquist, R. *J. Am. Chem. Soc. 1958*, *80*, 3908-3911.

[79] Lukyanov, S. M.; Bliznets, I. V.; Shorshnev, S. V.; Aleksandrov, G. G.; Stepanov, A. E.; Vasil'ev, A. A. *Tetrahedron 2006*, *62*, 1849-1863.

[80] Bliznets, I. V.; Vasil'ev, A. A.; Shorshnev, S. V.; Stepanova, A. E.; Lukyanov, S. M. *Tetrahedron Lett. 2004*, *45*, 2571-2573.

[81] Shie, J.-J.; Fang, J.-M. *J. Org. Chem. 2007*, *72*, 3141-3144.

[82] Schmidt, B.; Meid, D.; Kieser, D. *Tetrahedron 2007*, *63*, 492-496.

[83] Mukhopadhyay, S.; Lasri, J.; Janúario Charmier, M. A.; Guedes da Silva M. F. C.; Pombeiro, A. J. L. *J. Chem. Soc. Dalton Trans. 2007*, 5297-5304.

[84] Mukhopadhyay, S.; Mukhopadhyay, B. G.; Guedes da Silva, M. F. C.; Lasri, Januário Charmier, M. A.; Pombeiro, A. J. L. *Inorg. Chem. 2008*, *47*, 11334-11341.

[85] Mont, N.; Pravinchandra Mehta, V.; Appukkuttan, P.; Beryozkina, T.; Toppet, S.; Van Hecke, K.; Van Meervelt, L.; Voet, A.; DeMaeyer, M.; Van der Eycken, E. *J. Org. Chem. 2008*, *73*, 7509-7516.

[86] Beryozkina, T.; Appukkuttan, P.; Mont, N.; Van der Eycken, E. *Org. Lett. 2006, 8*, 487-490.

[87] Lietard, L.; Albert Meyer, A.; Vasseur, J.-J.; Morvan, F. *J. Org. Chem. 2008*, *73*, 191-200.

[88] Balducci, E.; Bellucci, L.; Petricci, E.; Taddei, M.; Tafi, A. *J. Org. Chem. 2009*, *74*, 1314-1321.

[89] Santagada, V; Perissutti, E.; Fiorino, F.; Vivenzio, B.; Caliendo, G. *Tetrahedron Lett. 2001*, *42*, 2397-2400.

2.2. Nitrones

Introduction

Huisgen classified nitrones as allyl-type dipoles with nitrogen function as middle center. The 1,3-dipolar cycloaddition reactions involving nitrones and alkenes produce isoxazolidine ring, which is important five-membered nitrogen heterocycle used as precursor to many natural and pharmaceutical compounds. [1] For the synthetic purposes, nitrones are either used as isolated stable reagents or thermally generated *in situ* from corresponding aldehydes and hydroxylamines.

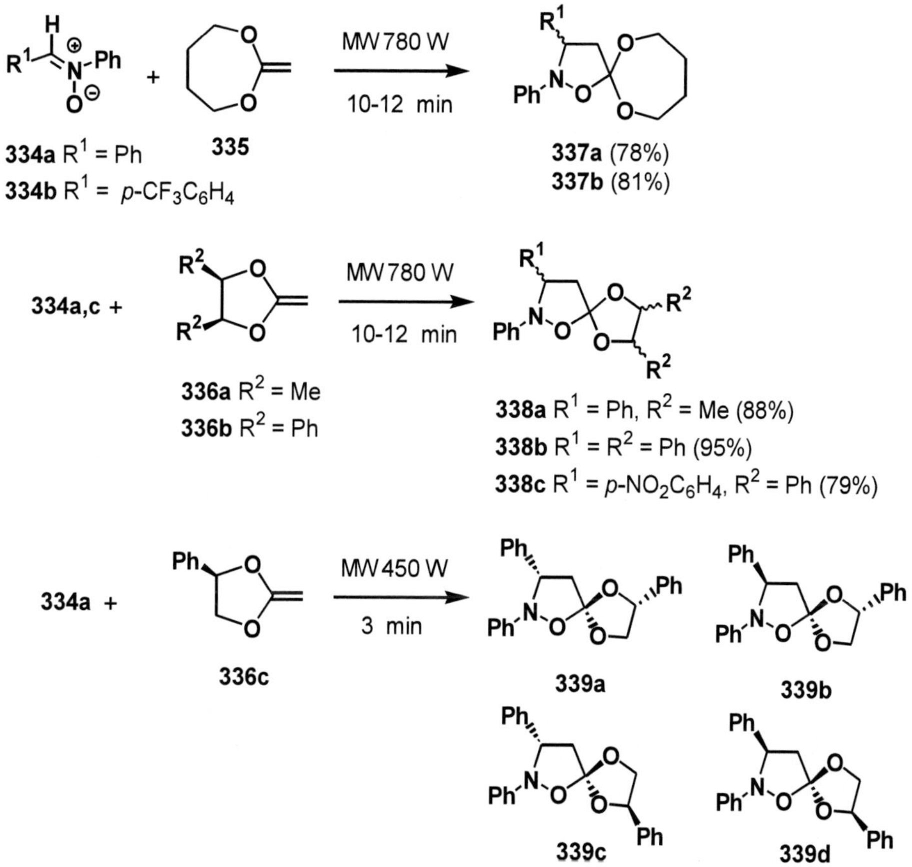

Scheme 72.

2.2.1. Intermolecular Reactions

The first account on the microwave assisted 1,3-dipolar cycloaddition of nitrones was published by Díaz-Ortiz. [2] The 1,3-dipolar cycloaddition of nitrones *334a-c* to ketene acetals *335* and *336a-c* under microwave irradiation in solvent-free conditions proceed in high yield and short reaction time to obtain spiro isoxazolidines *337a,b* and *338a-c* (Scheme

72). When *336c* was used as dipolarophile, complex mixture of stereoisomers *339a-d* was obtained. [3] It was found that the diastereomeric ratios were *339a/b*:38/36, and *339c/d*:18/8, with facial selectivity of 2.8:1. Significant changes in the facial selectivity were not observed either in thermal or under ultrasound irradiation.

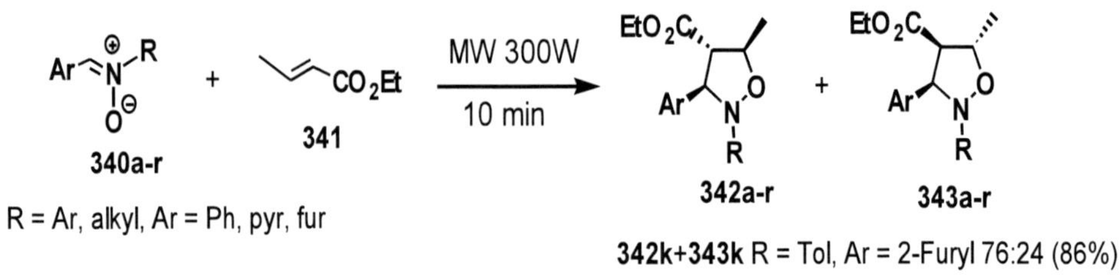

Scheme 73.

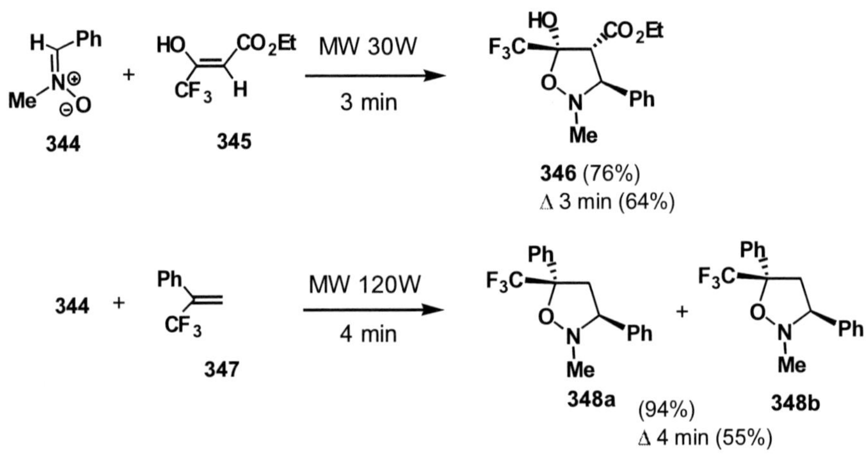

Scheme 74.

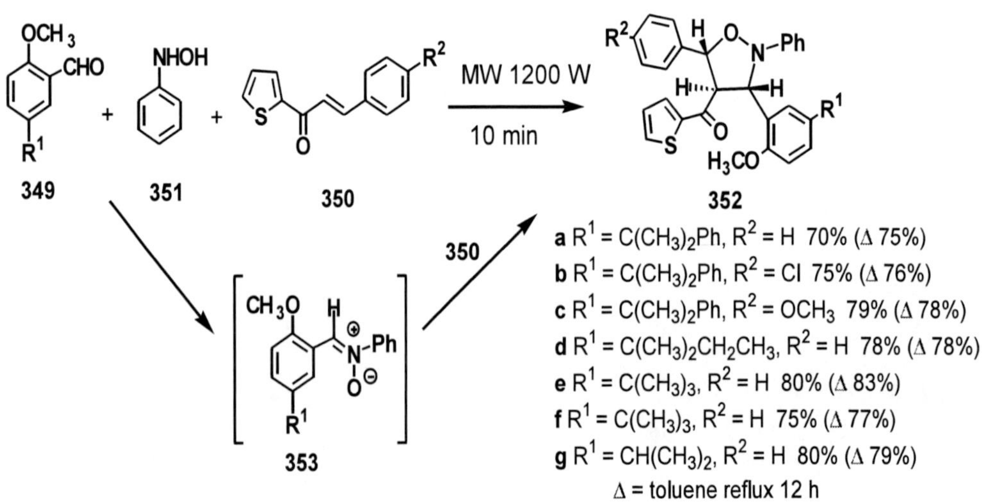

Scheme 75.

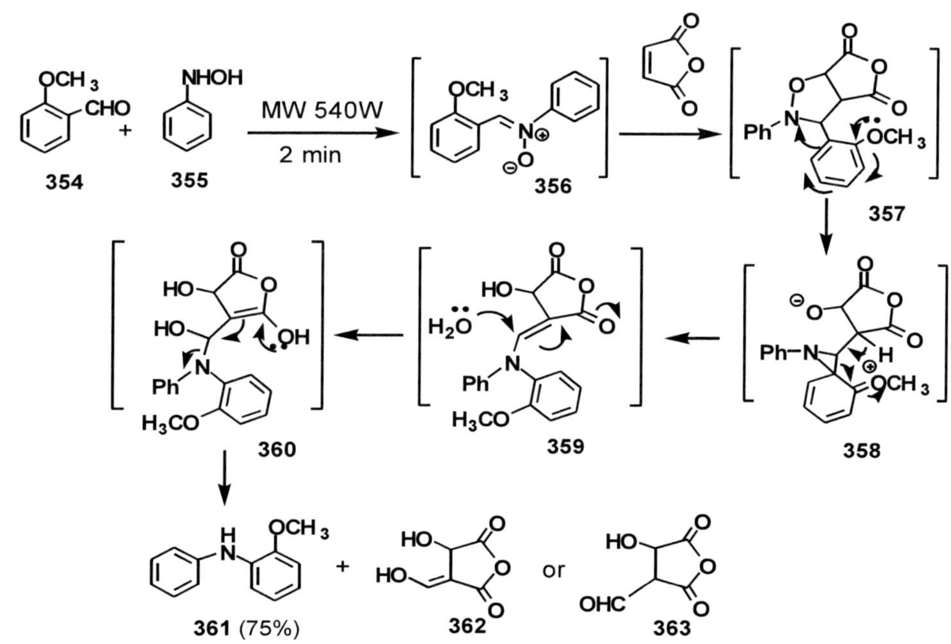

Scheme 76.

A library of *C*-aryl-*N*-alkyl and *C,N*-diarylnitrones *340a-r* was used in studies on their 1,3-dipolar cycloadditions to ethyl *trans*-crotonate *341* (Scheme 73). Andrade [4] carried out microwave reactions in the absence of solvent and completed within 10 minutes (75-98% yield). Cycloadditions were regioselective, with two diastereomers *342* and *343* obtained, in which *342* is dominant, regardless of substitution. Bulky R substituents diminish nitrone reactivity and require significantly longer reaction times (20-60 min) to obtain adducts in 26-87% yield.

Considerable improvement in yields and experimental conditions was obtained by Loupy [5] in reactions of nitrones to fluorinated dipolarophiles under solvent-free microwave activation, compared to classical reflux in toluene over 24-48 h. Thus, solvent-free microwave cycloaddition between *N*-(benzylidene)methylamine *N*-oxide *344* and ethyl trifluoracetoacetate *345* afforded the isoxazolidine *346* in 76% yield, regio- and stereo-selectively within only 3 min (Scheme 74). In the case of reaction *344* with α-trifluoromethylstyrene *347*, products *348a* and *348b* were isolated in 94% yield at irradiation power of 120 W. When performed under the same conditions of time and temperature, conventional heating led to only 64% and 55% yields, respectively.

Three-component stereoselective synthesis of tetra substituted isoxazolidines *352* from 5-substituted 2-methoxybenzaldehydes *349*, *N*-phenylhydroxylamine *351* and 1-(2-thienyl)-3-arylprop-2-en-1-ones *350* was achieved by solvent-free microwave enhanced reaction by Muthusubramanian [6] (Scheme 75). In the reaction conditions, aldehyde and *351* generate *in situ* corresponding nitrone *353*, which undergoes 1,3-dipolar reaction with dipolarophiles *350*. While the use of microwave conditions was advantageous in terms of significant shortening of the reaction times compared to classical reflux, the yields of the [3-(5-substituted-2-methoxyphenyl)-5-aryl-2-phenyltetrahydro-4-isoxazolyl](2-thienyl) methanone products were similar for both reaction conditions.

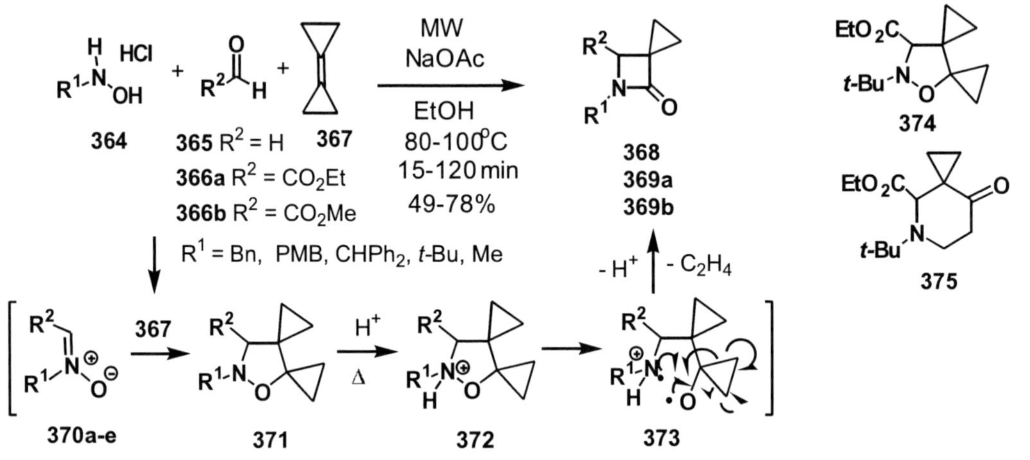

Scheme 77.

Sometimes in the reaction conditions employed, isoxazolidines obtained by cycloadditions undergo decomposition or rearrangements to other products. For instance, Muthusubramanian [7] has obtained unexpected multicomponent reaction of 2/4-methoxyarylaldehydes *354* with arylhydroxylamines *355* and maleic anhydride which led to unsymmetrical diarylamines *361* (Scheme 76). Microwave (2-3 min) or classical heating (4-5 h) conditions led to the formation of identical products, while microwave reactions gave on average 10% higher yields. A possible mechanism for the formation of *361* includes the expected isoxazolidine cycloadduct *357* as an intermediate. The role of the methoxy group in the migration of the aryl group from carbon to nitrogen must be important in mechanism. Intermediate *358* leads to the enone intermediate *359*. Nucleophilic attack of water, formed during nitrone formation, on the β-carbon of an enone intermediate obtained from *358*, then gives the diarylamine via intermediate *360*. Anhydrides *362* and *363* are the sideproducts of this process.

Three-component (one pot) cascade reaction was used by de Meijere [8] for direct conversion of alkylhydroxylamine hydrochlorides *364*, formaldehyde *365* (or an alkyl glyoxylate *366*) and bicyclopropylidene *367* (acting as dipolarophile) to yield 3-spirocyclopropanated β-lactams *368* (Scheme 77). Microwave heating of mixture of three components in the presence of sodium acetate in ethanol leads to initial formation of nitrone from *364* and *365*. Nitrone *370* undergoes 1,3-dipolar cycloaddition to *367*, and the resulting isoxazolidine *371*, under the slightly acidic conditions fragmented via intermediates *372* and *373* to ethylene and the β-lactam *368*. In the reaction of *N-tert*-butylhydroxylamine hydrochloride *364d*, ethyl glyoxylate *366a* and *367*, one of the proposed reaction intermediates was isolated, the unfragmented *N-tert*-butylisoxazolidine *374*, together with the thermal rearrangement product, the 3-spirocyclopropanated 4-oxopiperidine-2-carboxylate *375* (44%).

The synthesis of tiazofurin, a potent antitumor agent carried out by Merino [9] features 1,3-dipolar cycloaddition reaction of nitrones with allylic alcohol. Cycloaddition reaction of *C*-(2-thiazolyl)nitrones *376a* and *376b* with allylic alcohol provided mixture of *cis/trans* adducts *377a* and *44a* (or *377b* and *trans-378b*) (Scheme 78). Microwave heating without solvent produced low *cis/trans* ratio (60:40). The addition of $Zn(OTf)_2$ in dichloromethane was essential to obtain a complete *cis* selectivity. Selectivity was also achieved in the reactions of nitrone *376b*, although total yields were considerably lower. This is a good example of the superiority of microwave heating over classical heating. In the case of *376a*, $Zn(OTf)_2$ catalysed reaction (80°C, in dichloromethane) took 4 days to achieve identical yield and selectivity. To achieve cycloaddition of *376a* with (2*S*)-1,2-dihydroxy-3-butene *379* a CH_2Cl_2:EtOH mixture has to be used as a solvent, without $Zn(OTf)_2$ and the reaction afforded a 1:1:2:2 mixture of the four possible diastereomers. Here, incompatilibity of $Zn(OTf)_2$ catalyst with diol resulted in decomposition products only. On the other hand, cycloaddition between nitrone *376a* and the silylated compound *379b* under microwave irradiation in the presence of $Zn(OTf)_2$ afforded a 4:1 mixture of *cis* adducts *380b* and *381b*. A complete *cis* selectivity was observed, with diastereofacial selectivity (80% in favor of the (3*R*,5*S*)-isomer *381a*) rendered the process synthetically useful.

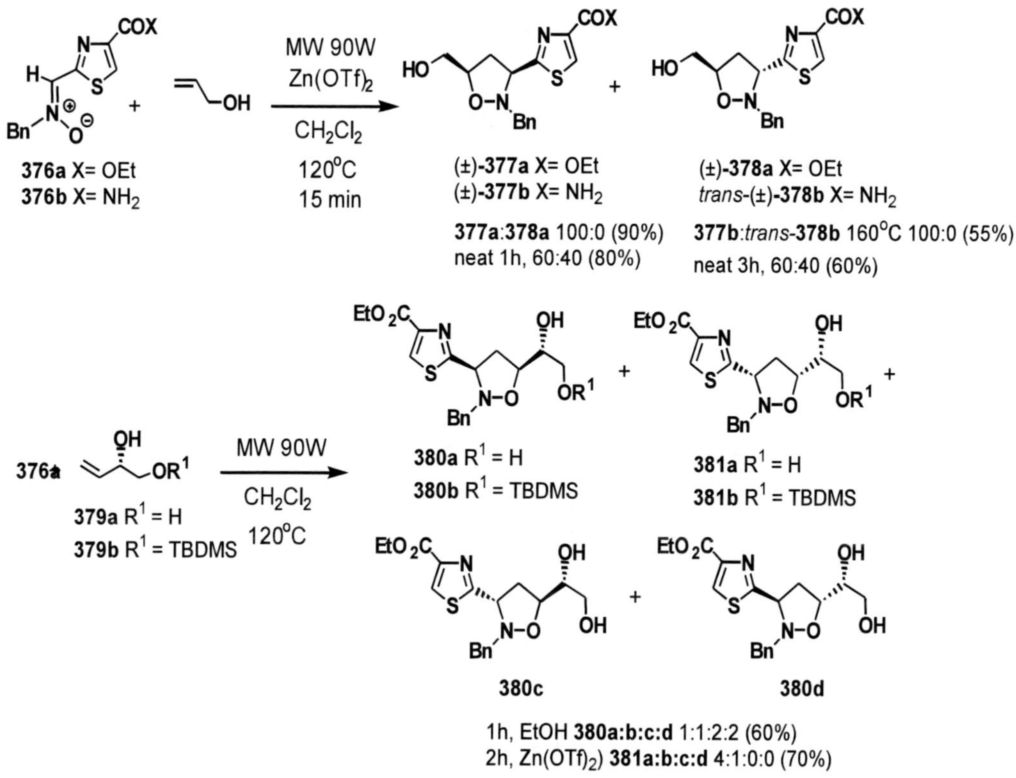

Scheme 78.

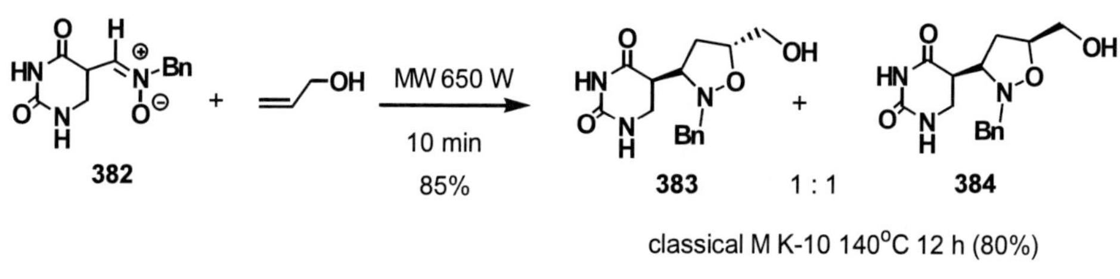

Scheme 79.

Chiacchio and Merino [10] applied microwave reactions of nitrone derivative of 5-formyluravil with allyl alcohol to obtain isoxazolidine analogues of pseudouridine, a new class of modified nucleosides. Since typical Lewis acids are not compatible with nitrone *382*, due to number of coordinating sites, several solid acids such as montmorillonite K-10 and Nafion (which are compatible of substrates with acidity) were tested in classical thermal reactions to obtain high yield. The advantage of microwave reaction was in the fact that cycloaddition could be carried out in high yield (85%), short reaction time (10 min), solvent-free and without the aid of an additive (Scheme 79). Hence, reaction of nitrone *382* with allyl alcohol in microwave conditions for 10 min yielded isoxazolidines *383* and *384* in 45 and 37% yields, respectively. The large excess of alcohol was used as reactant and solvent, which could be easily removed by distillation.

Nucleobases (4'-aza-2',3'-dideoxy nucleosides *386*: thymine, uracil, cytosine, 5-fluoro cytosine, adenine, 2-(*N*-trityl) guanine) could act as dipolarophiles for the synthesis of *N,O*-nucleosides. Bortolini [11] showed that solvent-free microwave cycloadditions proceed directly on unprotected vinyl nucleobases. Only exception being *N*-vinyl guanine for which only modest amounts of cycloadduct are found when employing the protected tritylated derivative *386f*. The unnecessary protection of the vinyl nucleobase, and consequent deprotection of the resulting *N,O*-nucleoside, represent an interesting improvement characterizing the overall synthetic protocol. The stereochemical outcome of the reaction showed a certain degree of control that appeared to be very highly regioselective with formation of the 1'-substituted isoxazolidines *387*. In addition, the *endo/exo* ratio is satisfactory, in the range 70:30 with a maximum value of 84:16 for the 5-fluoro cytosine derivative (Scheme 80).

Parmar [12] synthesized spiro-(indoline-isoxazolidines) by the cycloaddition reaction between ethyl (3-indolylidene)-acetate *388* and substituted α,*N*-diphenylnitrones *389a-j* (Scheme 81). Irradiation was carried out for 4-5 minutes, by 1 min at the time with 10 s interval. By conventional heating isoxazolidines *390a-j* and *391a-j* were prepared in combined yields of only 10-15%, after heating at 60°C in benzene for considerably longer time periods (150-180 h).

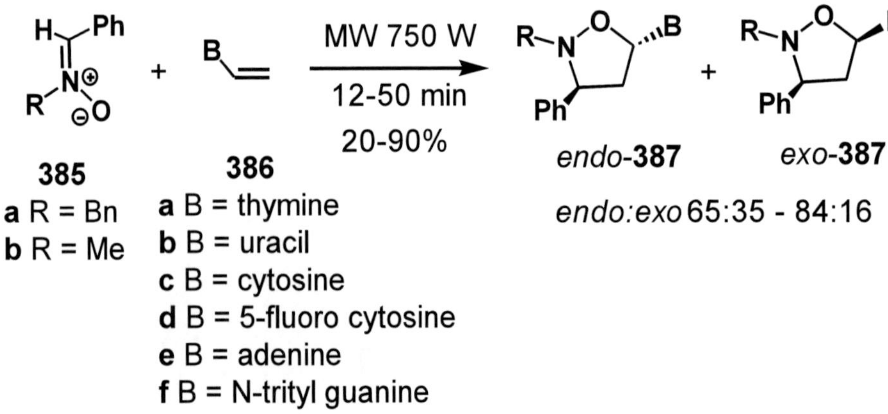

Scheme 80.

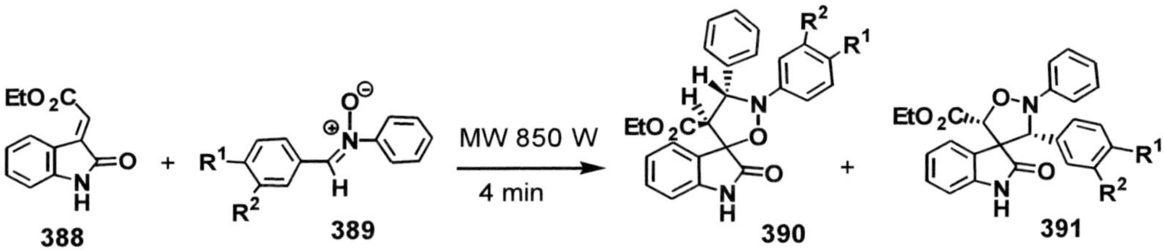

390d R^1 = F, R^2 = H (10%) **391d** (44%)

Scheme 81.

The use of carbohydrates as dipolarophiles in neat, microwave-assisted 1,3-dipolar cycloaddition was demonstrated by Barros. [13] Crotonyl sugars were reacted with *C,N*-diphenylnitrone *393* under microwave irradiation in neat conditions (as heterogeneous solid mixtures), producing optically active isoxazolidines. Cycloaddition to the crotonyl moiety located at the primary carbon position of glucose derivative *392* afforded two major diastereoisomeric products, *394a,b* and *394c,d* (90% yields). The cycloadducts *394a,b* show a 3,4-*trans*-4,5-*trans* relationship at the heterocyclic ring, whereas *3943c,d* are 3,4-*cis*-4,5-*trans* (Scheme 82). The observed *394a,b*/*394c,d* ratio was 87:13. Similar results were obtained at the α-anomeric position of glucose *395* with the ratio 3,4-*trans*-4,5-*trans*/3,4-*cis*-4,5-*trans* of 90:10. For this derivative, the ratio of the *endo* products *SRS/RSR* was 66:44. For xylose derivatives, different results were achieved when comparing both α and β anomers. In the case of β anomer *396*, the two diastereoisomers *SRS/RSR*, show a ratio of 57:43, denoting a minor facial selectivity when compared to the analogous anomer of glucose. However, when the α-anomer of xylose *397* was treated with nitrone *393*, only 3,4-*trans*-4,5-*trans* isoxazolidine isomers were formed, and the facial selectivity is improved to the *SRS/RSR* ratio of 81:19.

Microwave assisted 1,3-dipolar cycloaddition reactions between furanosyl *N*-benzyl-*C*-glycosyl nitrone *398* with an excess of methyl acrylate (which was also used as solvent) afforded mixtures of all possible 3,5-disubstituted isoxazolidines *400* (Scheme 83). Merino [14] reported that whereas the regioselectivity of this solvent free reaction was very high (the corresponding 3,4-regioisomers were not detected), both the *endo*/*exo* selectivity and the diastereofacial induction (*Re*/*Si*) were rather low. The cycloaddition is not dependent of the reaction conditions (solvent and temperature). Even the use of microwaves did not substantially change the ratio of the products, although it did accelerate the reaction rate considerably. In general, a slight trend towards *endo* adducts and *Re* attacks was observed.

The 1,3-dipolar cycloaddition of *Z*-α-phenyl-*N*-methylnitrone *401* with racemic methyl 2-[fluoro-(phenyl)methyl]acrylate *402* was used as a stereoselective route to enantiopure fluorine-containing isoxazolidines *403a-c* (Scheme 84). For this purpose, Comes-Franchini [15] used catalytic amounts of Lewis acid (triflates, 5 mol %). Neither the yields nor the *exo*/*endo* ratio could be substantially improved by running the reaction in different media (*403a*/*403b*/*403c* ratio was in all experiments 10:60:30). From the environmental point of view interesting are reactions carried in water which gave yields identical to those in toluene, while the use of ionic liquids was not beneficial (Table 2). A remarkably shorter reaction time was obtained under microwave irradiation without affecting the yields. The optimal reaction conditions were achieved when the reaction of *401* with *402* was performed under microwave activation (100°C, 15 min) without solvent in the presence of a catalytic amount of metal triflates.

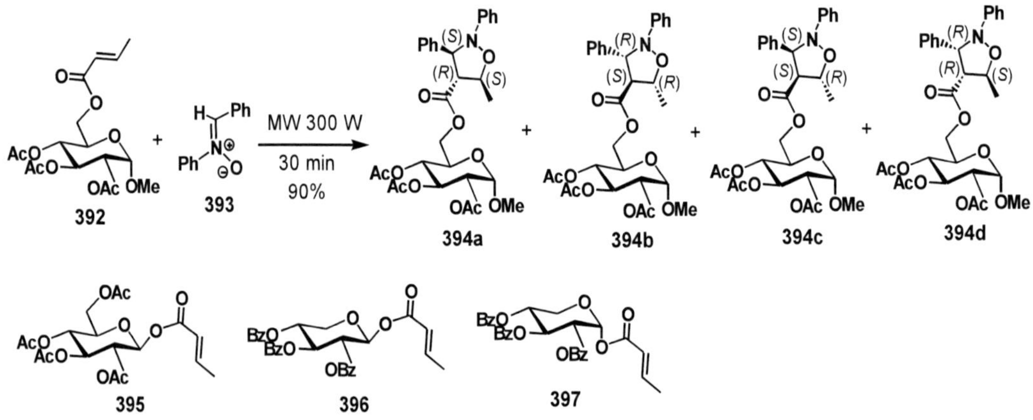

Scheme 82.

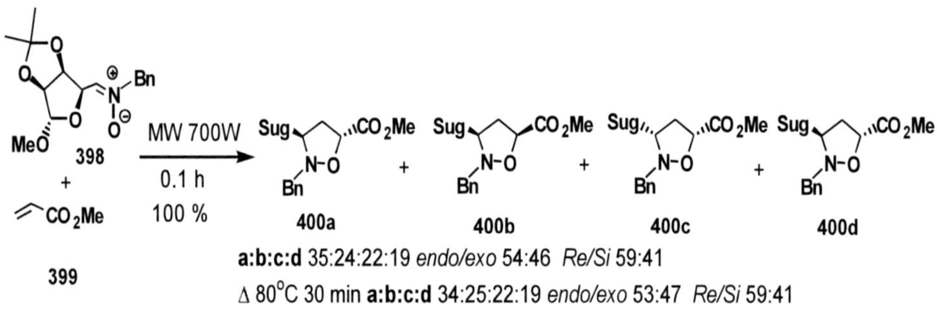

Scheme 83.

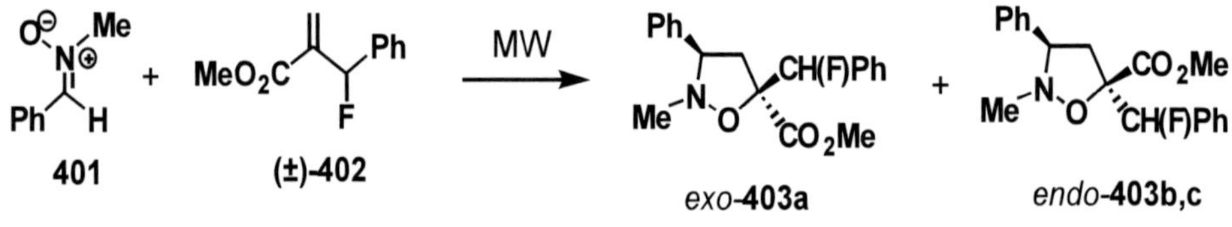

Scheme 84.

Taillefumier [16] found that cycloaddition reactions of activated exo-glycals such as *404* and nitrones *405-406* proceeded only under microwave activation, with excellent facial selectivities on furanoglycosylidenes and good stereocontrol producing only two diastereomeric spiroisoxazolidines. The *exo*-glycal *404* under microwave activation ensured a clean reaction of nitrone *405*, giving a mixture of two isomers in a 4:1 ratio and 50% yield (Scheme 85). Attack of the nitrone to *404* takes place from the less hindered β face. The C-4 configuration in the isoxazolidine ring is dictated by the geometry of the starting olefin, the two compounds obtained were the C-3 epimers *407* and *408* resulting from *endo/exo* attacks of the nitrone. Other nitrone *406* exhibited the same stereocontrol. Reactions to pyranose *exo-*

glycals such as *411* behave differently. The cycloaddition reaction on *411* (pure *Z*-isomer) with nitrone *406* under microwave activation yielded four cycloadducts *412-415* in a 1.3:8.1:3.5:1 ratio in 90% yield. The four isomers obtained are diastereoisomers at C-3 and/or C-5 (*trans/cis* ratio 2:1).

Table 2. Optimisation of conditions for reaction of 401 with 402

Solvent	MW (W)	T (°C)	Time	Yield (%)	Lewis acid
Toluene	-	110	24 h	46	$In(OTf)_3$
Toluene	120	110	10 min	48	$In(OTf)_3$
H_2O	-	rt	72 h	50	$Yb(OTf)_3$
H_2O	90	100	7 min	49	$Yb(OTf)_3$
$BimimBF_4$	-	90	16 h	19	$In(OTf)_3$
Toluene/$BimimBF_4$	90	165	10 min	39	$In(OTf)_3$
None	60	100	15 min	73	$Sc(OTf)_3$
None	60	100	15 min	75	$In(OTf)_3$

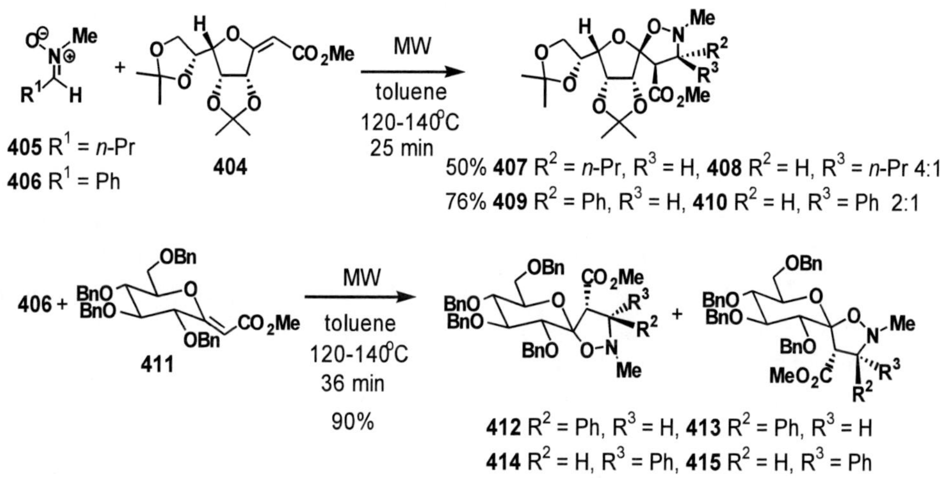

Scheme 85.

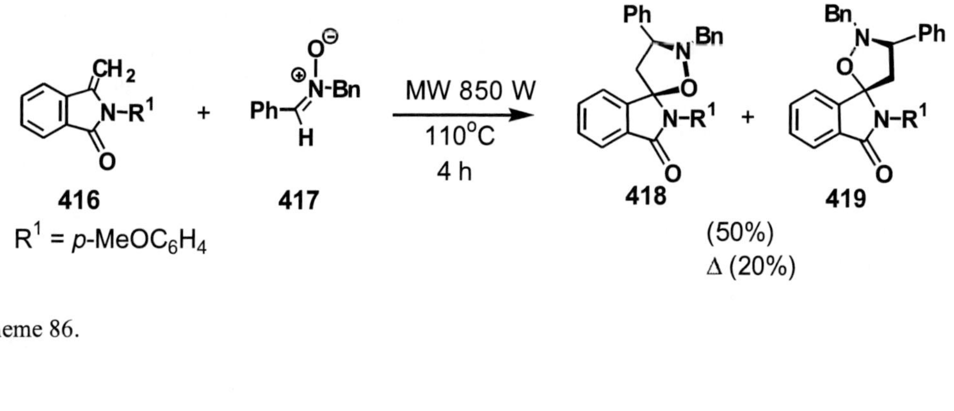

Scheme 86.

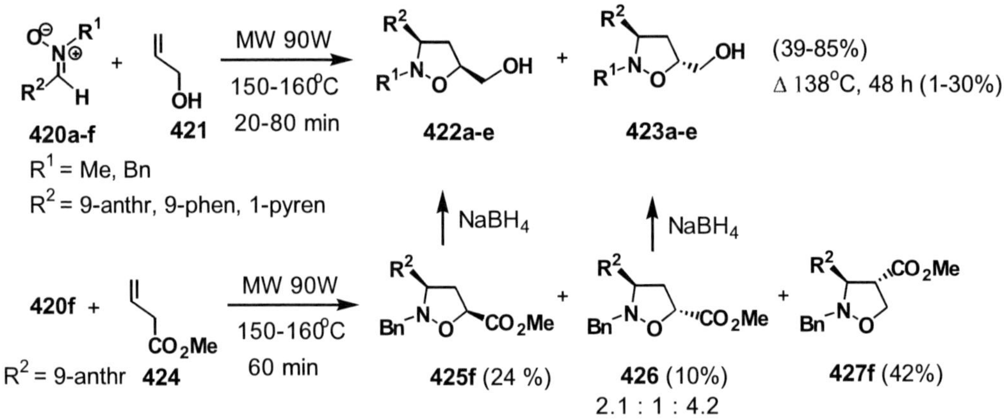

Scheme 87.

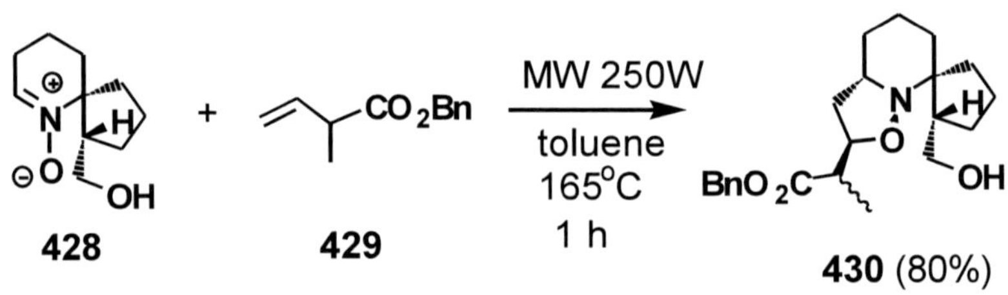

Scheme 88.

Regiospecific reaction of 2,3-dihydro-3-methylene isoindolones *416* with nitrones conducted by Vébrel [17] afforded good yields of diastereoisomeric cycloadducts. In this synthesis, spiro[1*H*-isoindol-1,5'(4'*H*)-isoxazole]-7(6*H*)-ones *418* and *419* were obtained with *N*-aryl nitrones such as *417* (Scheme 86). Reaction conducted with the aid of microwave irradiation are superior to classically heated reaction in terms of reaction yields. The slight stereoselectivity observed for cycloadditions with *N*-bezylnitrone *417* (35-60% de) is linked to secondary π-π orbital interactions.

Synthesis of isoxazolidines possessing polycyclic aromatic hydrocarbon substituents as potential DNA intercalators by Rescifina [18] was carried out by the cycloaddition process under microwave irradiation in solvent free conditions using a large excess of dipolarophile as solvent. The reaction of nitrones *420a-f* with allyl alcohol, under microwave irradiation for 20-80 min, afforded the 1,3-dipolar cycloadducts *422a-e* and *423a-e* in relative ratio from ca. 1.1:1 to 1.6:1, with a total yield in the range 39-85% (Scheme 87). The investigated reaction was found to be regiospecific, affording a mixture of *cis* and *trans* 5-substituted isoxazolidines *422* and *423*. The pericyclic reaction showed a low level of *cis/trans* stereoselectivity with the *cis* isomers *422* as the major product. These cycloadditions are markedly better than classical heating in a sealed tube, however, by this method, no reaction was observed for nitrone *420f* (Table 29). Isoxazolidines *422f* and *423f* have been prepared by an alternative procedure. Nitrone *420f* reacted with more reactive dipolarophile, methyl acrylate, followed by reduction with $NaBH_4$, to afford the target compounds in good yields. The cycloadducts *425f-427f* were obtained as pure compounds in a 2.4:1:4.2 ratio, with a 4/5 regioselectivity of 1.2:1 and the *trans/cis* ratio is 1.15:1. It was observed that the more reactive diene/dienophile components led to the decrease in product regioselectivity.

1,3-Dipolar cycloaddition of alkene *429* with nitrone *428* proceeded poorly under conventional thermal conditions using a variety of solvents and temperatures. It was found by Caprio [19] that the rate and yield can be much enhanced under microwave irradiation. Heating of a mixture of a three-fold excess of *429* and *428* in a microwave reactor for one hour resulted in clean cycloaddition to give isoxazolidine *430* as a mixture of two diastereomers in 80% yield (Scheme 88). The diastereomeric mixture *430* (d.r. = 1:1) was then further elaborated to the pinnaic acid, an alkaloid possessing spirobicyclic core structure.

For synthesis of functional spiro-oxazinanone-*C*-glycosides Li [20] applied the 1,3-dipolar cycloaddition of *exo*-glucal *431* with nitrones *432a-i* (Scheme 89). Reaction proceeds very efficiently in diglyme solution under microwave irradiation at 200°C for 2 min in a sealed pressure vial and stereoselectively afforded the corresponding adducts of spiro-isoxazolidine glycosides as the α- and β-anomeric mixture of *433a-i* and *434a-i* in 30-80% yield (*433:434* = 1.2-12.3:1). The the α- anomers *433a-i* are preferred, except for *para*-tolyl substituent, where preference was not observed.

Fišera [21] used chiral nitrone *435* for reaction with β-hydroxy-α-methylene esters *436*. Microwave reactions proceeded with complete regioselectivity, leading to the formation of diastereomeric 3,5,5-trisubstituted isoxazolidines *437a-f* as a mixture in a good overall yield (Scheme 90). Microwave conditions gave modest yields, the best one was obtained by conventional heating of *435* and *436* at 50°C, 2h, neat (92%), with *trans*:*cis* ratio 96:4 and also different *437a:b:c:d:e* ratio being 76:20:0:4:0. The replacement of *i*-Pr substituent in *436* with more bulky CH(OH)Ph group does not influence the preferred formation of *trans* isomer, and *437a* isomer is still dominant cycloadduct, regardless of heating source. Furthermore, it was found that the change of Bn substituent on the nitrone moiety with methyl group retains high *trans*:*cis* ratio (97:3), but diastereoisomer *437a(Me)* is increasingly more dominant in the mixture (*437a(Me)*:*b(Me)*:*c(Me)*:*d(Me)* ratio 87:9:3:1). [22]

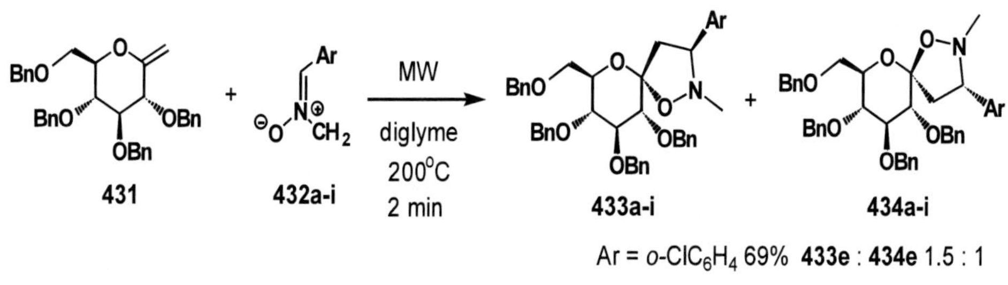

Scheme 89.

Stereoselectivity of cycloaddition of *N*-benzyl-*C*-ethoxycarbonyl nitrone *438* to (*S*)-5-hydroxymethyl-2(5*H*)-furanone *439* as chiral dienophile was studied by Fišera [23] (Scheme 91). The reaction proceeds in a highly face-selective manner, which affords the products *440-442* resulting from *anti* approach to the hydroxymethyl group of the dipolarophile. *Endo-exo* diastereoselectivity is significantly affected by the solvent. Microwave irradiation strongly accelerates the reaction with little effect on the diastereoselectivity, where small amounts of the *syn* adduct *443* were also obtained in addition to *anti* cycloadducts *440-442*. In microwave conditions, the highest proportion of adduct *440* was isolated, while thermal reactions (reflux, benzene, DMSO, dichloroethylene, 80°C) gave different proportions of *440-442* (*440* 30-53%, *108* 32-56%, *442* 10-18%).

The diastereomerically pure (*Z*)-nitrones *444* and *450* derived from *D*-erythrose- and *D*-threose were subjected to 1,3-dipolar cycloaddition reactions with *N*-phenylmaleimide *445* in boiling toluene (11 h). There are four possible isoxazolidine products; *cis*- and *trans*-isomers from *anti*- and *syn*-attack. In the case of D-*erythro*-derived nitrones *444* Fišera [24] isolated only two diastereoisomers, *erythro-trans 446* and *threo-trans 447* (74 % yield, *trans/cis* ratio 95:5, Scheme 92). Microwave irradiation dramatically decreased the reaction times of the cycloadditions. In the case of the cycloaddition of nitrone *444* with dipolarophile *445*, the reaction time decreased from 11 h to 8 min. Moreover, microwave irradiation could even reverse the ratio of *erythro-trans* / *threo-trans* adducts from 63:37 to 39:55 for *444* (*trans/cis* ratio 94:6). This reversal of stereoselectivity of the nitrone cycloaddition using microwave irradiation instead of conventional heating is a very rare phenomenon. On the other hand, in the case of the cycloaddition of nitrone *450* to imide *445* microwave irradiation only slightly changed the ratio of diastereoisomers while the reaction time decreased from 3 h to 10 min.

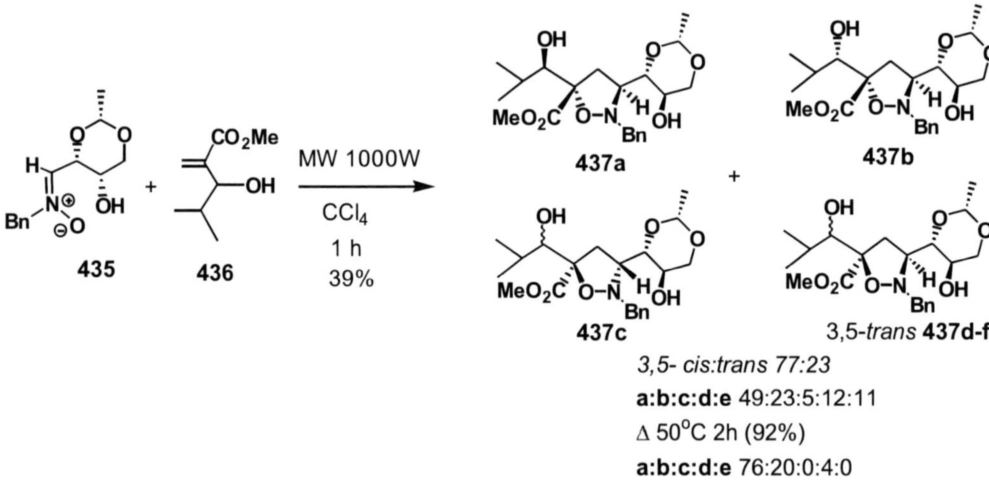

Scheme 90.

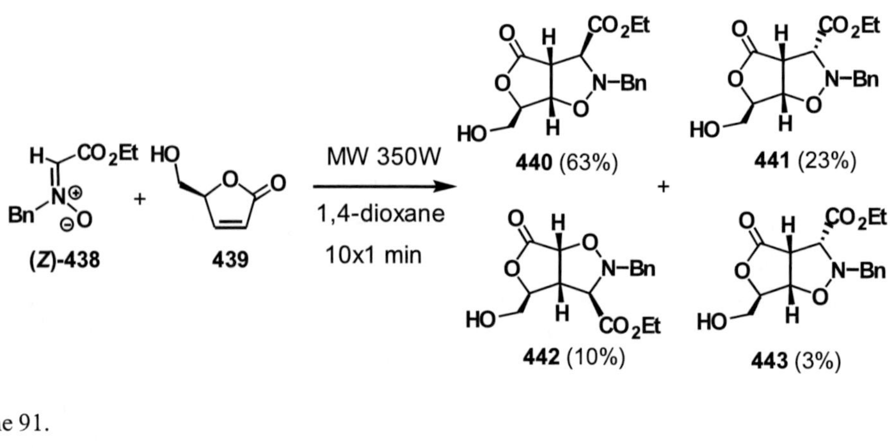

Scheme 91.

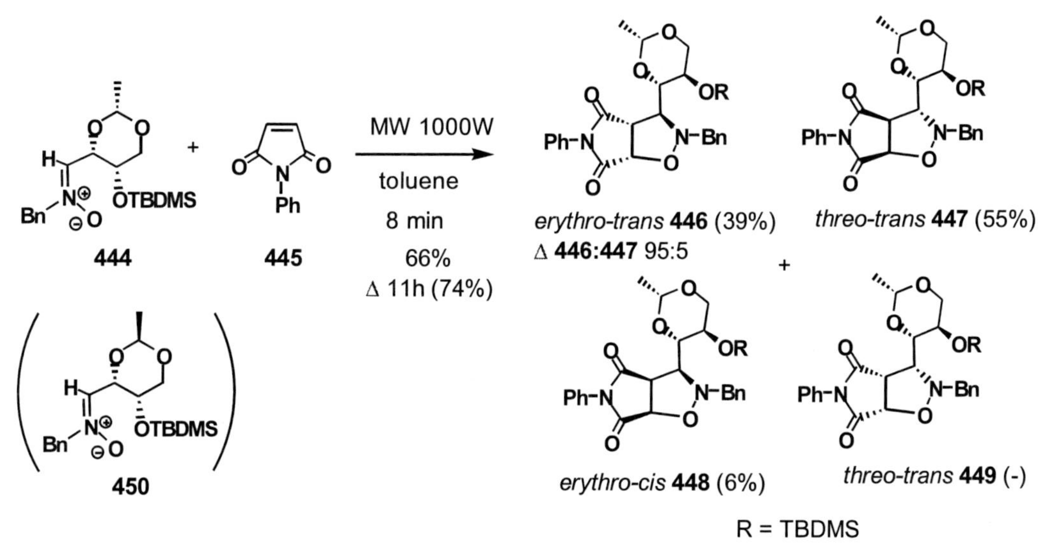

Scheme 92.

Bottino [25] used polyhedral oligomeric silsesquioxane *452a,b* which is externally covered by organic substituents as dipolarophilic component in 1,3-dipolar cycloadditions. The cycloaddition of *N*-methyl-C-ethoxycarbonyl nitrone *451* with vinyl-POSS *452a* in toluene at 130 °C afforded the 1,3-dipolar cycloadducts *453a*, *454a* and *455a* with a total yield of 20%. The conversion of vinyl- and styryl-POSS into isoxazole derivatives by the exploitation of microwave assisted cycloaddition was more successful. A sharp improvement of the cycloaddition process has been obtained under microwave irradiation within 90 min affording the *453a*, *454a* and *455a* in 3.5:19.8:1 relative ratio (same ratio was obtained in thermal conditions), but with a total yield of 73% (Scheme 93). The pericyclic reaction showed a good control of regio- and stereoselectivity with the *trans* isomer *454a*, as the major product.

Methyleneisoxazolidine nucleoside analogues were synthetised by Rescifina [26] *via* nitrone cycloaddition where allene was acting as dipolarophilic component. The microwave-assisted cycloaddition reaction of *C*-ethoxycarbonyl-*N*-methylnitrone *456* with protected thymallene *457* proceeded with high C-2–C-3 site selectivity and regioselectivity affording a diastereomeric mixture of isoxazolidine compounds *458* and *459* in a 1:1 ratio (Scheme 94). In thermal reaction, small quantity of water causes the formation of *461* and isoxazolidine derivative *462*. On the other hand, the cycloaddition reaction of nitrone *456* with *N*-acetylcytallene *463* has different siteselectivity, with the attack of the dipole occurring at C-1–C-2 of the dipolarophile. A possible explanation of the difference is that these compounds *458* and *459* are actually derived from cycloadducts *460*, which undergo a facile thermally promoted rearrangement, where the presence of a lone pair of electrons on the aza group of the pyrimidine nucleobase facilitates a [1,3] sigmatropic shift involving ring opening and ring closing. Further evidence of this mechanism is the reaction of products *464* and *465*, which on the prolongued heating in ethanol gave *466a/b*.

The use of heterodipolarophiles in 1,3-dipolar cycloaddition is uncommon due to their low reactivity. The cycloadditions of nitrones to nitriles usually require very drastic conditions, high pressure, or heating for long times at high temperatures. However, the employment of microwave irradiation has been beneficial for this reaction. Díaz-Ortiz [27] has found that under solvent-free conditions within short time, the corresponding 2,3-dihydro-1,2,4-oxadiazoles (Δ^3-1,3,4-oxadiazolines) *469a-g* were obtained from nitrones *467a-d* and nitriles *468a-c* (Scheme 95).

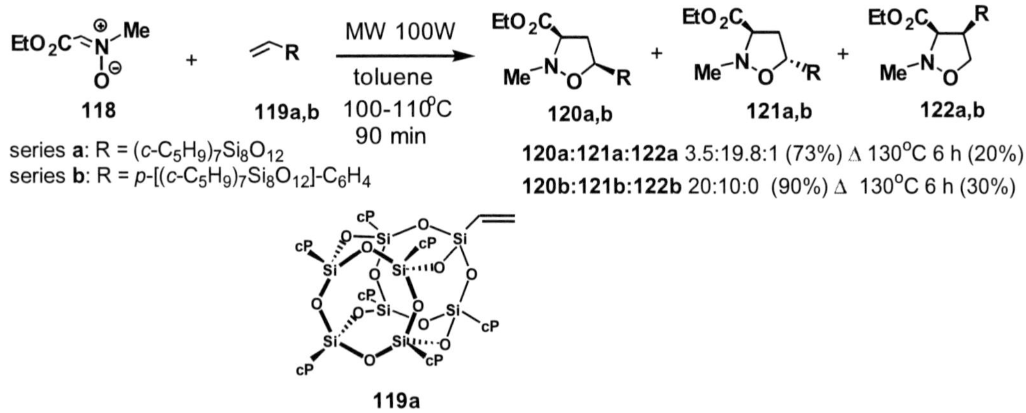

Scheme 93.

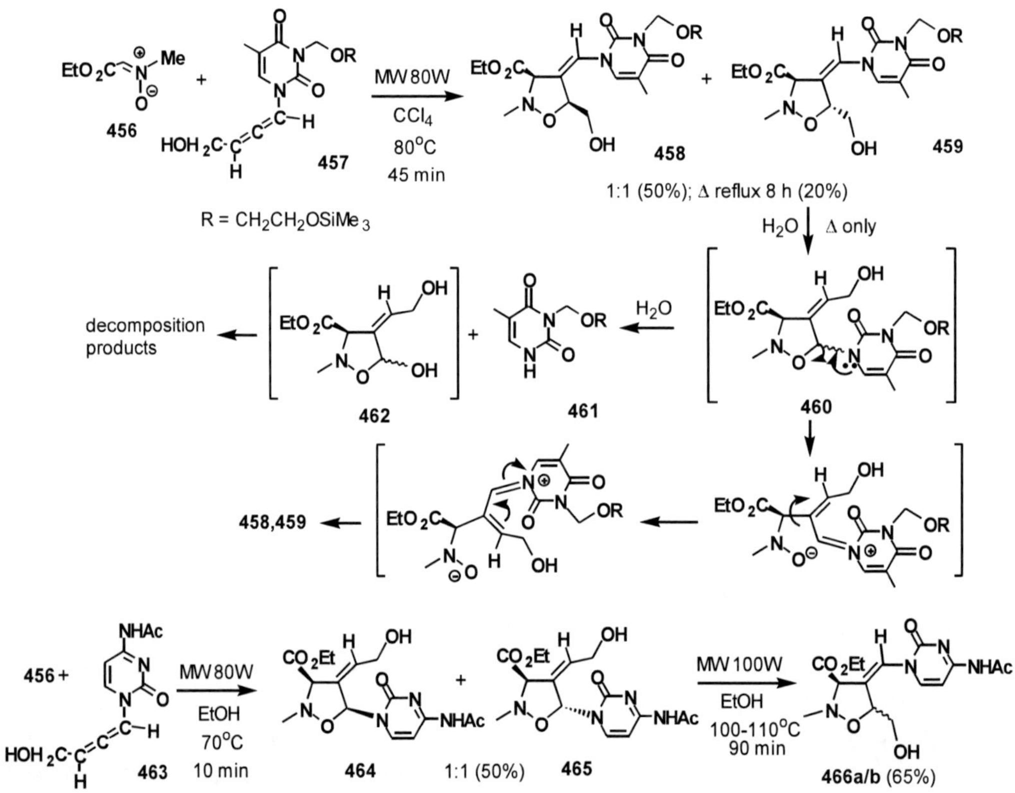

Scheme 94.

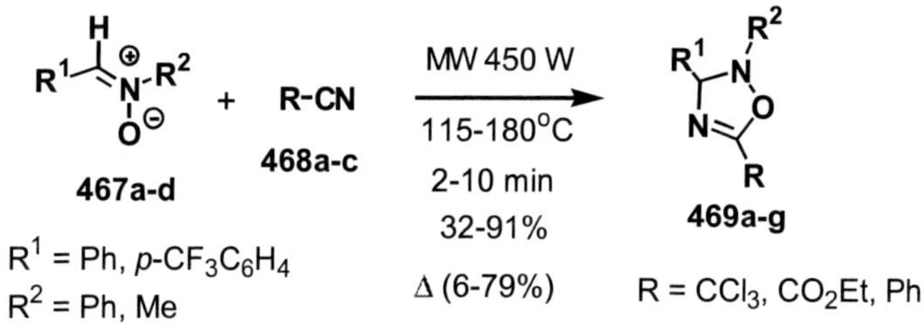

Scheme 95.

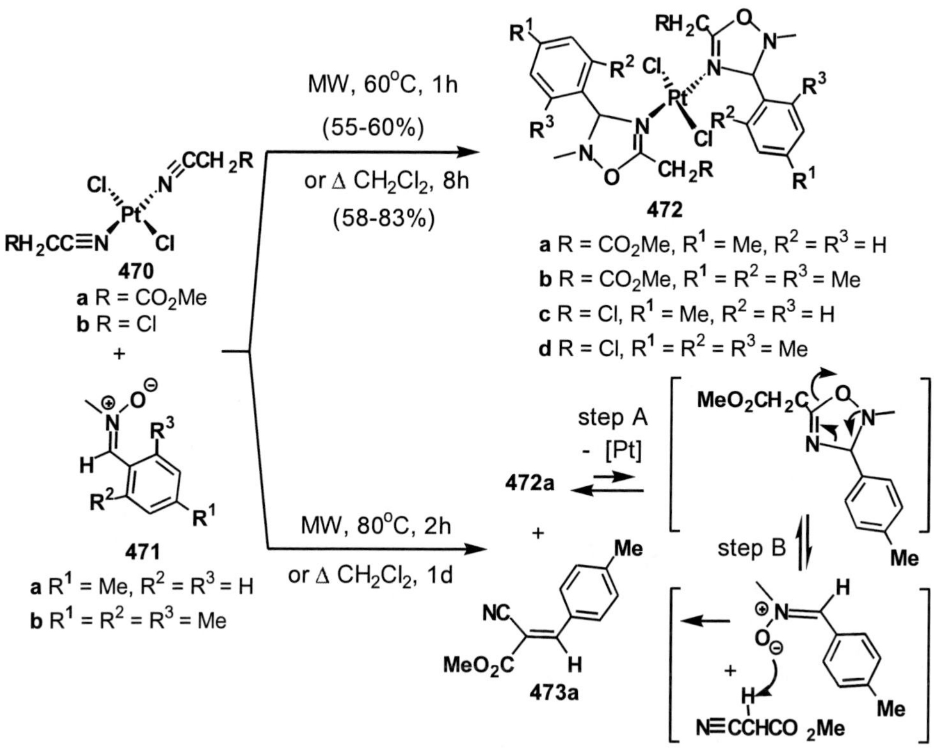

Scheme 96.

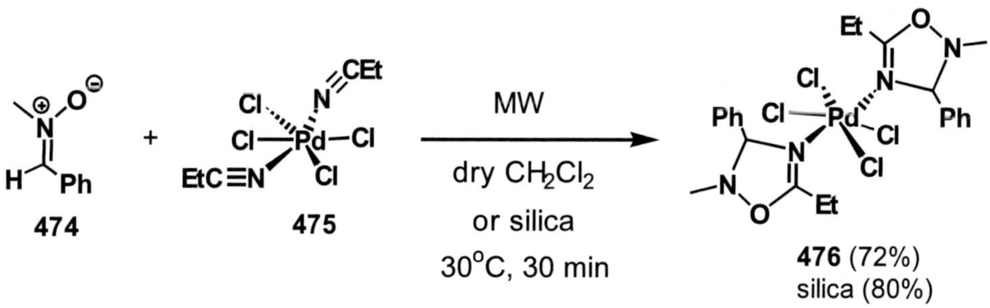

Scheme 97.

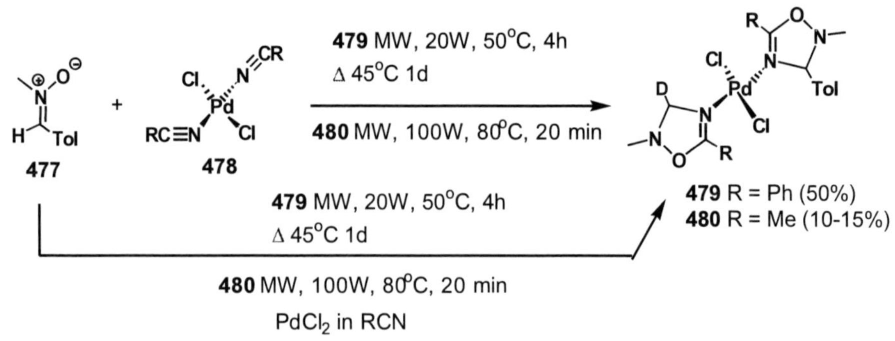

Scheme 98.

Pombeiro [28] has published the first example of Pt^{II}- complex promoted [3+2] cycloaddition of nitrones to nitriles. [29] The coordination of the nitrile to the Pt^{II} center activates the cyano group to such an extent that it becomes more reactive toward the nitrone in contrast to what was observed for the free nitrile. Therefore, treatment of *trans*-nitriles *470a,b* with nitrones *471a,b* upon conventional heating gives a 1:1 diastereomeric mixture of the corresponding *trans*-(Δ^4-1,2,4-oxadiazoline)-Pt^{II} complexes *472a-d* in moderate yields (Scheme 96). The reaction time is drastically reduced with microwave irradiation giving complexes *472* in comparable yields. The oxadiazoline-Pt^{II} complex *472a* also gives the corresponding cyano-alkene *473a* if the reaction is undertaken under more drastic conditions, then a mixture of *472a* and the cyano-alkene *473a* was obtained. Conversion of complex *472a* into the corresponding alkene *473a* possibly involves a 1,3-dipolar cycloreversion process. Liberation of the oxadiazoline ligand (step A) and dissociation of this species into the starting nitrile and nitrone components (step B) is followed by proton abstraction from the methylene group of the nitrile by the nitrone.

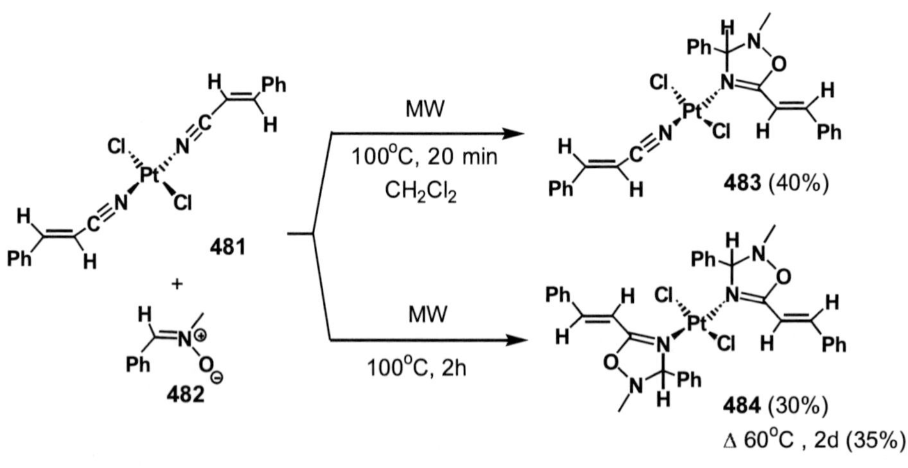

Scheme 99.

Except for Pt^{II} complexes dipolar cycloaddition of nitrones also proceeds smoothly with Pt^{IV} complexes, as shown by Pombeiro [30] (Scheme 97). The reaction of *trans*-$[PtCl_4(EtCN)_2]$ *475* (with a Pt^{IV} site) and nitrone *474* is greatly accelerated under focused microwave irradiation and similar or better isolated yields (in comparison with those for the conventional procedure) of the corresponding product *476* are obtained with only 30 min reaction times at 25-30°C in dry dichloromethane or in the solid state (dry silica gel).

The reaction between the nitrone and nitrile mediated by a Pd^{II} center could be performed by two synthetic routes (Scheme 98). As reported by Pombeiro, [31] in the first route, both $PdCl_2$ and *477* were reacted in a suspension of neat PhCN, while in the second route the nitrile complex *trans*-[$PdCl_2(PhCN)_2$] *478* dissolved in the corresponding nitrile was treated with the nitrone *477*. In both routes, the conventional reaction proceeds at 45 °C for 1 d (or for 4 h under microwave irradiation) and the isolated yield of the Δ^4-1,2,4-oxadiazoline complex *479*, prepared by the two methods, is almost the same (ca. 50%) irrespective of whether conventional or microwave conditions are applied. In the case of acetonitrile microwave irradiation results in the formation of a mixture of products, where the complex *480*, derived from the 1,3-dipolar cycloaddition is formed in 10-15% yield.

Cinnamonitrile is significantly less reactive dienophile than benzonitrile in analogous Pt^{II} complexes. As shown by Wagner, [32] if the solvent-free cycloaddition to $PtCl_2$(cinnamonitrile)$_2$ *481* is performed under microwave irradiation, the formation of the bis-oxadiazoline complex *484* is complete within 2 h (Scheme 99). Although this reaction time might appear unusually long for microwave reactions which often take place within a few minutes, the improvement relative to the thermal conditions (2 d under comparable conditions) is considerable. The product is obtained in similar yield and with an unchanged diastereomeric ratio of 1:1. Short-term microwave irradiation of nitrone *482* and complex *481* in dichloromethane results in selective formation of the mono-cycloaddition product *483*, independently whether one equivalent or an excess of nitrone is used.

Similar results were obtained by cycloaddition to *cis* and *trans*-[$PtCl_2(PhCN)_2$] complexes as dipolarophiles with nitrone *482*. [33] [2+3] cycloaddition of nitrones to the nitrile ligands in the *cis*- or *trans*-[$PtCl_2(PhCN)_2$] occurs under ligand differentiation and allows selective synthesis of complexes of the type *cis*- or *trans*-[$PtCl_2$(oxadiazoline)-(PhCN)]. Microwave irradiation (60°C, 20 min) favors the selectivity towards the mono-cycloadduct with respect to thermal conditions. Reaction of the *trans*-substituted mono-oxadiazoline complexes with different nitrone (60°C, 2.5 h) gives access to mixed bis-oxadiazoline compounds *trans*-[$PtCl_2$(oxadiazoline$_1$)(oxadiazoline$_2$)]. The corresponding *cis*-configured complexes, however, do not undergo further cycloaddition.

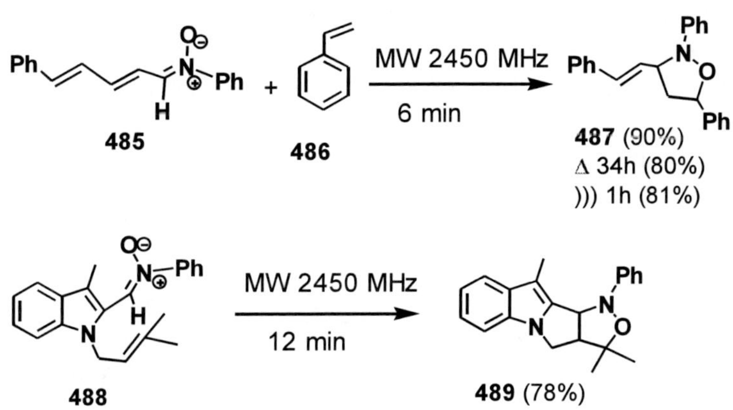

Scheme 100.

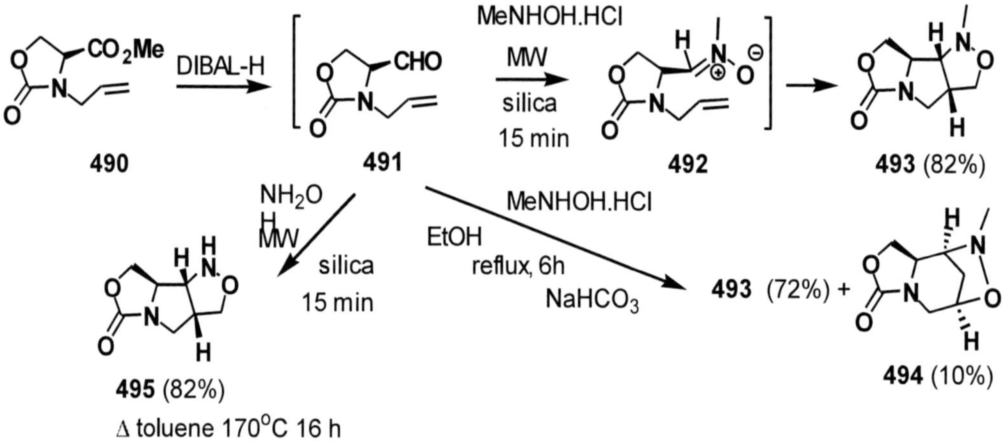

Scheme 101.

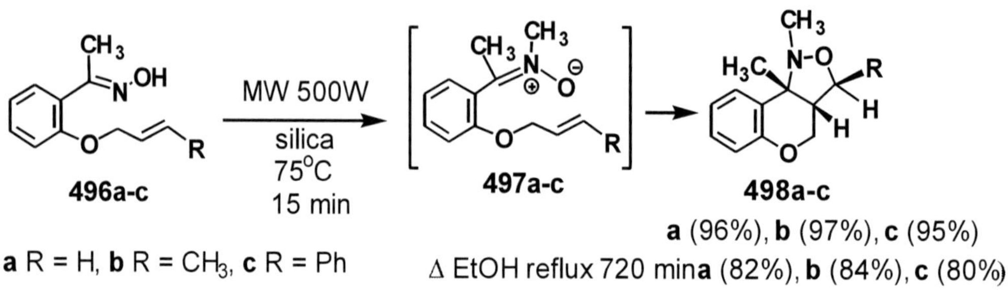

Scheme 102.

2.2.2. Intramolecular Reactions

The 1,3-dipolar cycloaddition of unreactive nitrones with unactivated alkenes proceeds rapidly within minutes and in the high yield, when microwave irradiation was applied in the absence of solvent. For instance, Sandhu [34] reported that reaction of α-cinnamyl-*N*-phenylnitrone *485* and styrene *486* within 6 min of irradiation produces *487* in 90% yield (Scheme 100). Classical heating requires 34 h to obtain 80% yield, while sonochemical reaction affords 81% yield after 1 h. Equally well proceeded intramolecular 1,3-dipolar cycloaddition reactions, such as the reaction of 3-methyl-1-(3'-methylbut-2'-enyl)indole *488* with methylhydroxylamine under microwave irradiation give 1,3,3,10-tetramethyl-1,3a,4,10b-tetrahydro-3*H*-isoxazolo [3',4':3,4]pyrrolo[1,2-a]indole *489* in 80% yield.

Intramolecular dipolar cycloaddition of nitrones is the key step in Cheng's [35] chiral synthesis of functionalized nitrogen heterocycles. Reduction of ester *490* with DIBAL-H gave the intermediate aldehyde *491*, from which non-stabilized nitrone *492* was generated by treatment with *N*-methylhydroxylamine (Scheme 101). Nitrone *492 in situ* underwent the intramolecular 1,3-dipolar cycloaddition to furnish an isoxazolidine *493* in 72% yield along with a bridged isoxazolidine *494* in 10% yield. However, when the intermediate *491* and *N*-methylhydroxylamine were mixed with silica gel and irradiated by microwaves for 15 min, it provided only the pyrrolidine derivative *493*, in 82% yield. It may be proposed that the selectivity in the microwave-induced reactions results from a faster transformation that leads to the kinetically controlled product. In addition, oxime *491* with hydroxylamine undergoes the intramolecular 1,3-dipolar reaction, which requires heating at 170°C in toluene in a sealed tube for 16 h to form an isoxazolidine *495*. Mixing of the oxime *491* with silica gel and irradiated by microwaves for 15 min affords isoxazolidine *495* in 82% yield.

Another example of intramolecular 1,3-dipolar cycloaddition of nitrones reported by Ishar [36, 37] features the oxygen tethered dipolarophile to the aromatic backbone. In a solvent-free reactions oximes *496a-c* were adsorbed on silica gel and exposed to microwave heating (500 W, 75°C) to obtain cyclic products *498a-c* in yields superior to classical heating in refluxing ethanol (Scheme 102).

References

[1] Katritzky, A. R.; Rees, C. W. (Eds.) *Comprehensive Heterocyclic Chemistry*, Pegamon Press, Oxford *1984*.

[2] Díaz-Ortiz, A.; Díez-Barra, E.; de la Hoz, A.; Moreno, A.; Prieto, P. *J. Chem. Soc. Perkin Trans. 1 1994*, 3595-3598.

[3] Díaz-Ortiz, A.; Díez-Barra, E.; de la Hoz, A.; Prieto, P.; Moreno, A.; Langa, F.; Prangé, T.; Neuman, A. *J. Org. Chem. 1995*, *60*, 4160-4165.

[4] Andrade, M. M.; Barros, M. T.; Pinto, R. C. *Tetrahedron 2008*, *64*, 10521-10530.

[5] Loupy, A.; Petit, A.; Bonnet-Delpon, D. *J. Fluor. Chem. 1995*, *75*, 215-216.

[6] Sridharan, V.; Saravanakumar, S. P.; Muthusubramanian, S. *J. Heterocycl. Chem. 2005*, *42*, 515-518.

[7] Sridharan, V.; Karthikeyan, K.; Muthusubramanian, S. *Tetrahedron Lett. 2006*, *47*, 4221-4223.

[8] Zanobini, A.; Brandi, A.; de Meijere, A. *Eur. J. Org. Chem. 2006*, 1251-1255.

[9] Chiacchio, U.; Rescifina, A.; Saita, M. G.; Iannazzo, D.; Romeo, G.; Mates, J. A.; Tejero, T.; Merino, P. *J. Org. Chem. 2005*, *70*, 8991-9001.

[10] Chiacchio, U.; Corsaro, A.; Mates, J.; Merino, P.; Piperno, A.; Rescifina, A.; Romeo, G.; Romeo, R.; Tejero, T. *Tetrahedron 2003*, *59*, 4733-4738.

[11] Bortolini, O.; D'Agostino, M.; De Nino, A.; Maiuolo, L.; Nardi, M.; Sindona, G. *Tetrahedron 2008*, *64*, 8078-8081

[12] Raunak; Kumar, V.; Mukherjee, S.; Poonam; Prasad, A. K.; Olsen, C. E.; Schäffer, S. J. C.; Sharma, S. K.; Watterson, A. C.; Errington, W.; Parmar, V. S. *Tetrahedron 2005*, *61*, 5687-5697.

[13] Andrade, M. M.; Barros, M. T.; Pinto, R. C. *Synth. Commun. 2010*, *40*, 855-860.

[14] Merino, P.; Franco, F.; Merchan, F. L.; Romero, P.; Tejero, T.; Uriel, S. *Tetrahedron Asymm. 2003*, *14*, 3731-3743.

[15] Bernardi, L.; Bonini, B. F.; Comes-Franchini, M.; Fochi, M.; Folegatti, M.; Grilli, S.; Mazzanti, A.; Ricci, A. *Tetrahedron Asymm. 2004*, *15*, 245-250.

[16] Enderlin, G.; Taillefumier, C.; Didierjean, C.; Chapleur, Y. *Tetrahedron Asymm. 2005*, *16*, 2459-2474.

[17] Rigolet, S.; Goncalo, P.; Mélot, P.; Vébrel, J. *J. Chem. Res. Synop. 1998*, 686-687.

[18] Rescifina, A.; Chiacchio, M. A.; Corsaro, A.; De Clercq, E.; Iannazzo, D.; Mastino, A.; Piperno, A.; Romeo, G.; Romeo, R.; Valveri, V. *J. Med. Chem. 2006*, *49*, 709-715.

[19] Yang, S.-H.; Caprio, V. *Synlett 2007*, 1219-1222.
[20] Li, X.; Wang, R.; Wang, Y.; Chen, H.; Li, Z.; Ba, C.; Zhang, J. *Tetrahedron 2008*, *64*, 9911-9920.
[21] Dugovič, B.; Fišera, L.; Hametner, C.; Cyrański, M. K.; Prónayová, N. *Monatsheft Chem. 2004*, *135*, 685-696.
[22] Dugovič, B.; Fišera, L.; Hametner, C.; Prónayová, N. *Arkivoc 2003*, *xiv*, 162-169.
[23] Ondruš, V.; Orsag, M.; Fišera, L.; Prónayová, N. *Tetrahedron 1999*, *55*, 10425-10436.
[24] Blanáriková, I.; Dugovič, B.; Fišera, L.; Hametner, C.; Prónayová, N. *Arkivoc 2001*, *ii*, 109-121.
[25] Chiacchio, M. A.; Borrello, L.; Di Pasquale, G.; Pollicino, A.; Bottino, F. A.; Rescifina, A. *Tetrahedron 2005*, *61*, 7986-7993.
[26] Chiacchio, U.; Corsaro, A.; Iannazzo, D.; Piperno, A.; Romeo, G.; Romeo, R.; Saita, M. G.; Rescifina, A. *Eur. J. Org. Chem. 2007*, 4758-4764.
[27] Díaz-Ortiz, A.; Díez-Barra, E.; de la Hoz, A.; Moreno, A.; Gómez-Escalonilla, M. J.; Loupy, A. *Heterocycles 1996*, *43*, 1021-1030.
[28] Lasri, J.; Charmier, M. A. J.; Haukka, M.; Pombeiro, A. J. L. *J. Org. Chem. 2007*, *72*, 750-755.
[29] These Pt and Pd activation of nitriles are covered in recent review on nitrone dipolar cycloadditions: Bokach, N. A.; Kukushkin, V. Yu. *Rus. Chem. Bull. Int. Ed. 2006*, *55*, 1869-1882.
[30] Charmier, M. A. J.; Kukushkin, V. Y.; Pombeiro, A. J. L. *Dalton Trans. 2003*, 2540-2543.
[31] Bokach, N. A; Krokhin, A. A.; Nazarov, A. A.; Kukushkin, V. Y.; Haukka, M.; da Silva, J. R. F.; Pombeiro, A. J. L. *Eur. J. Inorg. Chem. 2005*, 3042-3048.
[32] Desai, B.; Danks, T. N.; Wagner, G. *Dalton Trans. 2003*, 2544-2549.
[33] Desai, B.; Danks, T. N.; Wagner, G. *Dalton Trans. 2004*, 166-171.
[34] Baruah, B.; Prajapati, D.; Baruah, A.; Sandhu, J. S. *Synth. Commun. 1997*, *27*, 2563-2567.
[35] Cheng, Q.; Zhang, W.; Tagami, Y.; Oritani, T. *J. Chem. Soc., Perkin Trans. 1*, *2001*, 452-456.
[36] Singh, S.; Ishar, M. P. S.; Singh, R. S. *Can. J. Chem. 2005*, *83*, 260-265.
[37] Ishar, M. P. S. *Can. J. Chem. 2006*, *84*, 1015-1018.

2.3. Nitrile Oxides

According to Huisgen's classification of 1,3-dipoles, nitrile oxides belong to nitrilium betaines. Cycloaddition of nitrile oxides to acetylenic dipolarophiles forms isoxazole ring, while addition to alkene C=C bond generates isoxazoline ring. Isoxazolines constitute an important class of heterocyclic compounds which exhibit important biological activities, but they are also widely used as versatile synthons for isoxazoles and other important heterocycles.

The first example of 1,3-dipolar cycloaddition of nitrile oxides enhanced by microwave irradiation was published by Hamelin. [1] Simple synthesis of isoxazoles *501* and isoxazolines *502* was achieved by 1,3-dipolar cycloaddition of various dipolarophiles and nitrile oxide *500* generated *in situ* from *499* by dechlorination in dry media over basic alumina under microwave irradiation (Scheme 103). This synthetic protocol was more successful than classical heating, or microwave generation of nitrile oxides from methyl nitroacetate over acidic support (alumina, K-10 clay, SiO_2, MS, PTSA).

Further variation of this method was the use of *N*-methylmorpholine for dechlorination of phenyl (hydroxyimino) acetate *503* (NMM, xylene, 30 min, 130°C). [2] Analogous isoxazoline products were obtained with alkene dipolarophiles, while in the case of alkynes isoxazole products *504* and *505* were accompanied with enamine *506* and ether *507*, indicating that the NMM as a base is not a better way to generate 1,3-dipoles in the presence of alkynes (Scheme 104).

Hamelin [3] developed improved procedure for synthesis of isoxazolines and isoxazoles by simple mixing of oximes *508a,b*, *N*-chlorosuccinimide, Al_2O_3, and a dipolarophile without solvent followed by microwave irradiation in a single-mode apparatus (Scheme 105). *N*-phenylmaleimide, DMAD, methyl acrylate and methyl propiolate were used as dipolarophiles. For instance, methyl propiolate produced mixture of two regioisomers *511* and *512* after 10 min in 55% yield.

For the synthesis of isoxazoles of type *517* Müller [4] used one-pot, three component reaction. Synthesis starts with Sonogashira coupling of acid chlorides *513* and acetylenes *514*, followed by thermal cycloaddition of the *in situ* formed alkynones *515* with *in situ* generated nitrile oxides from hydroximinoyl chloride *516* (Scheme 106). Cycloaddition under conductive heating proved to be time-consuming (2-4 days) and were often inefficient, conventional heating for 12 hours (85°C) of the alkynone gave rise to formation of isoxazoles in low yield. One major drawback of the extended reaction times was that a side reaction of the *in situ* generated nitrile oxides gave rise to the formation of furoxan oxides *518*. Furoxane *518*, for instance, was formed under conductive heating conditions in 32% yield, whereas dielectric heating diminished the amount of this side product to 14% and formation of isoxazoles was in high yields.

The *N*-allyl and *N*-propargyl derivatives of saccharin were used as dipolarophiles by Bourgin [5] in cycloaddition with nitrile oxides. Nitrile oxides were generated *in situ* in solvent free conditions by the reaction of arylaldoximes *519* using NCS impregnated on Al_2O_3 as base. Thus, the microwave cycloaddition involving dipolarophiles *520* or *521* and arylnitrile oxides *522* within 3 minutes furnished the isoxazoline *523* or isoxazole *524*, respectively, in high yields (Scheme 107). Final reaction temperatures varied from 110-150°C. The procedure was greatly improved when compared to the conventional method and the reaction time was considerably decreased 100-fold. When the same reaction conditions were used in classical heating (3 min, final temperature, and atmospheric pressure), yields of only 0 and 4% were obtained. Classical heating for 5 hours with arylaldoxime, NaOCl in DCM gave 20 and 14% yield, respectively. A complete conversion and formation of only one regioisomer without any post-isomerization are another advantage compared to the classical conditions.

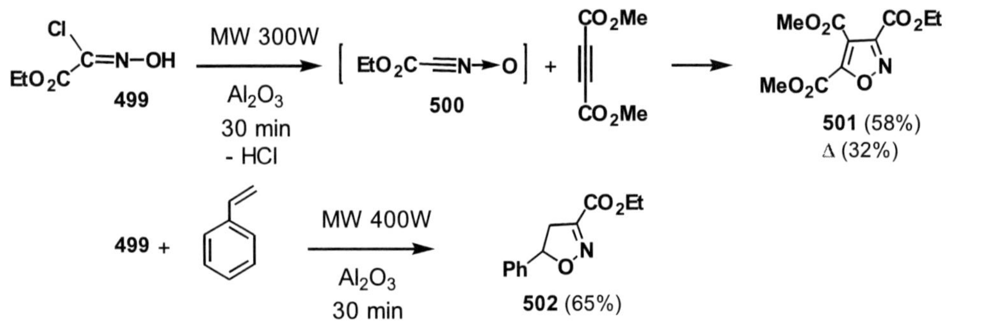

Scheme 103.

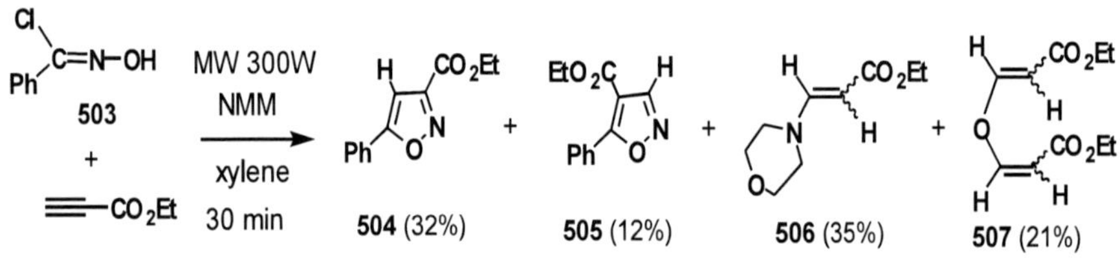

Scheme 104.

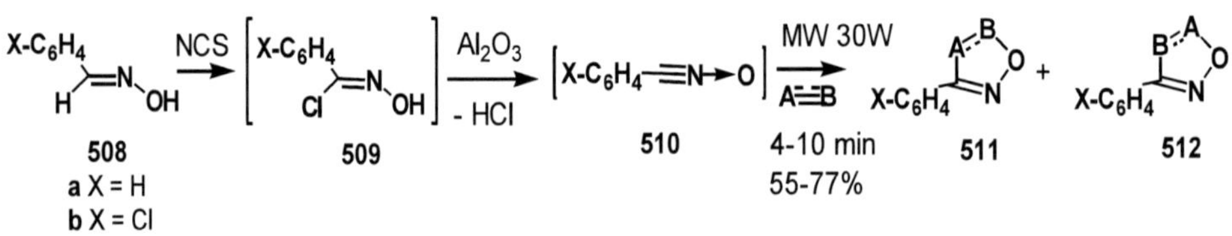

Scheme 105.

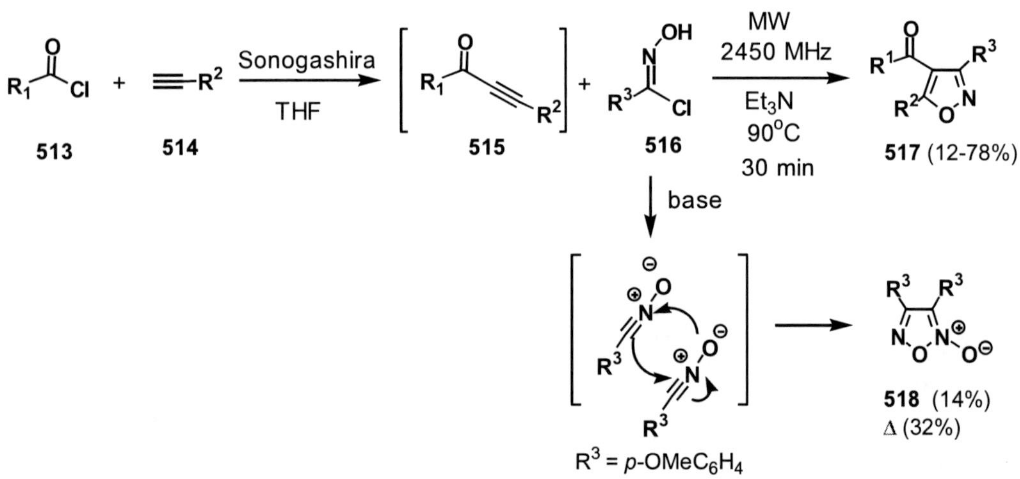

Scheme 106.

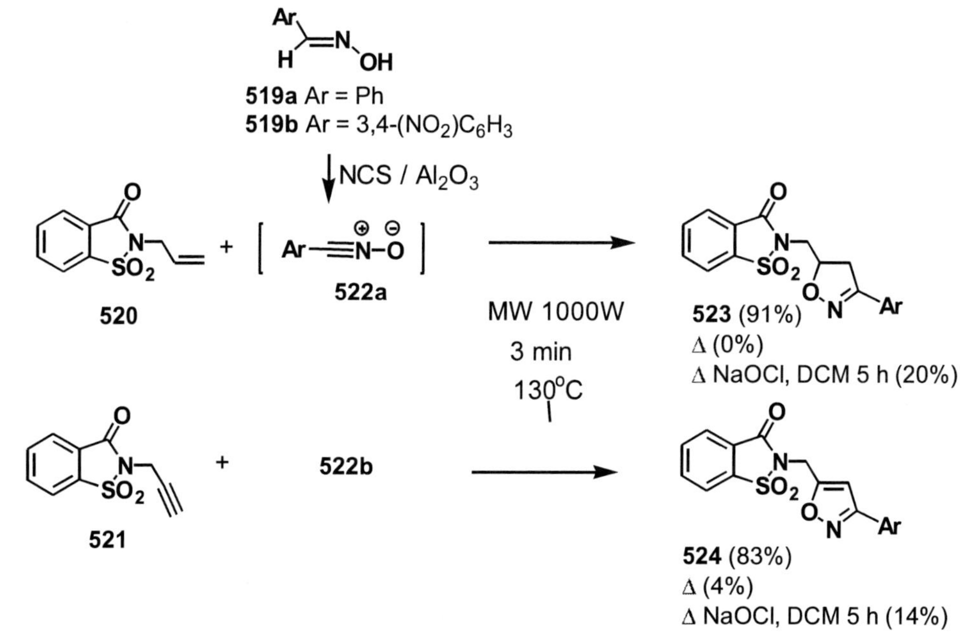

Scheme 107.

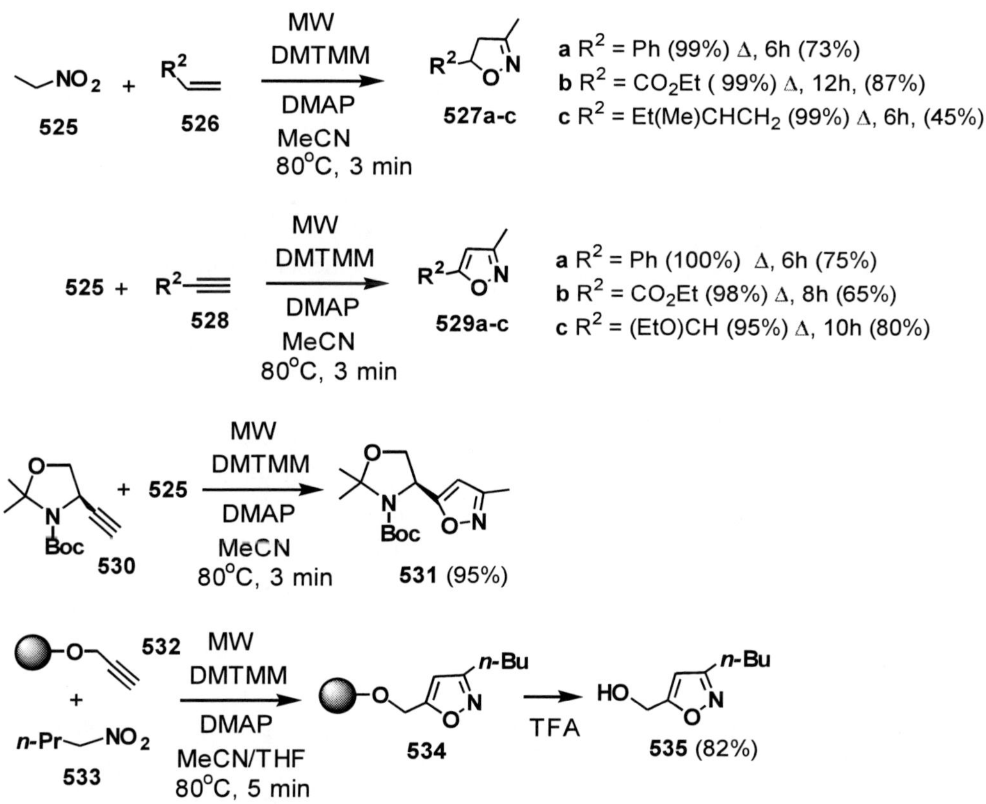

Scheme 108.

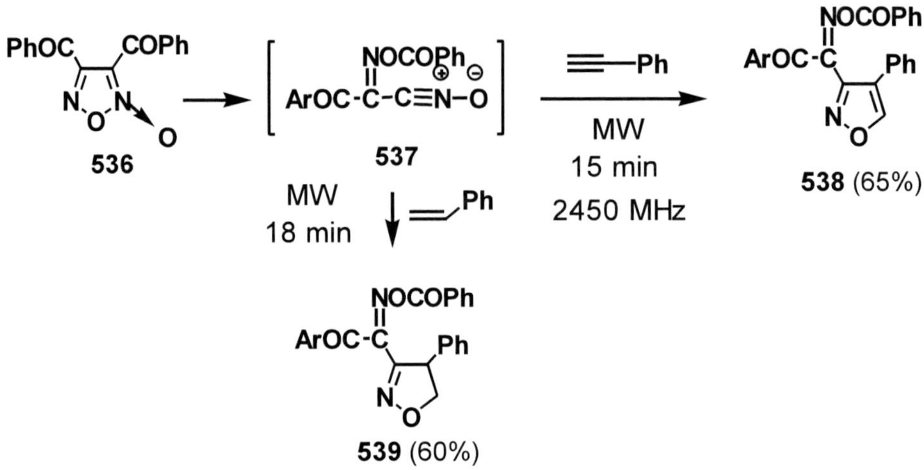

Scheme 109.

A mixture of 4-(4,6-dimethoxy[1,3,5]triazin-2-yl)-4-methylmorpholinium chloride (DMTMM) in the presence of DMAP as catalyst could be used for conversion the nitroalkanes to nitrile oxides. Reaction was consequently exploited by Giacomelli [6] for the transformation of nitroalkanes into isoxazolines *527a-c* or isoxazoles *529a-c* under very mild conditions and in good yields. Carrying out reactions using microwave irradiation conditions, as opposed to conventional heating, has the advantage to shorten the reaction times and increase yields (Scheme 108). The reaction conditions were also used for the synthesis of α-aminoacids containing the isoxazole moiety. Thus, a sample of (*R*)-4-ethynyl-2,2-dimethyloxazolidine-3-carboxylic acid tert-butyl ester *530* was treated with nitroethane. After 3 min of microwave irradiation, (*S*)-3-methyl-5-(2,2-dimethyl-3-tertbutoxycarbonyloxazolidin-4-yl) isoxazole *531* was obtained without loss of optical purity. Procedure can be applied to the synthesis on polymeric support: propargyl alcohol was anchored on trityl chloride resin and the mixture irradiated for 5 min. The isoxazole (3-methylisoxazol-5-yl)methanol *534* was cleaved off the resin in 82% yield.

Sandhu [7] described different way of *in situ* generation of aroyl nitrile oxide *537* by microwave heating of 3,4-dibenzoyl furoxan *536*. The transient nitrile oxides were trapped with acetylenic dipolarophiles to afford isoxazoles *533* or isoxazolines *539* when styrene was used as dipolarophile (Scheme 109). The advantages of this synthetic procedure are solvent free conditions, regioselectivity, and high yields.

Benzyne could be also regarded as acetylenic dipolarophile. Moses [8] developed an efficient route to 1,2-benzisoxazoles *542 via* 1,3-dipolar cycloaddition of *in situ* generated nitrile oxides and benzyne. The optimal results were obtained using K_2CO_3 as a base for nitrile oxide generation (1.5 equiv), while *540* (2.0 equiv.) and *t*-BuONO (2.0 equiv.) were required for benzyne formation. Varying the temperature relative to 120°C resulted in lower yields, whereas change of the solvent from MeCN did not offer any benefit (Scheme 110). This method was found to be amenable to substrates bearing two hydroximoyl chloride functionalities. In the case when benzyne dipolarophile was not present in the reaction mixture, furoxan *543* was the sole product arising from the 1,3-dipolar cycloaddition reaction of two molecules of *in situ* formed nitrile oxide.

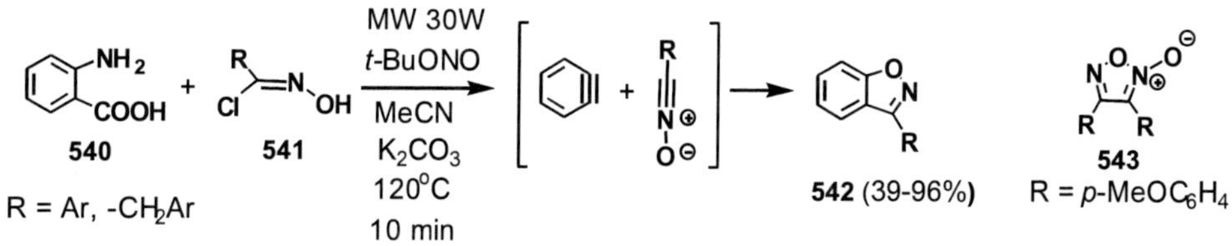

Scheme 110.

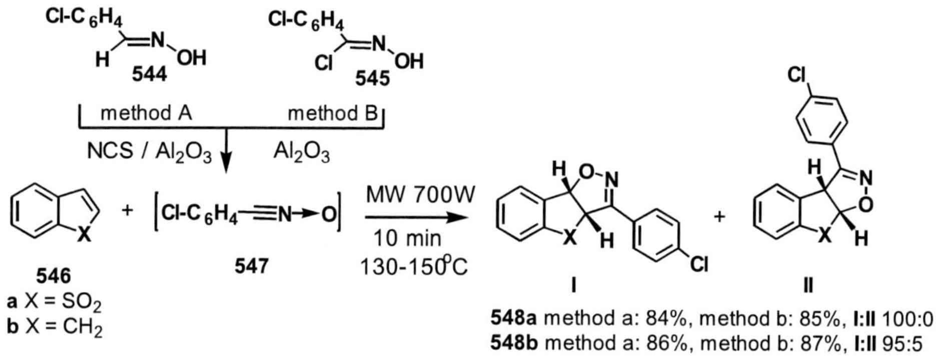

Scheme 111.

Soufiaoui [9] used two methods to generate nitrile oxide *547* required for isoxazoline preparation. Nitrile oxide *547* was generated *in situ*: A) from aldoxime *544*/NCS and B) from benzohydroximinoyl chloride *545*, with Al_2O_3 as base. The 1,3-dipolar cycloaddition of *547* with sulfone *546a* or indene *546b* as dipolarophiles effectively produced isoxazoline cycloadducts *548a* and *548b* (Scheme 111) and two synthetic routes seem to be equally efficient. Short reaction times are accompanied with great regioselectivity of cycloaddition process.

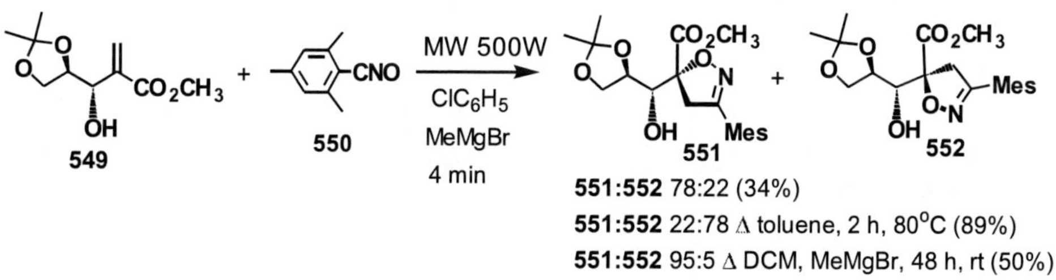

Scheme 112.

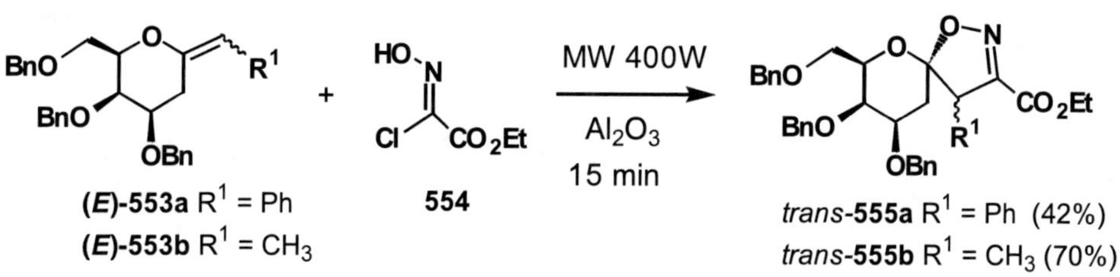

Scheme 113.

Thermal cycloadditions of mesitonitrile oxide *550* with electron deficient dipolarophiles such as *549* are completely regioselective with the 5-substituted isoxazolines *551* and *552* being isolated. Fišera [10] showed that changing the solvent, reaction temperature or microwave irradiation has no influence on the regioselectivity of the reaction and isoxazolines *552* were formed as a main product (Scheme 112). Formed mixtures of diastereoisomers have *de* ranging from 56 to 4%, while ratio of diastereoselective isoxazolines *551*:*552* is 22:78. [11] The addition of a Grignard reagent (MeMgBr) as a Lewis acid affects and even reverses the sense of induced stereoselectivity: >95:<5 for *551*. The reversal of the stereoselectivity presumably results from the imposition of a chelated transition state with geometry different from a `non-chelated' transition state. Acceleration of the cycloaddition by microwave irradiation was evident. Commonly used dichloromethane or toluene solvents were changed to chlorobenzene, as the higher boiling point was required in microwave conditions. In the case of chelated cycloaddition under the microwave irradiation, reaction time was reduced. For example, in the case of dipolarophile *549*, the reaction time decreased from 48 h to 4 min, but diastereomeric excess also dropped from >90 to 56%. However, depending on the dipolarophilic substrate, diastereomeric excess was in some instances increased, or remained unchanged.

Solvent-free conditions coupled with microwave irradiation have proven useful for the generation of nitrile oxides and good yields were found even with sluggish dipolarophile. Lieberknecht [12] used low reactive 1-phenylenitols (*E*)- and (*Z*)-*553a*, in the reaction with the ethoxycarbonylnitrile oxide, which was generated from *554 in situ* using Al_2O_3 as base (Scheme 113). The conventionally heated reactions with the phenyl-substituted enol ethers *553* afforded lower yields than the ones with the methyl substituent, but the yields and/or reaction times were improved with the aid of microwave irradiation (d.r.>95:5). The use of higher power (700 W, 15 min), or prolonged heating (400 W, 30 min) in the reaction of (*E*)-*553a* resulted in the decrease of the reaction yield of *trans-555a* to 35 and 40%, respectively.

Corsaro [13] has shown that dipolar cycloaddition of nitrile oxide to polycyclic aromatic hydrocarbons under microwave (in solvent or solvent-free) conditions was advantageous over thermally conducted reactions. For instance, microwave reaction of 2,4,6-trimethyl-benzonitrile oxide *557* and anthracene *556* produced the mixture of 1:1 cycloadduct *558*, and regioisomeric 2:1 cycloadducts *559* and *560* (Scheme 114). These yields are better than the ones obtained by the solvent-free heating at 180°C, and by far better than these obtained by reflux in benzene or toluene for 24 h.

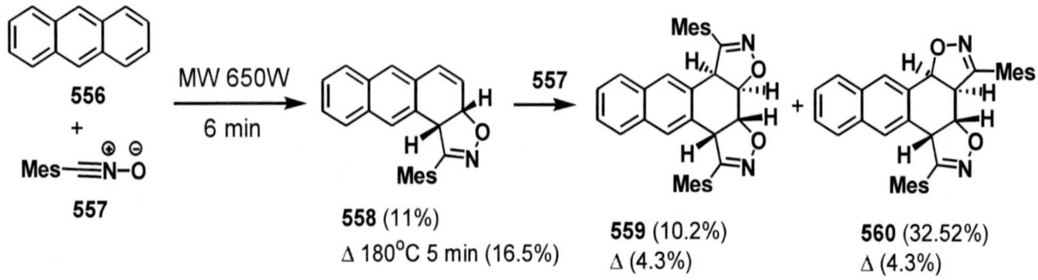

Scheme 114.

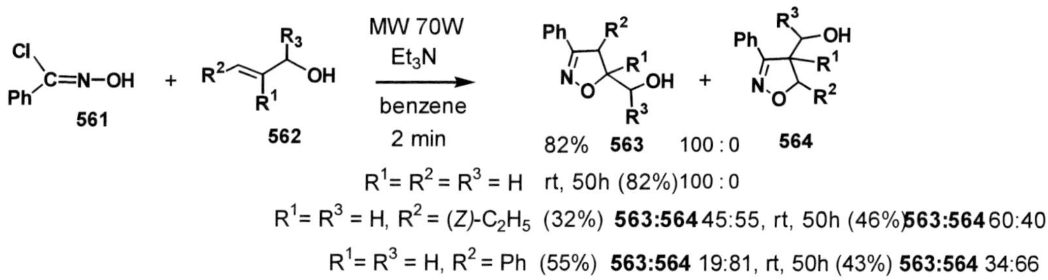

Scheme 115.

Cycloaddition of benzonitrile oxide generated *in situ* from *561* and triethyl amine as base with allyl alcohols produces mixture of two regioisomeric substituted Δ^2-isoxazolines *563* and *564* (Scheme 115). Lu [14] demonstrated that regioselectivity depends on the substituents positioned on allylic alcohol. More interestingly, microwave irradiation alters the regioselectivity of the cycloaddition reaction favoring the non-hydrogen-bond directed cycloadduct, isoxazoline *564*. The percentage of the product *564* increases as the size of the R^2 substituent increases.

Díaz-Ortiz [15] carried out 1,3-dipolar cycloaddition of nitrile oxides *565a,b* to ketene acetal *566* under microwave irradiation in solvent-free conditions which proceeded in high yield to obtain spiro isoxazolidines *567a,b* (Scheme 116).

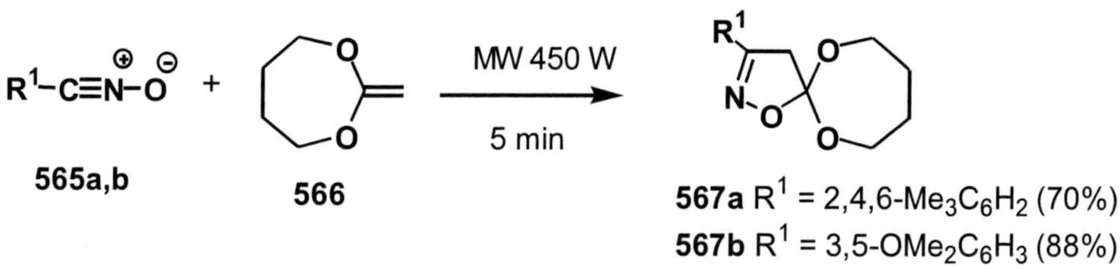

Scheme 116.

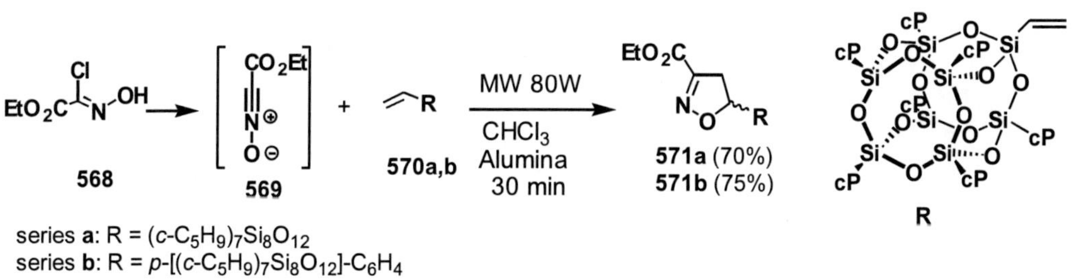

Scheme 117.

Bottino [16] used polyhedral oligomeric silsesquioxanes (POSS) which are externally covered by organic substituents as dipolarophiles. The conversion of vinyl- and styryl-POSS into 4,5-dihydroisoxazole derivatives was effectively achieved by microwave assisted 1,3-dipolar cycloaddition of nitrile oxide. Synthesis of 3-ethoxycarbonyl functionalized 4,5-dihydroisoxazoles *571* employed 1,3-dipolar cycloaddition of nitrile oxide *569* (generated *in situ* from chloro oxime *568* in the presence of Al_2O_3) with vinyl- *570a* and styryl-POSS *570b* as dipolarophiles under microwave assisted conditions. The 1,3-dipolar cycloaddition proceeds regiospecifically and always affords 5-substituted isoxazolines. Under microwave irradiation the reaction time was dramatically reduced compared to conventional heating and the yield increased to 70-75% (Scheme 117).

Díaz-Ortiz [17] used nitriles as dipolarophile in solvent free microwave reactions with nitrile oxide *573* to afford 1,2,4-oxadiazoles *575a-e* (Scheme 118). Reaction times are dramatically reduced, and reaction conditions are milder than other described for these cycloadditions. The differences of yields between microwaves and classical heating were not large when the cycloaddition was a facile process resulting in stable cycloadducts. However, when yields of classical heating were moderate, microwave irradiation obtained better yields of cycloadducts. Since nitrile oxides are less stable than nitrones, they are more reactive dipoles and their resulting 1,2,4-oxadiazole adducts are more stable than their 2,3-dihydro derivatives.

Imine was used as dipolarophile by Kumar [18] in 1,3-dipolar cycloaddition with nitrile oxide. The cycloaddition of nitrile oxide, generated *in situ* from benzohydroximinoyl chloride *577* and triethylamine, with benzothiazepines *576* affords stereoselectively 1,2,4-oxadiazolo benzothiazepine *77* (Scheme 119). Classical reaction conditions require prolongued heating to obtain less amount of product. The cycloaddition of nitrile oxide to *576* is also chemoselective, as it was found that further cycloaddition of nitrile oxide to the benzylidene of *578* does not occur, even with the presence of an excess of nitrile oxide. This may be ascribed to the steric hindrance for the cycloaddition on both the faces of the benzylidene C=C bond of *578*.

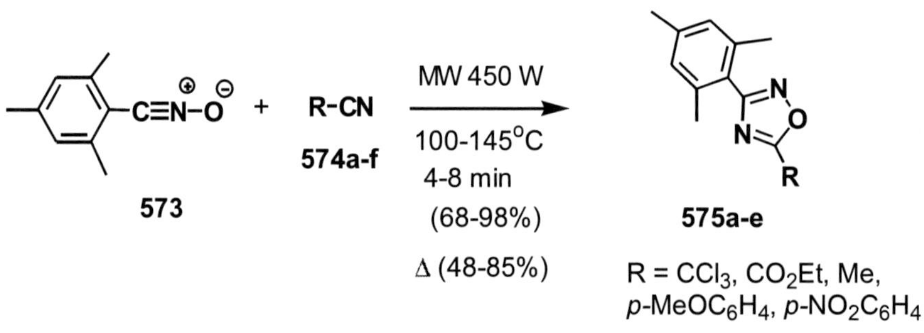

Scheme 118.

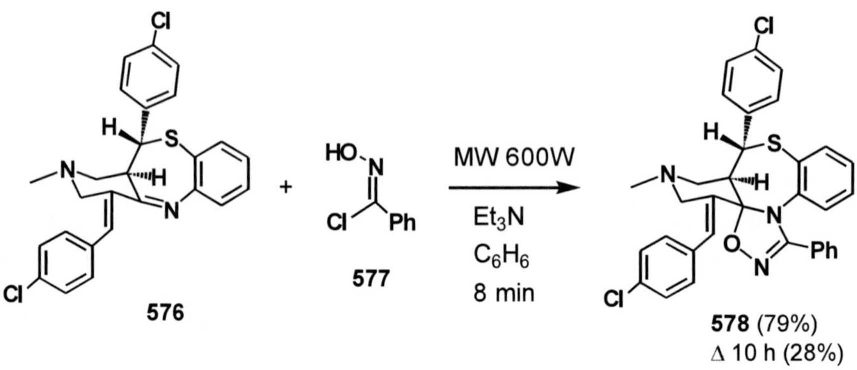

Scheme 119.

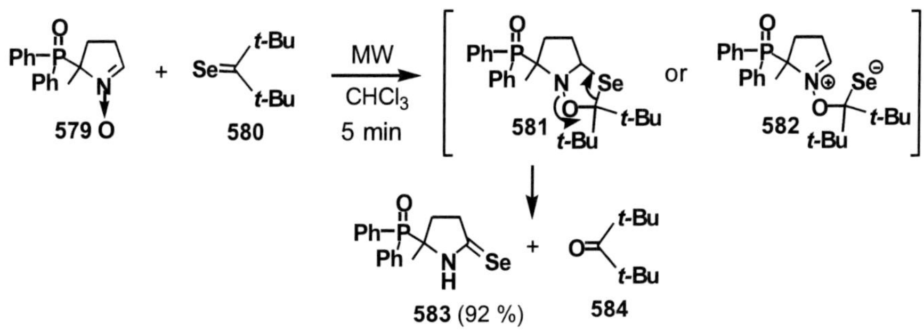

Scheme 120.

The use of selenoketone as dipolarophile was studied by Shioji. [19] Cycloaddition of 3,4-dihydro-2H-pyrrole *N*-oxide *579* with di-*tert*-butyl selenoketone *580* did not proceed at room temperature. However, under microwave irradiation, the cycloaddition of *579* with *580* afforded selenolactam *583* in 92% yield along with the di-*tert*-butyl ketone *584* (Scheme 120). These results suggest that the initial adduct of the reaction was oxaselenazolidine *581* or betaine *582*, in which rearrangement and subsequent elimination took place under these conditions.

References

[1] Touaux, B.; Klein, B.; Texier-Boullet, F.; Hamelin, J. *J. Chem. Res. Synop. 1994*, 116-117.

[2] Kaddar, H.; Hamelin, J. Benhaoua, H. *J. Chem. Res. Synop. 1996*, 718-719.

[3] Touaux, B.; Texier-Boullet, F.; Hamelin, J. *Heteroat. Chem. 1998*, *9*, 351-354.

[4] Willy, B.; Rominger, F.; Müller, T. J. J. *Synthesis 2008*, 293-303.

[5] Mabrour, M.; Bourgin, K.; Benhida, R.; Loupy, A.; Soufiaoui, M. *Tetrahedron Lett. 2007*, *48*, 443-447.

[6] Giacomelli, G.; De Lucca, L.; Porcheddu, A. *Tetrahedron 2003*, *59*, 5437-5440.

[7] Karmakar, D.; Prajapat, D.; Sandhu, J. S. *Synth. Commun. 1998*, *13*, 2415-2420.

[8] Spiteri, C.; Mason, C.; Zhang, F.; Ritson, D. J.; Sharma, P.; Keeling, S.; Moses, J. E. *Org. Biomol. Chem. 2010*, *8*, 2537-2542.

[9] Syassi, B.; Bourgin, K.; Soufiaoui, M. *Tetrahedron Lett. 1997*, *38*, 8855-8858.

[10] Mičúch, P.; Fišera, L.; Cyrański, M. K.; Krygowski, T. M.; Krajčik, J. *Tetrahedron 2000*, *56*, 5465-5472.

[11] Mičúch, P.; Fišera, L.; Cyrański, M. K.; Krygowski, T. M. *Tetrahedron Lett. 1999*, *40*, 167-170.

[12] Colinas, P. A.; Jäger, V.; Lieberknecht, A.; Bravo, R. D. *Tetrahedron Lett. 2003*, *44*, 1071-1074.

[13] Corsaro, A.; Chiacchio, U.; Librando, V.; Fisichella, S.; Pistarh, V. *Heterocycles 1997*, *45*, 1567-1572.

[14] Lu, T.-J.; Tzeng, G.-M. *J. Chin. Chem. Soc. (Taipei) 2000*, *47*, 189-196.

[15] Díaz-Ortiz, A.; Díez-Barra, E.; de la Hoz, A.; Moreno, A.; Prieto, P. *J. Chem. Soc. Perkin Trans. 1 1994*, 3595-3598.

[16] Chiacchio, M. A.; Borrello, L.; Di Pasquale, G.; Pollicino, A.; Bottino, F. A.; Rescifina, A. *Tetrahedron 2005*, *61*, 7986-7993.
[17] Díaz-Ortiz, A.; Díez-Barra, E.; de la Hoz, A.; Moreno, A.; Gómez-Escalonilla, M. J.; Loupy, A. *Heterocycles 1996*, *43*, 1021-1030.
[18] Kumar, R. R.; Perumal, S. *Tetrahedron 2007*, *63*, 7850-7857.
[19] Shioji, K.; Matsumoto, A.; Takao, M.; Kurauchi, Y.; Shigetomi, T.; Yokomori, Y.; Okuma, K. *Bull. Chem. Soc. Jpn. 2007*, *80*, 743-746.

2.4. Nitrile Imines

Introduction

In Huisgen's nomenclature, nitrile imines are classified as nitrilium betaines (1,3-dipoles of propargyl-allenyl type). When nitrile imines and alkenes are involved in 1,3-dipolar cycloaddition reaction, pyrazoline ring is delivered, while addition of nitrile imines to imine C=N bond forms 1,2,4-triazoline ring. In view of the immense biological importance of pyrazolines, various synthetic studies were carried out aiming to increase reaction efficiency and microwave irradiation represensts an important improvement. The 1*H*-1,2,4-triazole compounds are considered interesting heterocycles since they possess important pharmacological activities such as antifungal and antiviral activities [1] and they could be efficiently obtained by microwave assisted dipolar cycloaddition of nitrile imines to imine bond.

2.4.1. Synthesis of Pyrazolines

First paper on the 1,3-dipolar cycloaddition reactions of nitrile imines under microwave irradiation was published by Loupy and Soufiaoui. [2] Impregnated indene dipolarophile *585* and hydrazonyl chloride *586* on KF-Al_2O_3 support when heated with microwaves (multimode) in solvent-free conditions afforded pyrazoline products in 96% yield (as a 93:7 mixture of two regioisomers *587a* and *587b*) (Scheme 121). The best results were achieved on mineral support in a monomode reactor (96%). Under solid-liquid phase transfer conditions yield was lower, due to the partial decomposition of Aliquat-336, however much better than under classical heating conditions.

In addition, Soufiaoui [3] found that montmorillonite-K10 clay is also efficient as a base for solvent free 1,3-dipolar cycloaddition reactions of nitrilimines under microwave irradiation. Microwave synthesis of *591* requires significantly lower temperature and irradiation power than reaction carried out without clay (Scheme 122). Furthermore, when imine *589* was employed as dipolarophile, in the course of reaction, the morpholino group was eliminated forming the corresponding pyrazole *592*.

Synthesis of spiro-rhodanine-pyrazolines by 1,3-dipolar addition of diphenylnitrilimine *594* to 5-arylidenerhodanines *593* on solid inorganic support under microwave irradiation was carried out by Soufiaoui [4] (Scheme 123). Diphenylnitrilimine *594* was prepared *in situ* by dehydrohalogenation of hydrazonyl chloride *590* over alumina, which were trapped by *593* in high yields. Thermal effects of microwaves are displayed by comparison with classical heating, showing that just traces of cycloadducts *596* were formed by conventional heating at 135°C for 10 min.

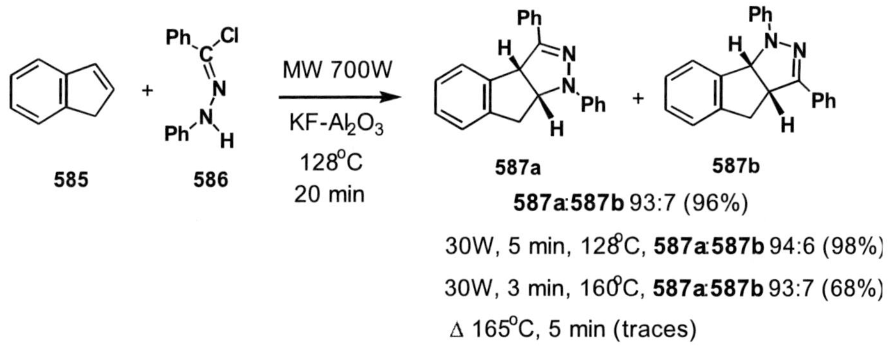

Scheme 121.

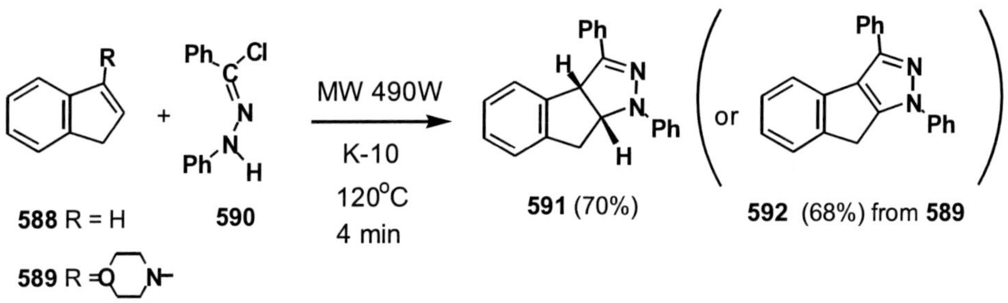

Scheme 122.

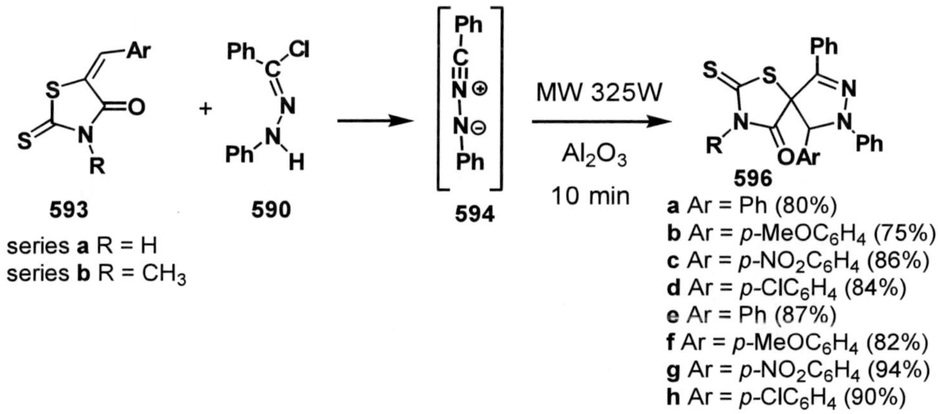

Scheme 123.

Later these authors [5] found that porous calcium hydroxyapatite (p-HAP) is highly efficient as a base for solvent free 1,3-dipolar cycloaddition reactions of nitrilimine and indene under microwave irradiation. Several supports were tested, amongst them p-HAP prepared by heating at three different temperatures (p-HAP100 (basic), p-HAP300 (basic), and p-HAP800 (low basic)), KF/NP (basic), KF/γ-Al_2O_3 (high basic), and α- Al_2O_3 (acid).

Amongst them, the use of p-HAP300 catalyst yielded a cycloadduct *599* in 98% yield (Scheme 124). Catalyst can be regenerated by heating at 300°C, and after ten successive runs, pyrazoline product *599* was obtained with the same yield. Reactions with other dipolarophiles produced corresponding pyrazolines in 87-97% yields at reaction temperatures varying from 104-145°C.

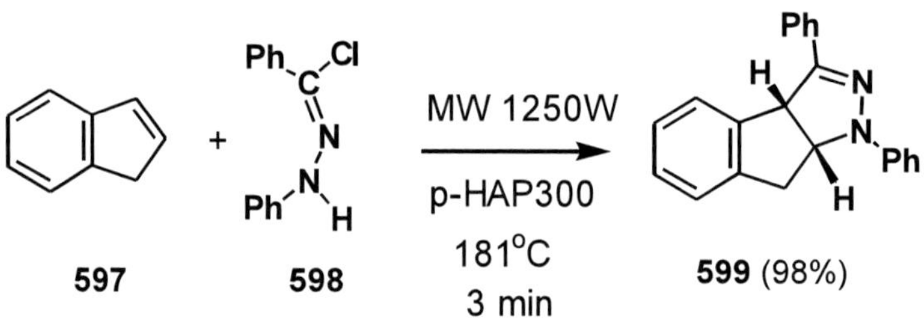

Scheme 124.

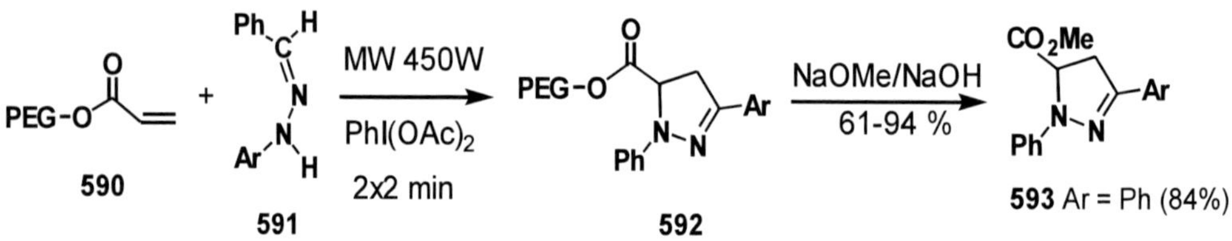

Scheme 125.

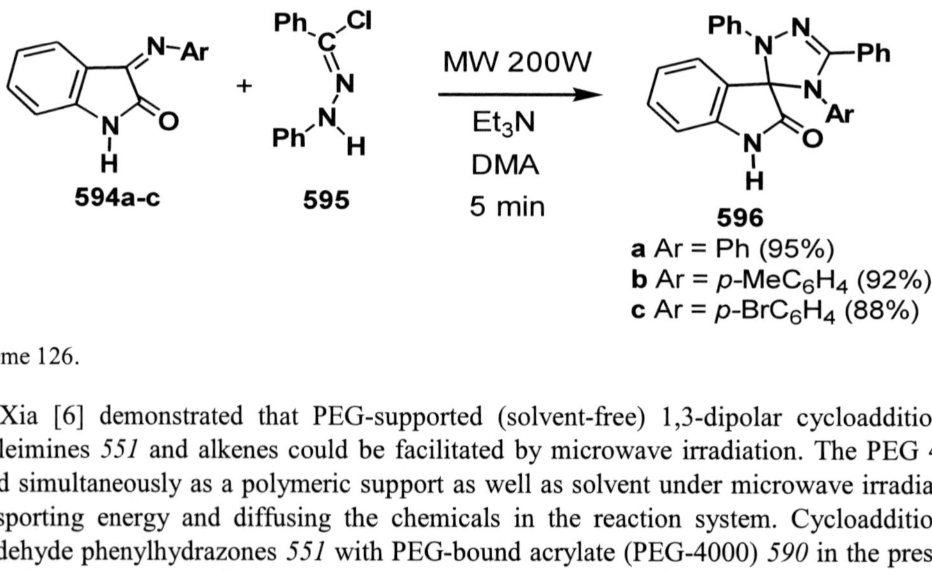

Scheme 126.

Xia [6] demonstrated that PEG-supported (solvent-free) 1,3-dipolar cycloaddition of nitrileimines *551* and alkenes could be facilitated by microwave irradiation. The PEG 4000 acted simultaneously as a polymeric support as well as solvent under microwave irradiation, transporting energy and diffusing the chemicals in the reaction system. Cycloaddition of aryldehyde phenylhydrazones *551* with PEG-bound acrylate (PEG-4000) *590* in the presence of $PhI(OAc)_2$ as an oxidant obtained 1-phenyl-3-substituted-2-pyrazolinyl-5-carboxylates *592*. Reaction was regioselective, and pyrazoline products *593* were completely released from the support in NaOMe/MeOH solution (Scheme 125). Owing to the excellent precipitation of PEG in ether, the PEG-bound pyrazoline intermediates *592* could be obtained in excellent purities after simple filtration and washing, avoiding the complicated isolation and separation procedure.

2.4.2. Synthesis of 1,2,4-triazolines

Azizian [7] developed an efficient synthesis of 1,2,4-triazolines with the aid of microwave irradiation. In this synthesis a mixture of isatin imines *594-c*, triethylamine, and hydrazonyl chloride *595* in *N,N*-dimethylacetamide was irradiated in the microwave oven at 200W for 5 min. *In situ* generated nitrilimine reacted efficiently and regiospecifically across C=N bond to obtain spiroindoles *596a-c* (Scheme 126). In thermal conditions, corresponding yields of 1,2,4-triazolines were significantly lower 20-30%, after 6-10 h of heating.

References

[1] Al-Masoudi, I. A.; Al-Soud, Y. A.; Al-Salihi, N. J.; Al-Masoudi, N. A. *Chem. Heterocycl. Comp. 2006*, *42*, 1377-1403.
[2] Bougrin, K.; Soufiaoui, M.; Loupy, A.; Jaquault, P. *New J. Chem. 1994*, *19*, 213-219.
[3] Syassi, B.; Bougrin, K.; Lamiri, M.; Soufiaoui, M. *New J. Chem. 1998*, 1545-1548.
[4] Ben-Alloum, A.; Bakkas, S.; Bougrin, K.; Soufiaoui, M. *New J. Chem. 1998*, 809-812.
[5] Atir, R.; Mallouk, S.; Bougrin, K.; Soufiaoui, M.; Laghzizil, A. *Synth. Commun. 2006*, *36*, 111-120.
[6] Xia, M.; Pan, X.-J. *Synth. Commun. 2004*, *34*, 3521-3528.
[7] Azizian, J.; Soozangarzadeh, S.; Jadidi, K. *Synth. Commun. 2001*, *31*, 1069-1073.

2.5. Diazoalkanes

Introduction

Microwave irradiation was rarely used to promote 1,3-dipolar cycloaddition reactions of diazo compounds. These reactions produce pyrazoline ring, which often rearranges by dinitrogen elimination to more stable products, such as cyclopropane. With acetylenic dienophiles, pyrazoles are formed in the course of reaction. Huisgen classified diazoalkanes as diazonium betaine 1,3-dipoles of propargyl-allenyl type.

2.5.1. Intermolecular Reactions

Hamelin [1] investigated reactivity of ethyldiazoacetate *597* towards alkenes under microwave irradiation. Comparison with literature shows that microwave irradiation, without the use of solvent could accelerate the process and even enhance the reaction yields. The addition of *597* to methyl benzylidene malonate *598* without solvent leads *via* the unstable 1-pyrazoline *599* to the cyclopropane *600* and 2-pyrazoline *601*, depending on the reaction conditions (Scheme 127). Below 100 W irradiation power the formation of *601* is exclusive, however, when the irradiation power is raised, the cyclopropane *600* appears and the yield goes up to 27% at a power of 150 W. When $Cu(acac)_2$ catalyst was used for the reaction of *597* with alkenes, diastereoisomeric cyclopropanation *E/Z* products *603-604* were obtained exclusively. The use of solvent is essential, since without the presence of solvent, only methyl maleate and fumarate were obtained (dimerisation products of *597*), while the use of alkene as solvent caused polymerization of the substrate and low reaction yields.

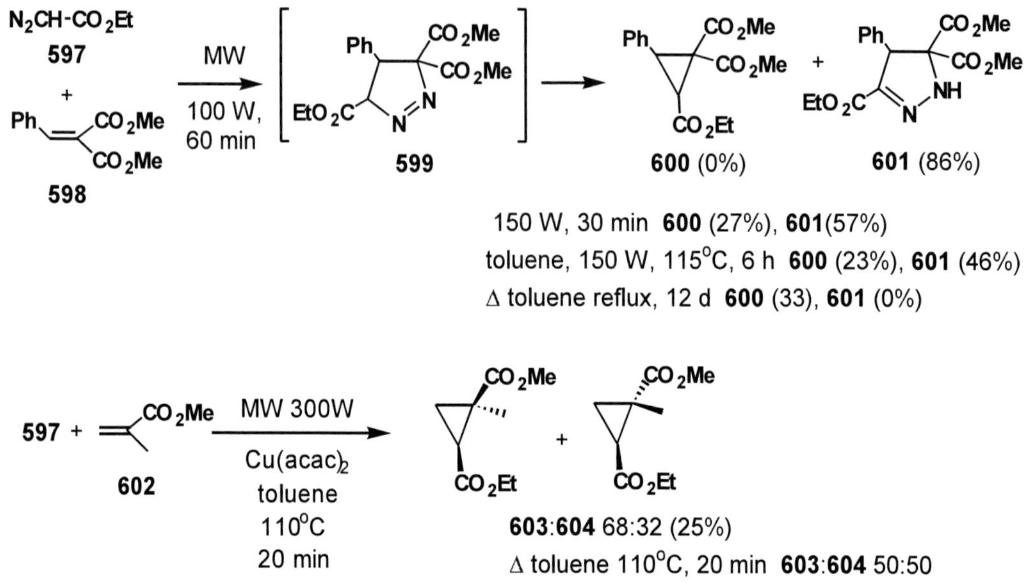

Scheme 127.

When acetylenes *605a-d* were used as dipolarophiles by Eckert-Maksić [2] in the solvent-free 1,3-dipolar cycloaddition reaction with ethyldiazoacetate *606*, or (trimethylsilyl) diazomethane *607*, pyrazoles *608-610* were obtained (Scheme 128). In the case of cycloaddition with *606*, regioisomer *608* dominates or represents the exclusive isomer formed. To afford products *608a* and *609a* in classical conditions, mixture of *605* and *606* was heated for 48h at 140°C with only 38% yield. Microwave assisted reactions of *607* using the same acetylenic compounds *605a-d* were carried out in 2M solution of *607* in hexane producing 3-pyrazoles *610*, as the consequence of the subsequent 1,5-Me_3Si shift in the initial cycloadduct.

2.5.2. Intramolecular Reactions

Intramolecular addition of diazo compounds employing *trans*-4-phenyl-3-buten-2-one substrate was investigated by Rosa [3] in synthesis of pyrazoles. Compound *614a* has been obtained under microwave activation following two methodologies. The first synthetic approach to synthesize the pyrazole derivatives *614* was the 1,3-cycloaddition of the corresponding tosylhydrazone *611* using a base (Scheme 129). Reaction mechanism for intramolecular 1,3-dipolar cycloaddition of tosylhydrazones *611* (derived from acyclic α,β-unsaturated carbonyl compounds *615*) with a β-hydrogen includes the formation of intermediate *613*, which is aromatized by subsequent hydrogen loss. The best performing conditions resulted from neat reagents in the presence of anhydrous K_2CO_3 and a catalytic amount of *N,N*-dimethylformamide as an energy transfer improver. Powdered graphite was less efficient (62%). In a "one-pot" approach to the synthesis of *614a*, the tosylhydrazone formation step was avoided, by starting directly from the carbonyl precursor *615* and tosylhydrazine employing the same catalytic conditions. By this procedure *614a* was obtained in high yield.

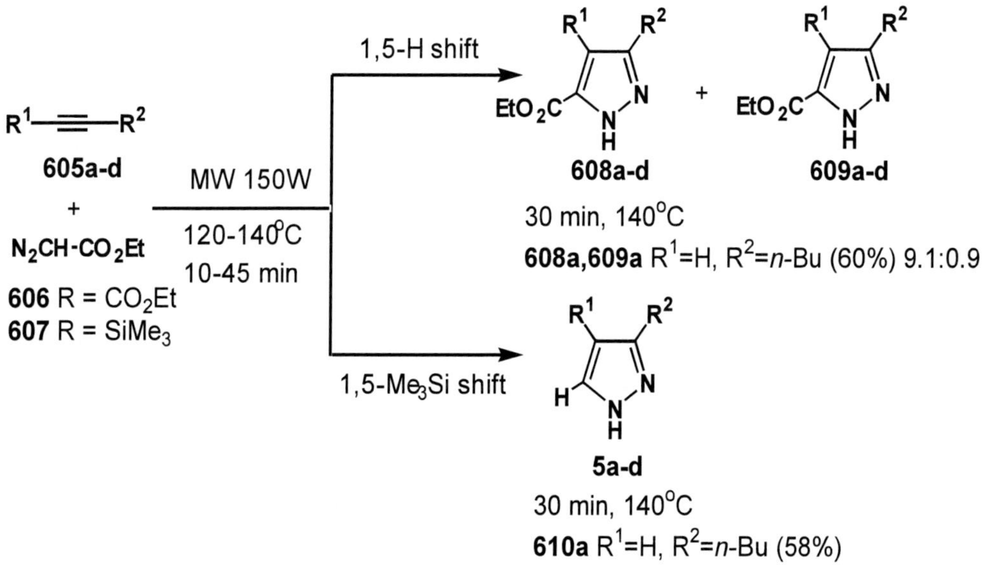

Scheme 128.

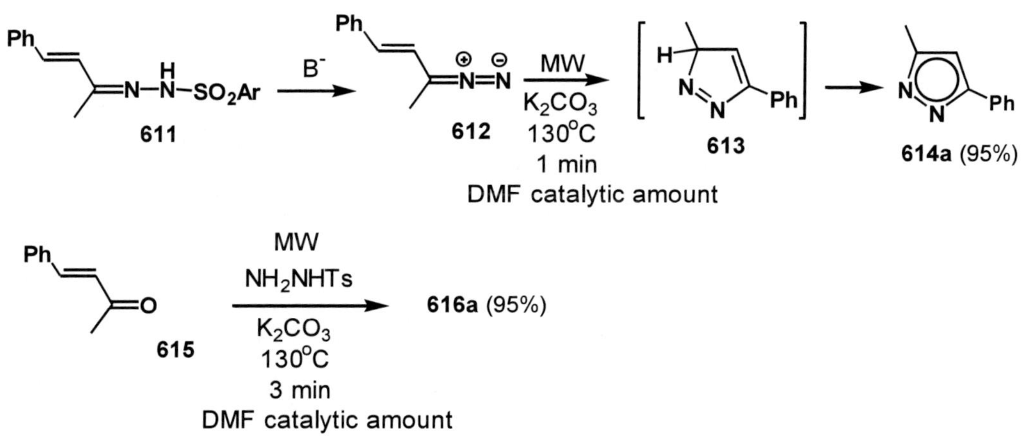

Scheme 129.

References

[1] Bendeddouche, K. C.; Rechteiner, B.; Texier-Boullet, F.; Hamelin, J.; Benhaoua, H. *J. Chem. Res. Synop. 2002*, 114.

[2] Zrinski, I.; Juribašić, M.; Eckert-Maksić, M. *Heterocycles 2006*, *68*, 1961-1967.

[3] Corradi, A.; Leonelli, C.; Rizzuti, A.; Rosa, R.; Veronesi, P.; Grandi, R.; Baldassari, S.; Villa, C. *Molecules 2007*, *12*, 1482-1495.

2.6. Azomethine Ylides

Introduction

Azomethine ylides are 1,3-dipoles of allyl type, with nitrogen function as middle center. In the course of the dipolar cycloaddition reaction with alkenes, five membered pyrrolidine ring is produced, which in the case of unstable products could be rearranged, or forms pyrrole by elimination and aromatization. Cycloadditions with acetylenes directly generate pyrroles, while additions to carbonyl double bond afford oxazolidine ring.

2.6.1. Intermolecular Reactions

The first report on 1,3-dipolar cycloadditions enhanced by microwave irradiation was published by Díaz-Ortiz [1] in 1994. It is also the first report on microwave 1,3-dipolar cycloadditions in solvent free conditions. Pyridinium azomethine ylide *616* (*N*-dicyanomethylide) reacted with ethyl phenylpropiolate *617a* and phenylacetylene *617b* to obtain two regioisomers *618a,b* (Scheme 130). Thermal reaction of *616* with *617a* in refluxing toluene afforded *618a* and *618b*, where isomer *618a* dominates. On the other hand, in these conditions, phenylacetylene as a very unreactive dienophile affords only traces of cycloadducts. Short reaction times and better yields were obtained when microwave solvent free method was used in comparison to classical heating, but the selectivity remained unchanged. Adsorption on solid supports affects the dipolarophile by increasing the polarity and changed the selectivity. Reactions with different supports were performed under microwave irradiation (K-10 Montmorillonite, bentonite, talc, celite, Hyflo superCell, basic alumina, florisil, and silicagel). The best selectivity was observed using a neutral bentonite (*618a*/*618b*=97:3, 69%), while the results with phenylacetylene were less satisfactory. Prolonged heating and increased temperatures cause the decomposition of ylide and decrease in reaction yields.

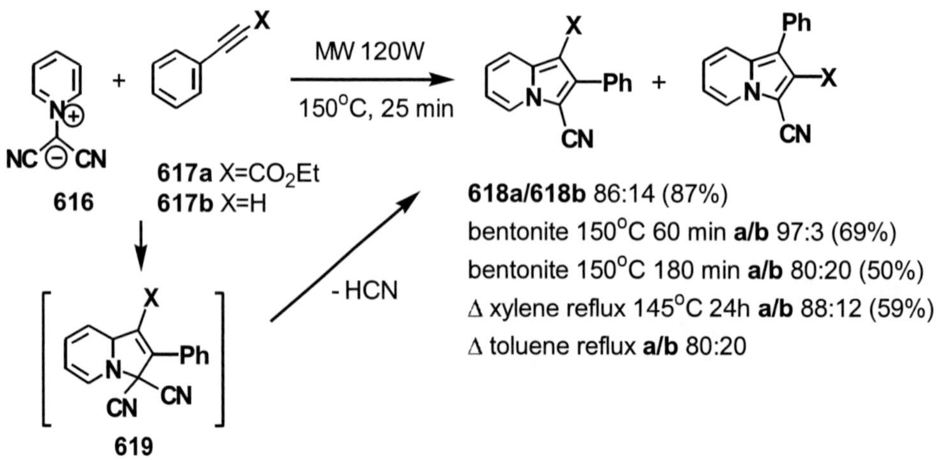

Scheme 130.

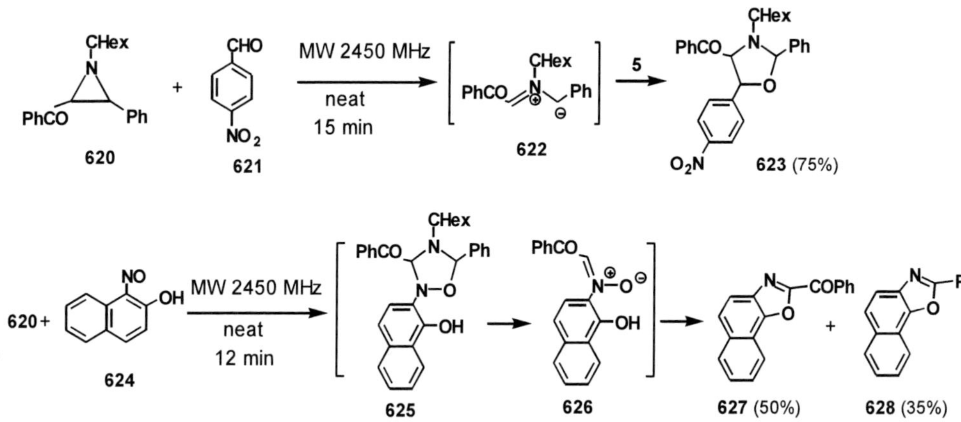

Scheme 131.

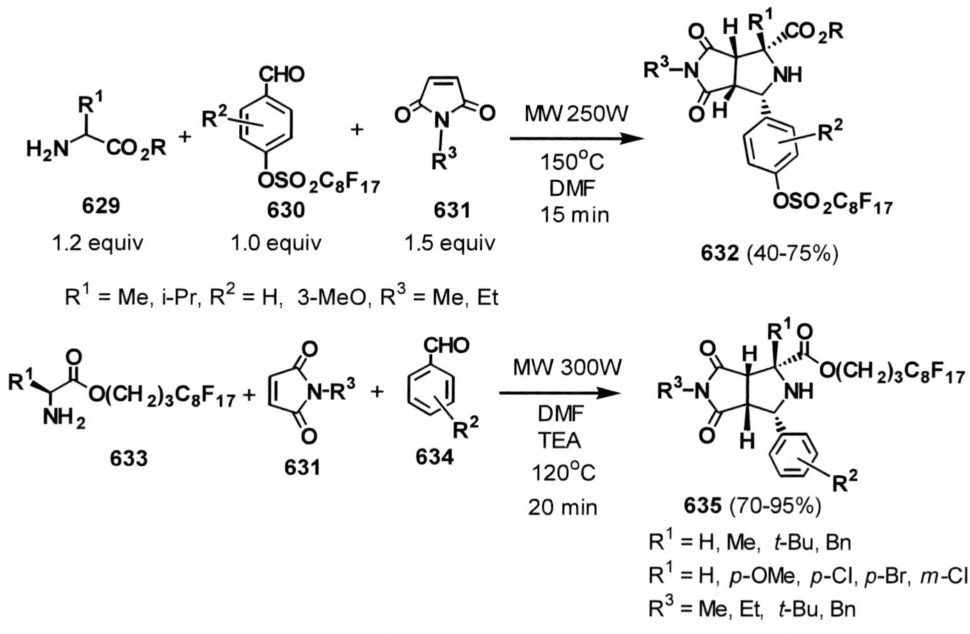

Scheme 132.

Another early example of 1,3-dipolar cycloaddition by Sandhu [2] deals with the addition to multiple bonds of azomethine ylide generated from 2-aroyl-aziridines *620*. When 2-benzoyl-1-cyclohexyl-3-phenylaziridine *620* and *p*-nitrobenzaldehyde *621* were subjected to microwave irradiation without solvent 4-benzoyl-3-cyclohexyl-c-5-(*p*-nitrophenyl)-r-2-phenyl oxazolidine *623* was obtained (Scheme 131). When 1-nitrosonaphth-2-ol *624* was used as dipolarophile, 1,3-dipolar cycloaddition to the nitrogen-oxygen bond gave oxadiazolidine *625*. Subsequent ring cleavage to a nitrone *626* and intramolecular cyclisation afforded mixture of 2-benzoyl-naphtho[1,2-d]oxazole *627* and 2-phenyl-naphtho[1,2-d]oxazole *628*.

The multicomponent reactions consisting of amino acid derivative, aldehyde and alkene dipolarophile afforded cycloadducts with the formation of new pyrrolidine ring. Perfluoroalkylsulfonyl protected hydroxy benzaldehydes were used by Zhang [3] for preparation of bicyclic proline analogs. In this multicomponent reaction, amino ester *629*, benzaldehyde, and maleimide were condensed under microwave irradiation to produce bicyclic proline *632* as a single diastereomer, where two ring-fused hydrogen atoms are *cis*- to the R^1 and *trans*- to the phenyl group (Scheme 132). The alternative synthetic route to perfluorinated pyrrolidine derivatives is the multicomponent reaction of a mixture of a fluorous amino ester, benzaldehyde, and *N*-alkylmaleimide. Zhang [4] showed that when mixture of these components, with triethylamine in DMF was heated under microwave irradiation at 130°C for 20 min proline derivatives *635* were obtained. Bicyclic prolines *635* with different R^1-R^3 substituents were synthesized in 75-90% yields. The stereochemistry of compound *635a* was established and there is no evidence for the racemization of chiral amino acid *633* during the cycloaddition.

Gallagher combined thermally induced formation of azomethine ylides, retro-Diels-Alder reaction and subsequent 1,3-dipolar cycloaddition in the single reaction pot. [5, 6] Dithioformates *640a-c* were released *in situ* from their cyclopentadiene cycloadducts *639a-c*, *via* [4+2] cycloreversion. At the same time, the key azomethine ylid intermediate *638* was released from oxazolidinone *636* to provide the racemic cycloadduct *641* as a 2.5:1 mixture of *exo* and *endo* isomers (Scheme 133). Classical heating in acetonitrile required 2 days, at 80 °C to obtain low yield (19%). Using toluene as a solvent, a 76% yield of the cycloadduct *641a* was isolated following microwave irradiation of *636* and *639*. Cycloadduct *641a* was obtained in 56% yield when the same reaction was carried out for 1 h using 10 mol % of ionic liquid (emimPF_6) as an additive. Interestingly, experiments in (microwave) open vessel (200 W, 5 h) (62%) or with emimPF_6 (10 mol %, 55 W, 4 h) gave lower product yields than pressure vessel (62 and 40%, respectively). More elaborated substrate, C(6)-O-silylated hydroxyethyl *β*-lactam-based oxazolidinone *637* also reacted with *639c* to give the corresponding dipolar cycloadduct *642c*. Although OTBS substituent survives the applied reaction conditions, significantly lower yield was obtained.

Solvent-free microwave-assisted synthesis of functionalized spiropyrrolidines and pyrrolizidines was achieved through 1,3-dipolar cycloaddition by Raghunathan. [7] Methyl 2-(2,3-dihydro-2-hydroxy-1,3-dioxo-1H-inden-2-yl) acrylate *643* was utilized as a dipolarophile in 1,3-dipolar reactions of azomethine ylides *645* generated *in situ* from mono-, di- and tri-ketones and sarcosine *644* or proline by the decarboxylative condensation. The follow-up 1,3-dipolar cycloaddition reaction afforded the novel spiro adducts *647* in high yields with a high regioselectivity. Three different sets of reaction conditions were compared: microwave/methanol, solvent-free K-10 Montmorillonite clay/microwave and methanol reflux. It was found that solvent-free conditions gave the best yields with high regio- and stereoselectivity, such as obtained for reaction with isatin *646* (Scheme 134).

Using the same method as described previously in Scheme 134, greener approach for the synthesis of 1-*N*-methyl-(spiro[2.3']oxindolespiro[3.2"] and spiro[2.3']indan-1,3-dionespiro[2.2"]) cyclopentanone-4-aryl pyrrolidines was developed by Raghunathan. [8] A range of dipolar cycloaddition reactions in solvent-free conditions combining supported reagents (on Bentonite clay) and microwave irradiation in a domestic microwave oven was carried out. The cycloaddition reaction of azomethine ylides *650* formed using the triketone ninhydrin *648*, with sarcosine *649* and 2-arylidene-1-cyclopentanones *651a-d* as dipolarophiles afforded a novel dispiro heterocycles *652a-d* (Scheme 135). Yields were much better in comparison with conventional refluxing. In a similar reaction, the condensation of isatin and sarcosine afforded corresponding dispiropyrrolidinyl oxindole derivatives.

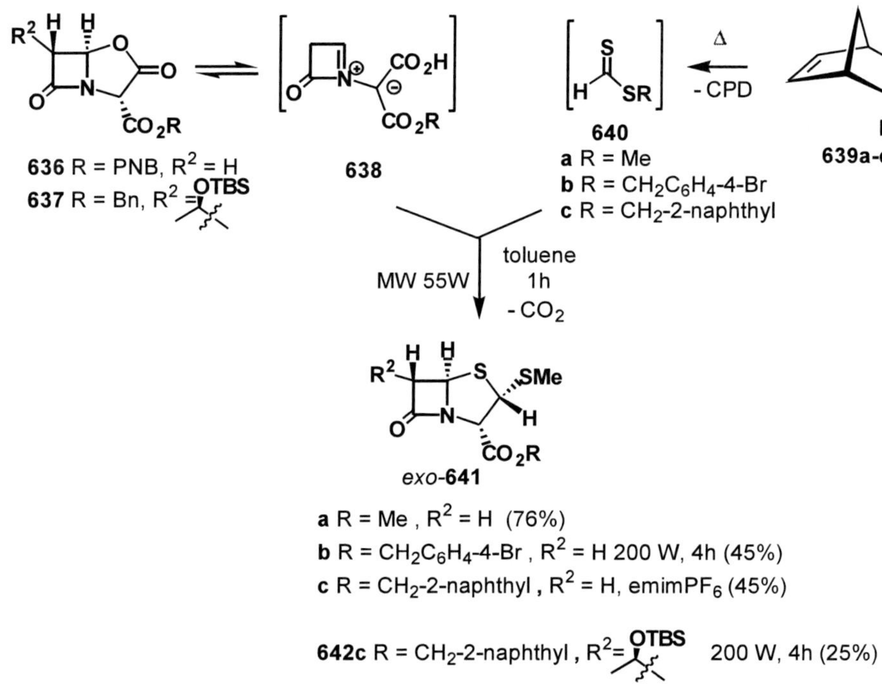

Scheme 133.

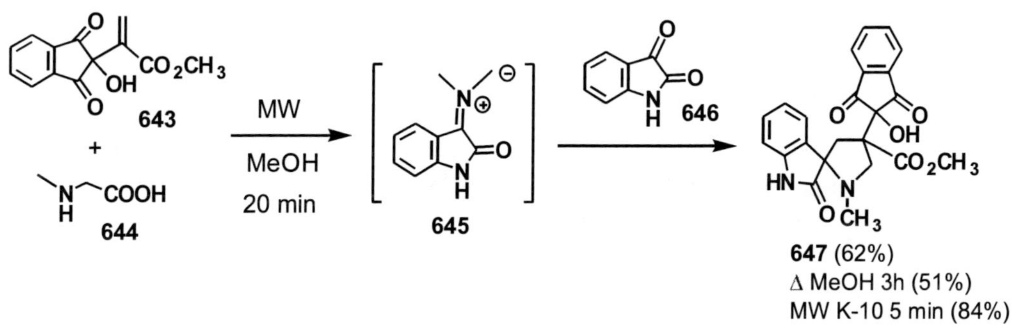

Scheme 134.

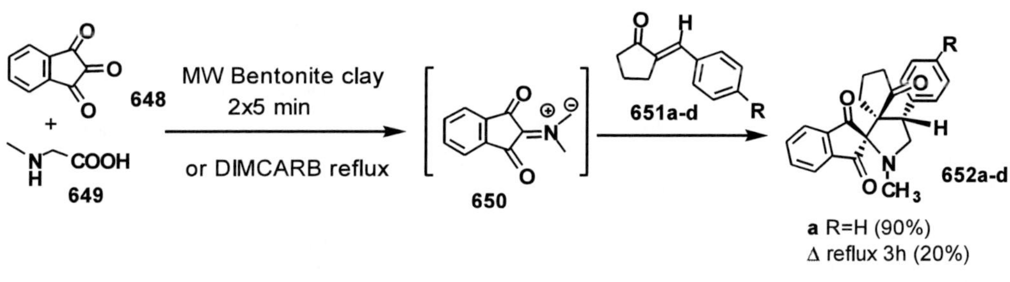

Scheme 135.

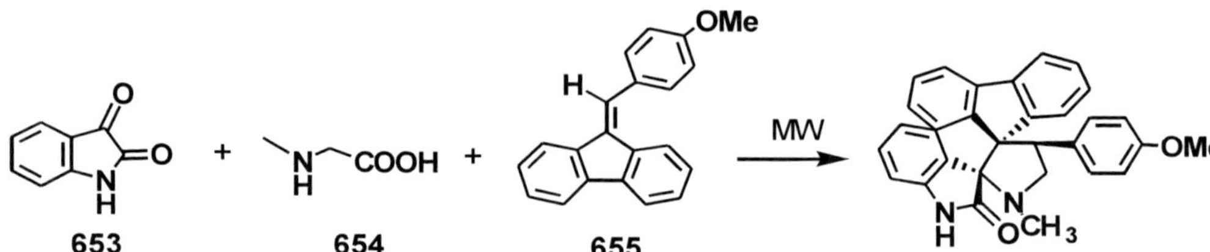

methanol 6.5 min (56%)
K-10 4.5 min (78%)
neat 4.5 min (79%)
Δ methanol reflux 6.5 h (58%)

Scheme 136.

Raghunathan [9] compared four different conditions used in multicomponent 1,3-dipolar cycloaddition reaction of azomethine ylides: conventional methanol/reflux; microwave/methanol; K-10 Montmorillonite/ microwave and microwave/neat. *In situ* generated 1,3-dipol via decarboxylative route from sarcosine/proline *654* and isatin *653* with 9-arylidine-fluorene *655* as the dipolarophile using a solvent-free microwave-assisted approach gave products *656* with the highest yields in a shortest time (Scheme 136). When ninhydrin was used instead of isatin, fluorenespiro[9.3']-(4'-aryl)pyrrolidine/ pyrrolizidine/tetrahydropyrrolo[1,2-c]thiazolespiro[2'.2'] indan-1",3"-dione were prepared. Again, neat microwave reactions afforded the highest yields. [10]

The same author [11] discovered that TiO_2 impregnated silica could be used as an efficient solid-support catalyst for this type of reactions. The synthesis of a series of novel dispiroheterocyclic systems *660* was achieved by the cycloaddition of an azomethine ylide generated by the decarboxylative route from tetrahydroisoquinoline-3-carboxylic acid *658* and acenaphthenequinone/isatin *657* with various dipolarophiles such as 2-arylidene-1,3-indanediones and (*E*)-2-oxoindolino-3-ylidene acetophenones *659* in a one pot three component tandem reaction (Scheme 137). The high regiochemistry and stereochemistry at the spiro centre is governed by the nature of transient *anti*-ylide *661* and steric repulsion for the formation of *syn*-ylide. TiO_2/silica was superior to $BiCl_3$/silica or silica supported three-component reactions.

Microwave assisted, solvent free reaction supported on clay was also successfully employed by Raghunathan [12] for 1,3-dipolar cycloadditions of dipole generated from benzylamine with isatin. Out of three sets of reaction conditions examined, solvent free reaction supported on K-10 Montmorillonite clay gave by far largest yields (Scheme 138). Condensation of benzylamine with isatin *662* give rise to two configurationally distinct azomethine ylides, *664a* and *664b*, favoring isomer *664a* due to the steric interactions. It reacts with bis-arylmethylidene cyclohexanones *665* to give a series of dispiro oxindole derivatives *666* in a regioselective manner. The reaction takes place at one of the exocyclic double bonds of *665*. Even with excess of 1,3-dipole and longer reaction time, the other double bond remained unaffected, which may be attributed to steric hindrance.

Synthesis of spiropyrrolizidines *671*, a potent antimicrobial agents was carried out employing the above described chemistry by Raghunathan. [13] This synthesis was accomplished by 1,3-dipolar cycloaddition reaction of azomethine ylide *669* (generated from proline and isatin *667*) with dipolarophilic (*E*)-2-arylidine-1-keto carbazoles such as *670*.

Classical reaction was carried out in dioxane:methanol, while the microwave irradiation in solvent free conditions, with K-10 Montmorillonite clay support greatly outperforms classical heating in terms of reaction times and yields (Scheme 139).

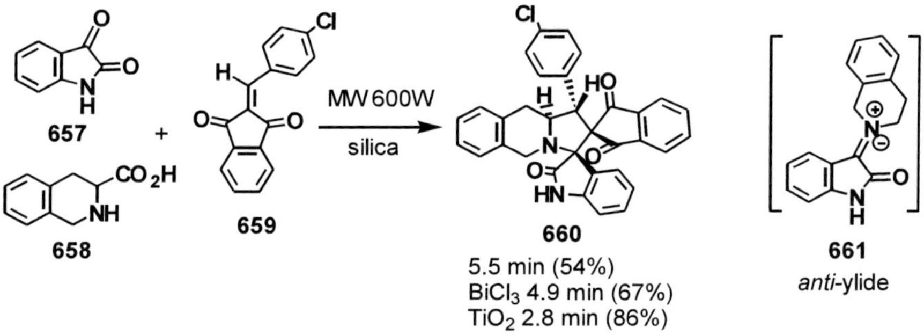

Scheme 137.

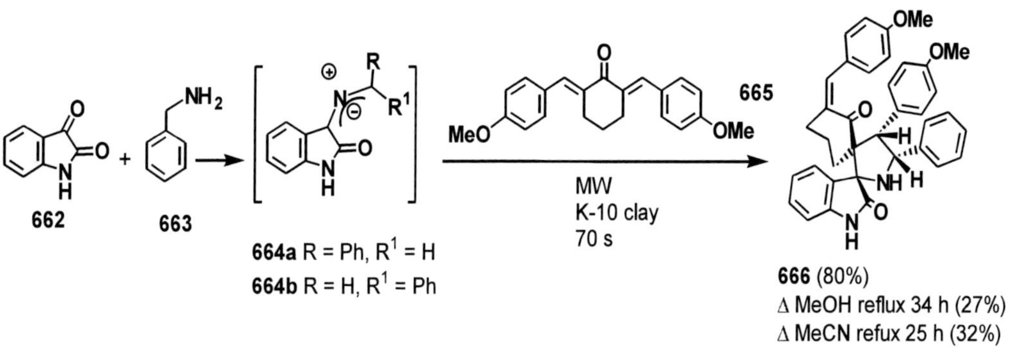

Scheme 138.

Dinica [14] reported that microwave enhanced solvent free reaction could be also carried out on Alumina as solid support. The synthesis of substituted 7,7'-bis-indolizines such as *675a* was achieved via two-fold 1,3-dipolar cycloaddition of non-stabilized 4,4'-bipyridinium ylide *673a* under microwave irradiation. Reactions were carried out by solvent free cycloaddition of reactants adsorbed on Alumina (Scheme 140). Reaction takes place via non-isolable intermediates *674*, which undergo dehydrogenation process.

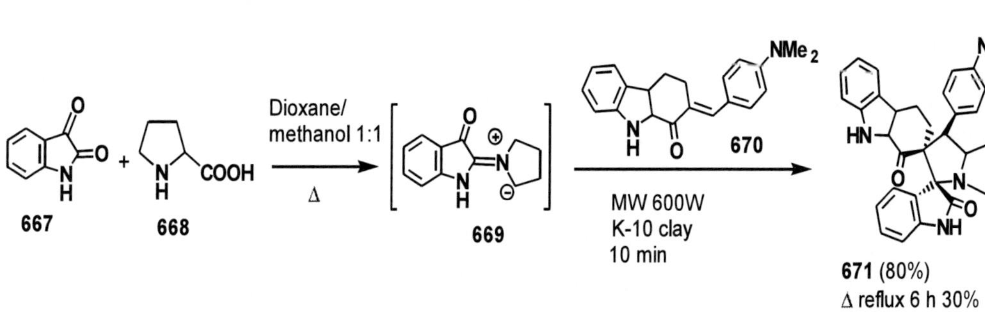

Scheme 139.

Interesting example of multicomponent one pot diastereoselective reaction was published by Azizian. [15] The mixture of of isatin *676*, proline, and *N*-substituted maleimides were heated by microwaves. Firstly condensation of isatin and proline affords the intermediate *678*. Stereoselective cycloreversion of *678* generates dipolar azomethine ylide *679* which undergoes 1,3-dipolar cycloaddition with maleimides *680* to obtain spiro products, 2-substituted-1,1',2,3,3aα,-3bβ,4,5,6,8,8aα-8-(spiro-3'-indolino-2'-one)'pyrrolo[3,4-a]-pyrrolizine-1,3-diones *681* in 71-94% yields (Scheme 141). Classical reflux in absolute ethanol required longer times and obtained yields were smaller.

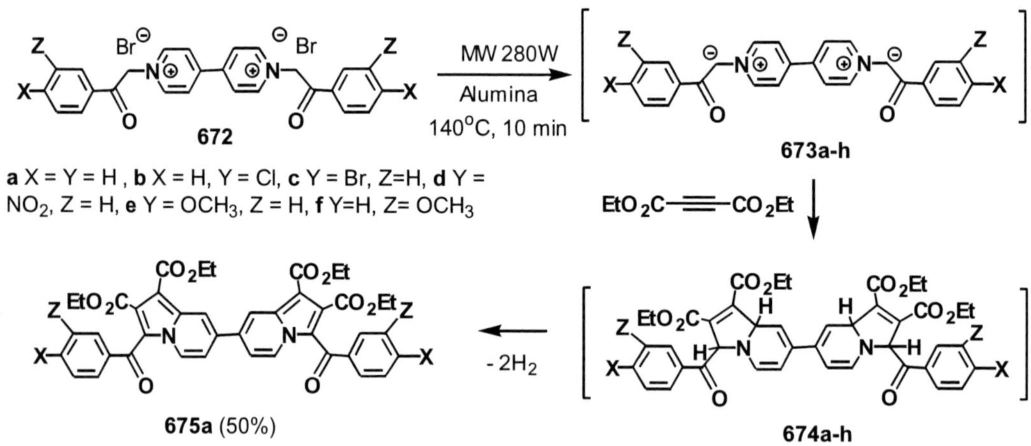

Scheme 140.

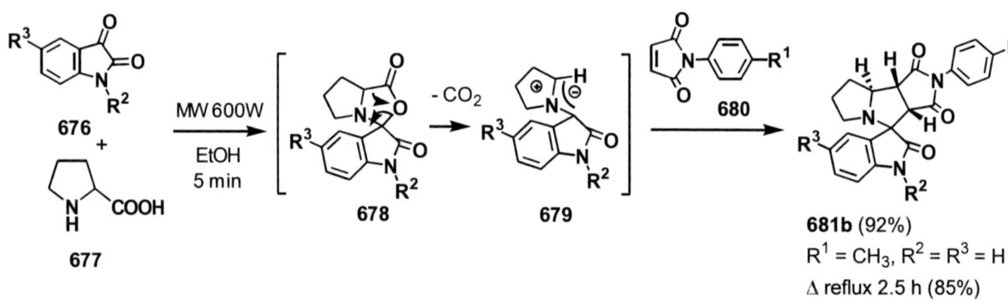

Scheme 141.

Imidates could be also employed as a synthetic source of azomethine ylides. Bazureau [16] has shown that imidate *682* is in equilibrium with azomethine ylide form *682b* by thermal protropy. 4-alkylidene-1*H*-imidayol-5(4*H*)-one derivatives *687* (aromatic, aliphatic and heteroaromatic) were obtained in good to excellent yield by 1,3-dipolar cycloaddition using solvent-free conditions under microwave irradiation from imidate *682*, aldehydes *683* and dry acetic acid as catalyst. Similar cycloadducts were obtained by cycloaddition from imidate *682* and aldimines *684* as dipolarophiles (Scheme 142).

The reactivity of these azomethine ylides could be further enhanced by addition of cyano group. [17] Azomethine ylide form *688b* of imidate *688* undergoes regioselective cycloadditions to aldehydes *689a-f* as dipolarophiles in solvent free microwave conditions in short reaction times. In the process, 4-cyano-2-oxazoline-4-carboxylates *691a-f* were obtained in good yields (81-98%) and moderate diastereoisomeric ratios (Scheme 143). No change in stereoselectivity was observed between microwave and thermal heating. If the cyano group of *688* is replaced by carbmethoxy substituent, due to symmetry, a single 2-oxazoline-4,4-dicarboxylate was obtained. [18] Using *N*-benzylidene methylamine *692* as dipolarophile, the mixture of diastereoisomeric ethyl 4-cyano-2-methyl-5-phenyl-2-imidazoline-4-carboxylate *693* was obtained.

Ionic liquids as environmentally suitable solvents could be used in 1,3-dipolar cycloadditions involving azomethine ylides. Bazureau [19] reported that imidate *694* and aldehyde *695* react in air stable ionic liquid 1-ethyl-3-methylimidazolium tetrafluoroborate ([emim][BF_4]) to obtain oxazolidine adduct *696* (Scheme 144). Classical heating in the same ionic liquid afforded *696* in 78% yield after 10 hours. Focused microwave irradiation with glacial acetic acid in [emim][BF_4] was suitable for increasing the reaction rate and the yield of isolated cycloadduct *696*. For comparison, heating in oil bath at 70°C for 2.5 h, afforded *696* in 52% yield.

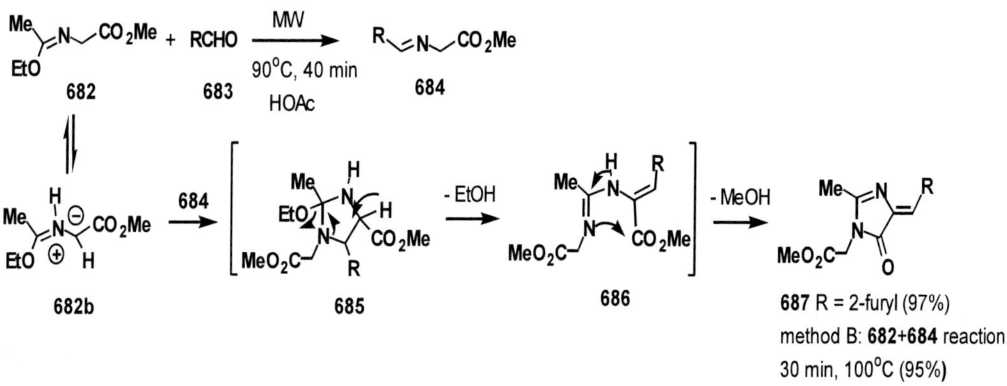

Scheme 142.

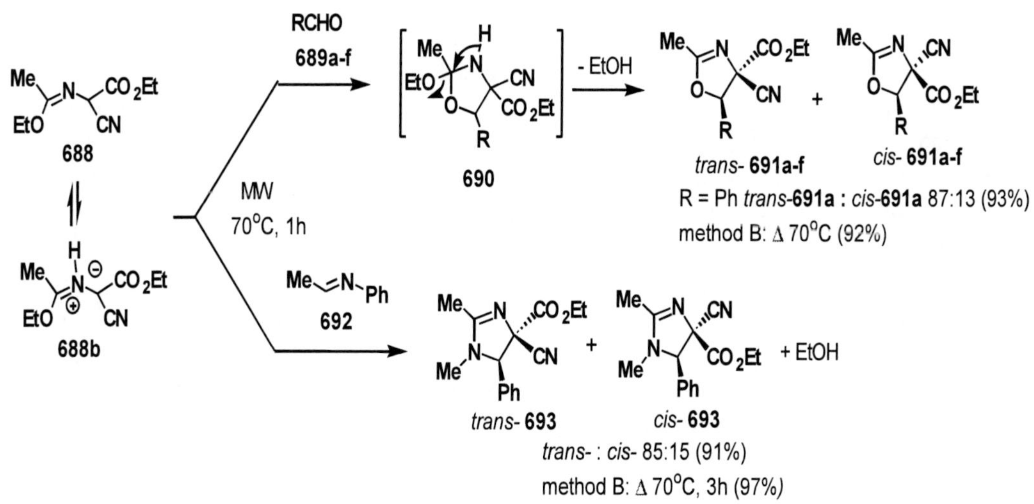

Scheme 143.

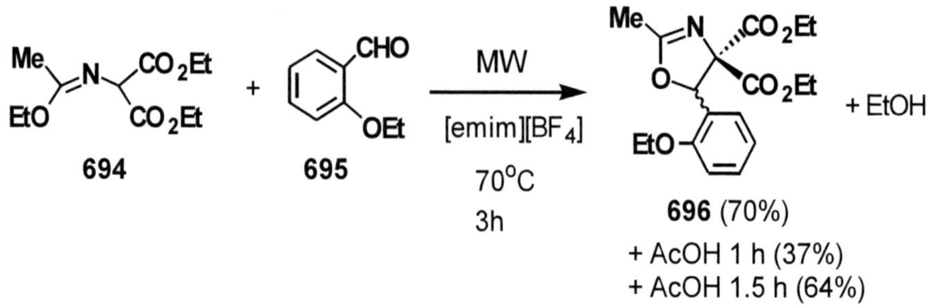

Scheme 144.

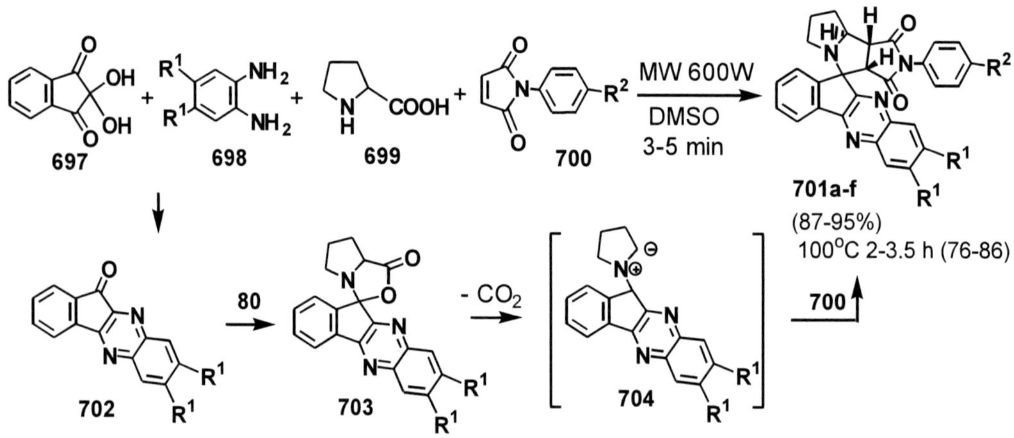

Scheme 145.

An interesting example of multicomponent reaction in which the key step is 1,3-dipolar cycloaddition of azomethine ylides enhanced by microwave irradiation was published by Azizian. [20] Novel spiro pyrrolizidine derivatives *701a-f* were obtained in four component synthesis through diastereoselective 1,3-dipolar cycloaddition of azomethine ylides with *N*-aryl maleimides *700* as electron deficient dipolarophiles (Scheme 145). Reaction proceeds through condensation of ninhydrin *697* and 1,2-phenylene diamines *698* to indenoquinoxaline-11-one *702*, which was subsequently condensed with *L*-proline to intermediate *703*. Generation of azomethine ylide *704* as the result of thermal decarboxylation of *703* is subsequently followed by cycloaddition reaction with aryl maleimide *700* as dipolarophile. Classical heating (2-3.5 h) gave yield inferior to those obtained by microwave.

Three sets of reaction conditions were used for the 1,3-dipolar cycloaddition reaction of dipolarophiles *707a-f* with azomethine ylides (generated by decarboxylative condensation of acenaphthoquinone *705* with proline *706*). Jadidi [21] obtained pyrrolizidines *708* in high yields using either classical heating, microwave irradiation or ionic liquid (Scheme 146). All the reactions were highly diastereoselective and regioselective. For this particular reaction, the use of microwave heating was beneficial in terms of reducing reaction times, while the yields were almost unaffected.

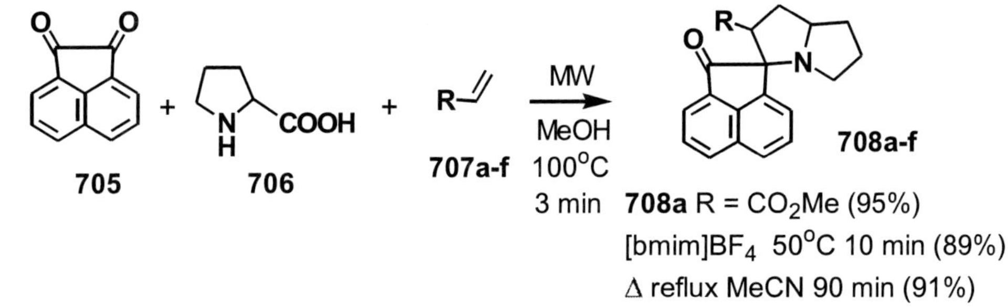

Scheme 146.

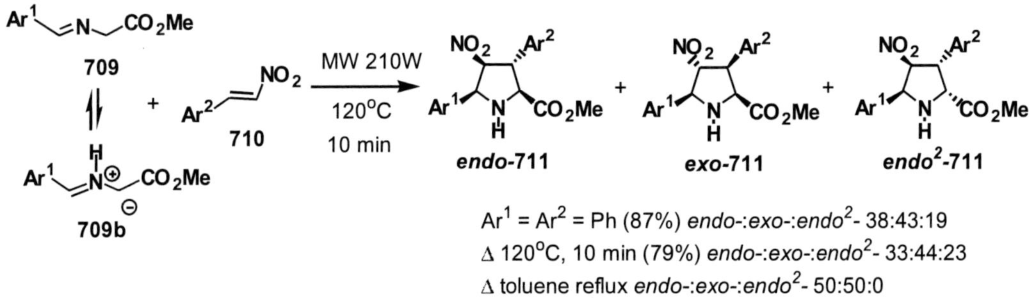

Scheme 147.

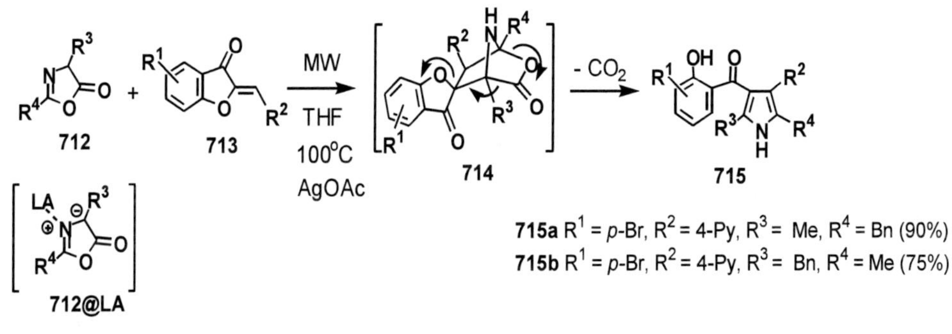

Scheme 148.

Employment of solvent free thermal reaction conditions could affect the stereochemical outcome of dipolar cycloaddition. As shown by Cossío and Díaz-Ortiz, [22] thermal and microwave-assisted [3+2] cycloaddition between azomethine ylides and nitrostyrenes *710* produced three stereoisomers (Scheme 147). In the absence of solvent, microwave irradiation induces the thermal isomerization of imines *709* to the corresponding azomethine ylides *709b*. The subsequent 1,3-dipolar cycloadditions with substituted *β*-nitrostyrenes *710*, which occur within 10-15 min, afford the corresponding pyrrolidines *711* in 81-87% yield. Three stereoisomers, *endo-711*, *exo-711*, and *endo^2-711*, were obtained in each cycloaddition. The *exo-711* is major isomer, followed by smaller proportion of *endo-711*, and even less *endo^2-711*. Formation of third stereoisomer involves an azomethine ylide resulting from isomerization of the initially formed one or from a double proton transfer between two starting imine molecules. In contrast to solvent-free conditions, the *endo^2-711* isomer was not found in the reaction carried out in the classical conditions in solvent (toluene, reflux), instead approximately 50:50 mixture of *endo-711* and *exo-711* was obtained. [23]

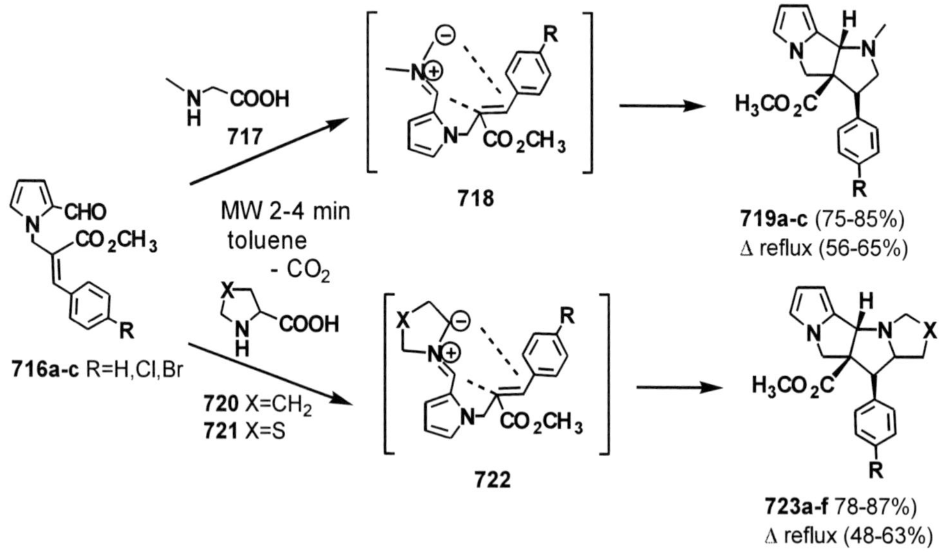

Scheme 149.

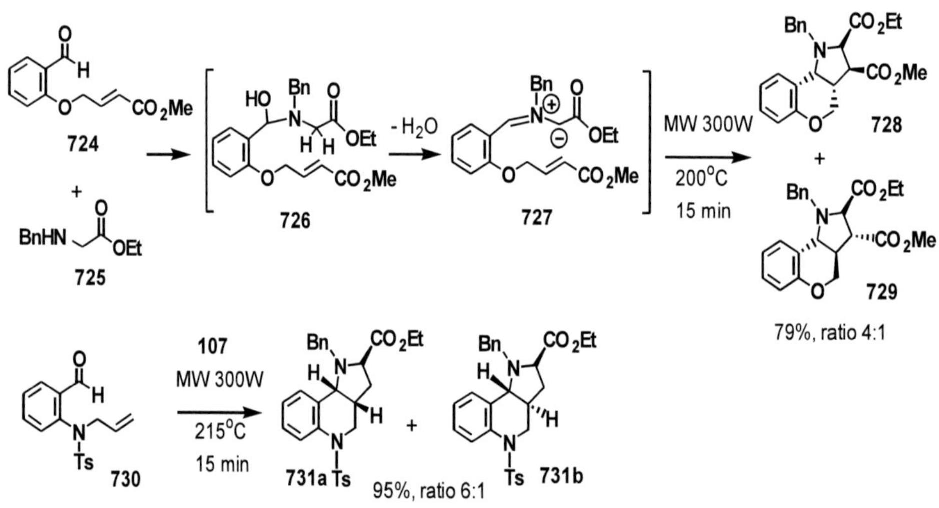

Scheme 150.

Regioselective synthesis of tetrasubstituted pyrroles *via* 1,3-dipolar cycloaddition of α,β-unsaturated benzofuran-3(2*H*)-one *713* and azlactones *712* described by Park [24] requires decarboxylation of the initially formed cycloadduct. In this process *713* serves as nonsymmetric dipolarophile leading to the formation of bridgehead intermediate *713* through the regioselective [3+2] cycloaddition with 1,3-dipole *712@LA* (formed by complexation of *712* with Lewis acid) (Scheme 148). The transient intermediate *714* was converted to pyrrole derivatives through spontaneous decarboxylation and subsequent ring opening of benzofuran-3(2*H*)-one, which is a key step for this transformation. A variety of Lewis acids [AgOAc,

AgOTf, Ag_2O, $Cu(OAc)_2$, Cu_2O] and bases [2,6-lutidine, TEA, DBU] was tested, and found that AgOAc (10 mol%) being most successful catalyst. The nearly complete control of the regiochemistry in tetrasubstituted pyrroles was achieved in high to excellent yields. The orthogonal synthesis of complementary regioisomers *715a* and *715b* was achieved by the use of different azlactones *712a* and *712b*, derivativs of *N*-acetylated phenylalanine and *N*-2-phenylacetylated alanine, respectively.

2.6.2. Intramolecular Reactions

Intramolecular cycloadditions of azomethine ylides often lead to products with increased complexity. For instance, Raghunathan [25] carried out reactions of *N*-alkenyl aldehydes *716a-c* (*E*-isomer), which were transformed to the unstable azomethine ylides by decarboxylative condensation with secondary amino acids. The condensations of *716a-c* and *717* generated azomethine ylides *718* that underwent intramolecular 1,3-dipolar cycloaddition yielding *cis*-adducts pyrrolo[2,3-a]pyrrolizine *719a-c* (Scheme 149). When the same reaction was extended with proline *720* and thiazolidine carboxylic acid *721*, a series of pyrrolizidine and thiopyrrolizidine heterocycles (pyrrolizine[2,3-a]pyrrolizine) *723a-f* was obtained with high regio- and stereoselectivity. There was an increase in the yields of the products with a decrease in reaction time compared to classical heating, when the same reactions were carried out under microwave irradiation in toluene.

The key reaction step in Potaček's [26] synthesis of compounds containing hexahydrochromeno[4,3-*b*]pyrrolidine skeletons is intramolecular 1,3-dipolar cycloaddition. Azomethine ylide *727* was derived from reaction of an aldehyde, *trans*-4-(2-formylphenyloxy)crotonate *724* and an α-amino acid ester *725* (*N*-benzyl ethylglycinate). Depending on the orientation of incoming reagents in the tansition state, two diastereoisomeric cycloadducts were formed, *728* and *729*, in a 4:1 (R^2 = benzyl, ethyl, and *n*-butyl) ratio (Scheme 150). Under classical reaction conditions long reaction time was required, to afford products in low yields and with low selectivity. Replacement of *724* with protected *ortho*-(3-alkenyl)amino benzaldehyde *730* led to the synthesis of ethyl hexahydro-1*H*-pyrrolo[3,2-*c*]quinoline-2-carboxylate skeleton, where *731a* is the major isomer. [27] Reaction carried under classical conditions (reflux, toluene, 1080 min), gave again the same major isomer, but reaction yield was significantly lower.

Interestingly, Potaček [28] discovered that amine *733* which brings additional acidic hydrogen into the reacting system, enables a competitive reaction to the formation of azomethine ylide *736* (Scheme 151). During the water elimination step, the hydrogen atom in intermediate *735* attached to the nitrogen atom competes with the hydrogen atom attached to the α-carbon next to the carbonyl group. Hence, the reaction could proceed *via* two different intermediates, *736* and *739*, leading to products *737*, *738* and *740*. Since an amide hydrogen atom is more acidic than the hydrogen atom situated on an α-carbon next to the amide group and due to the irreversibilty of the water elimination step under the reaction conditions, only the formation of intermediate *739* is preferred, and cyclic compound *740* was observed as the only product in 85% yield. No products of intermolecular reaction (dimers, trimers, ...) were detected.

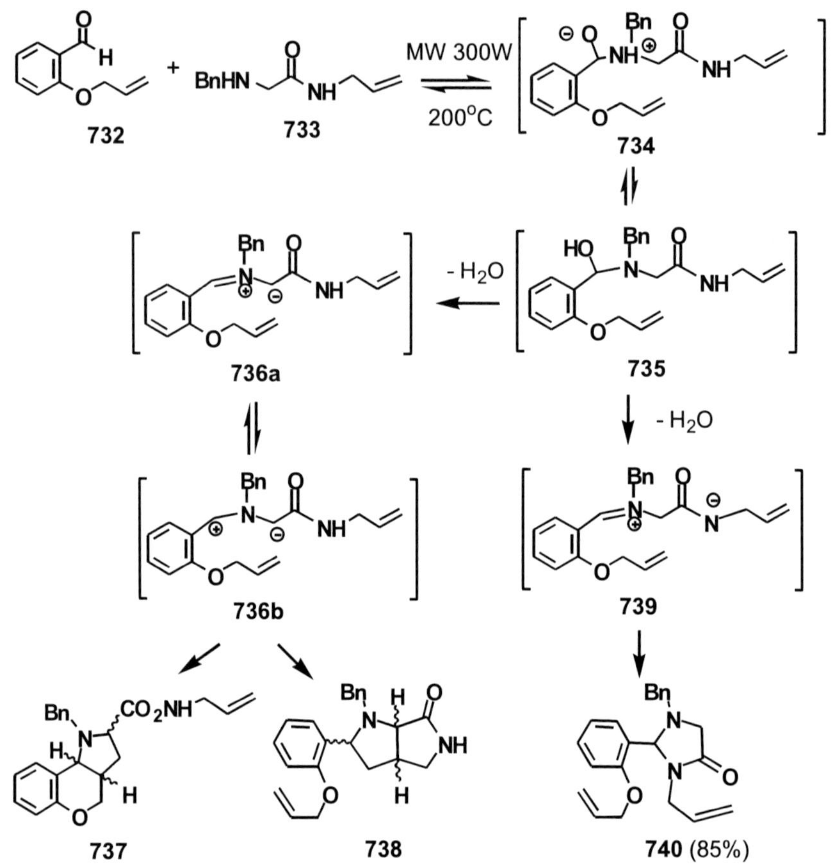

Scheme 151.

Bashiardes [29] investigated a series of substrates analogous to these in Scheme 151 and two sets of reaction conditions to provide benzopyrano-pyrrolidines. In the case of benzopyrano-pyrrolidines compound *743a* the all *cis* compound formed in the reaction was accompanied by approximately 5% of diasteroisomer in which the ring junction was *trans* (Scheme 152). Compound *743b* (with CO_2Me substituent at the terminal alkene *741*) was obtained as a 1:1 mixture of *cis*/*trans* diastereoisomers. There is a clear advantage in using microwave mode as compared to the thermal conditions: the yields under microwaves are often the highest. This reflects a cleaner reaction with fewer side-products. The use of an acetylenic substrate could be used for synthesis of pyrrole products. For this purpose, intramolecular 1,3-dipolar cycloaddition of azomethine ylide formed by condensation of *ω*-propargylic benzaldehyde *744* with ethyl *N*-benzylglycinate *745* (or methyl sarcosinate *742*), led to the intermediate dihydropyrrole cycloadduct *746*. Cycloaddition was followed by *in situ* oxidation by sulfur, which efficiently provides substituted benzopyrano-pyrroles *747*.

Following these findings Bashiardes [30] condensed *N*-propargylpyrroles *748* with with α-aminoacids, such as sarcosine or proline derivatives, to obtain azomethine ylides which undergo the intramolecular cycloaddition with the adjacent propargylic group. Hence, pyrrole carbaldehyde *748* condensed with methyl sarcosinate *749* in the presence of sulfur on heating in refluxing xylene provided tricyclic pyrrolizinopyrroles *750* in 60% yield (Scheme 153). The reaction of *N*-propargylic indole carbaldehyde *751* with methyl prolinate *752* was complete within 15 minutes under microwave conditions, providing the pentacyclic product *753* in 80% yield, while under conventional heating the time required was two hours and an inferior yield of 67% was obtained.

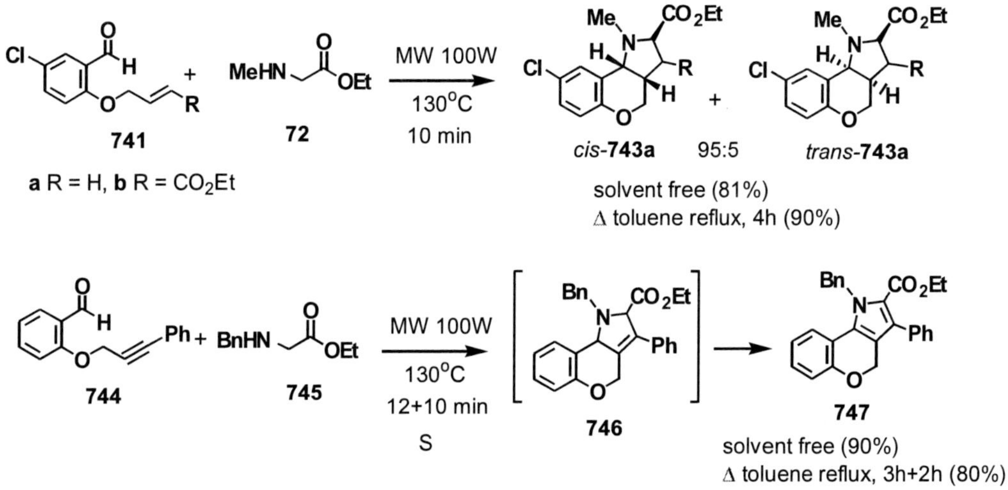

Scheme 152.

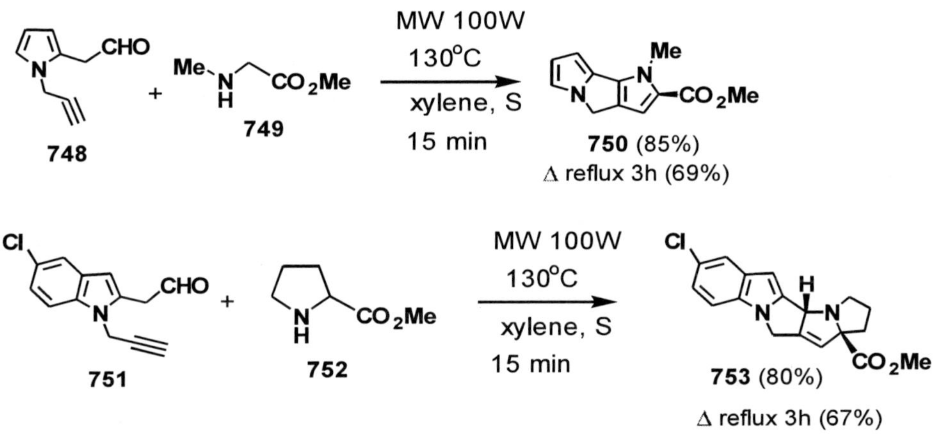

Scheme 153.

Sometimes initially formed cycloadducts, with the excess of one of the reagents could further react and deliver polycylic products. In one example by Zhang [31] methyl glycinate *755* with *O*-allyl salicylaldehyde *754* undergoes intramolecular 1,3-dipolar cycloaddition to form a tricyclic proline derivative *758*. In addition to product *758*, another major product *756* was formed (Scheme 154). The ratio of mono- and dicycloaddition products was 45:55. The formation of the dicycloaddition product was proportional to the number of equivalents of allylic salicylaldehyde *754*. When 1.25 equiv. of *754* was used, the *758*:*756* ratio was 96:4. The amount of the dicycloaddition product increased significantly with increasing equivalence of *754*. The *758*:*756* ratio was 80:20 when 4.0 equiv of *754* was used. A mechanism leading to formation of *756* starts with condensation to form azomethine ylide *757*, followed by intramolecular 1,3-dipolar cycloaddition. With an excess of *754*, monocycloaddition product *758* is being transformed to azomethine ylide *759*, which gives the final product *756 via* the second intramolecular 1,3-dipolar cycloaddition.

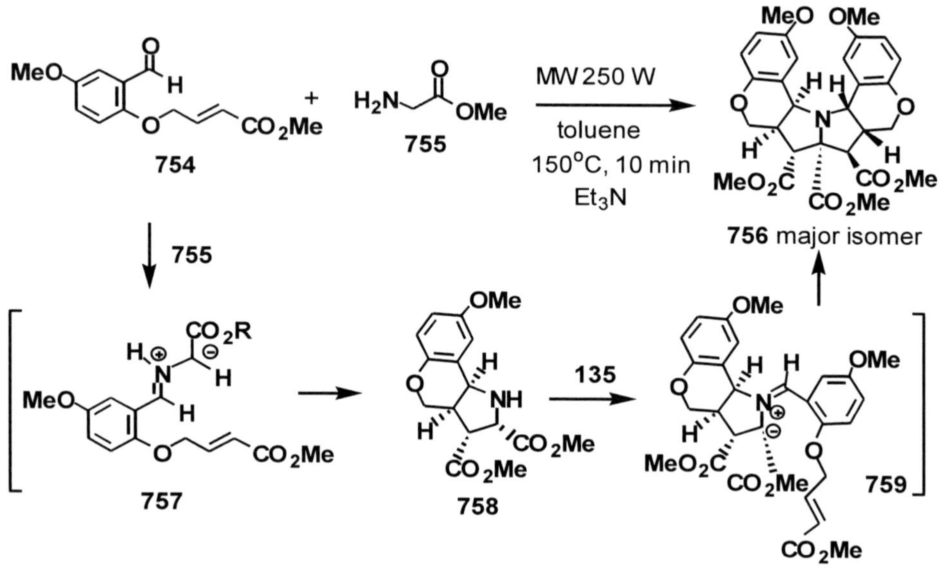

Scheme 154.

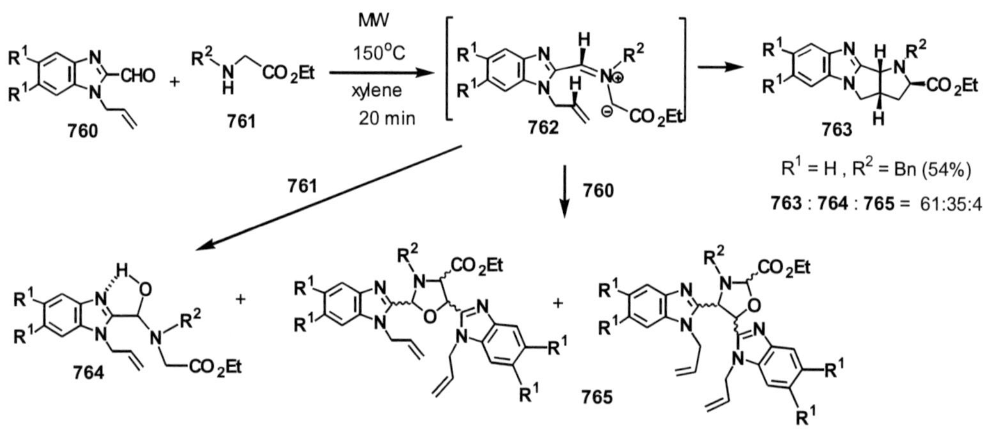

Scheme 155.

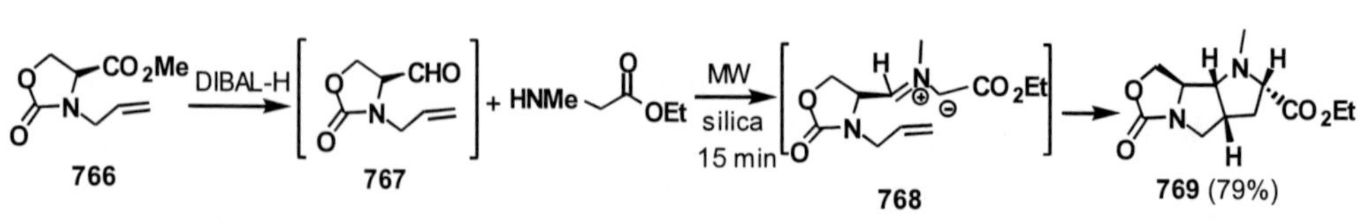

Scheme 156.

Construction of the polycyclic fused pyrrolidine ring systems by Kurth [32] employs the intramolecular cycloaddition of azomethine ylides as powerful tool for the synthesis of pyrrolidino[2',3':3,4]pyrrolidino[1,2-*a*]benzimidazole-2-carboxylates. The condensation of carbaldehyde *760* with a secondary amino ester *761* and the subsequent 1,3-dipolar cycloaddition of the ylide *762* was mediated by microwave irradiation in xylene and delivered polycyclic pyrrolidine *763* in highly regio- and diastereoselective manner (Scheme 155). The yield of this condensation-cycloaddition reaction improved with increasing substrate concentration; the better yield of *763* (54%) was obtained at a 1 M concentration in xylene at 150 °C under microwave irradiation for 20 min. In addition to the desired product, dimer and hemiaminal side products *764* and *765* can make up a significant portion of the reaction mixture. Dimer formation occurs when the azomethine ylide *762* undergoes an intermolecular addition to its precursor aldehyde *760* to form 3-(substituted)-oxazolidine-4-carboxylates *765* as a complex mixture of regio- and stereoisomers. Depending on the R^1 and R^2 substituents, the *763*:*764*:*765* ratio could be changed in favor of *763*.

The use of enantiomerically pure substrates often leads to chiral synthesis of functionalized nitrogen heterocycles. Cheng [33] used aldehyde *767*, which was obtained as an intermediate by the reduction of chiral ester *766* with DIBAL-H. From *767* non-stabilized azomethine ylide *768* was generated by treatment with *N*-substituted α-amino esters and subsequently trapped intramolecularly by olefinic bond. Reaction was carried out in solvent free conditions, where a mixture of the aldehyde *767* and *N*-methylglycine ethyl ester was deposited on the surface of silica gel. Irradiation under microwaves *in situ* smoothly afforded the cycloadduct *769* in 79% yield (Scheme 156).

References

[1] Díaz-Ortiz, A.; Díez-Barra, E.; de la Hoz, A.; Loupy, A.; Petit, A.; Sánchez, L. *Heterocycles 1994*, *38*, 785-792.

[2] Bourah, A.; Baruah, B.; Prajapati, D.; Sandhu, J. S.; Ghosh, A. C. *Tetrahedron Lett. 1996*, *37*, 4203-4204.

[3] Zhang W.; Chen, C. H.-C. *Tetrahedron Lett. 2005*, *46*, 1807-1910.

[4] Zhang, Z.; Lu, Y.; Hiu-Tung Chen, C.; Curran, D. P.; Geib, S. *Eur. J. Org. Chem. 2006*, 2055-2059.

[5] Sanchez, S.; Bateson, J. H.; O'Hanlon, P. J.; Gallagher, T. *Org. Lett. 2004*, *6*, 2781-2783.

[6] Gallagher, T.; Sanchez, S.; Bateson, J. H.; O'Hanlon, P. J. *Pure Appl. Chem. 2005*, *77*, 2033-2040.

[7] Ramesh, E.; Kathiresan, M.; Raghunathan, R. *Tetrahedron Lett. 2007*, *48*, 1835-1839.

[8] Sridhar, G.; Gunasundarib, T.; Raghunathan, R. *Tetrahedron Lett. 2007*, *48*, 319-322.

[9] Jayashankaran, J.; Manian, R. D. R. S.; Raghunathan, R. *Tetrahedron Lett. 2004*, *45*, 7303-7305.

[10] Manian, R. D. R. S.; Jayashankaran, J.; Raghunathan, R. *Tetrahedron 2006*, *62*, 12357-12362.

[11] Suresh Babu, A. R.; Raghunathan, R. *Tetrahedron 2007*, *63*, 8010-8016.

[12] Jayashankaran, J.; Manian, R. D. R. S.; Venkatesan, R; Raghunathan, R. *Tetrahedron 2005*, *61*, 5595-5598.

[13] Periyasami, G.; Raghunathan, R.; Surendiran, G.; Mathivanan, N. *Bioorg. Med. Chem. Lett. 2008*, *18*, 2342-2345.

[14] Dinica, R. M.; Druta, I. I.; Pettinari, C. *Synlett 2000*, 1013-1015.

[15] Azizian, J.; Asadi, A.; Jadidi, K. *Synth. Commun. 2001*, *31*, 2727-2733.

[16] Kerneur, G.; Lerestif, J. M.; Bazureau, J. P.; Hamelin, J. *Synthesis 1997*, 287-289.
[17] Fraga-Dubreuil, J.; Cherouvrier, J. R.; Bazureau, J. P. *Green Chem. 2000*, *2*, 226-229.
[18] Lerestif, J. M.; Feuillet, S.; Bazureau, J. P.; Hamelin, J. *J. Chem. Res. Synop. 1999*, 32-33.
[19] Dubreuil, J. F.; Bazureau, J. P. *Tetrahedron Lett. 2000*, *41*, 7351-7355.
[20] Azizian, J.; Karimi, A. R.; Mohammadi, A. A.; Mohammadizadeh, M. R. *Synthesis 2004*, 2263-2265.
[21] Jadidi, K.; Moghaddam, M. M.; Aghapoor, K.; Gharemanzadeh, R. *J. Chem. Res. Synop. 2007*, 71-73.
[22] Arrieta, A.; Otaegui, D.; Zubia, A.; Cossío, F. P.; Díaz-Ortiz, A.; de la Hoz, A.; M. Herrero, A.; Prieto, P.; Foces-Foces, C.; Pizarro, J. L.; Arriortua, M. I. *J. Org. Chem. 2007*, *72*, 4313-4322.
[23] Díaz-Ortiz, A.; de la Hoz, A.; M. Herrero, A.; Prieto, P.; Sánchez-Migallón, A.; Cossío, F. P.; Arrieta, A.; Vivanco, S.; Foces-Foces, C. *Mol. Diversity 2003*, *7*, 175-180.
[24] Kim, Y.; Kim, J.; Park, S. B. *Org. Lett. 2009*, *11*, 17-20.
[25] Kathiravan, S.; Ramesh, E.; Raghunathan, R *Tetrahedron Lett. 2009*, *50*, 2389-2391.
[26] Pospíšil, J.; Potaček, M. *Eur. J. Org. Chem. 2004*, 710-716.
[27] Neuschl, M.; Bogdal, D.; Potaček, M. *Molecules 2007*, *12*, 49-59.
[28] Pospíšil, J.; Potaček, M. *Tetrahedron 2007*, *63*, 337-346.
[29] Bashiardes, G.; Safir, I.; Mohamed, A. S.; Barbot, F.; Laduranty, J. *Org. Lett. 2003*, *5*, 4913-4918.
[30] Bashiardes, G.; Barbot, F. *Synlett 2007*, 1707-1711.
[31] Zhang, W.; Lu, Y.; Geib, S. *Org. Lett. 2005*, *7*, 2269-2272.
[32] Meng, L.; Fettinger, J. C.; Kurth, M. J. *Org. Lett. 2007*, *9*, 5055-5508.
[33] Cheng, Q.; Zhang, W.; Tagami, Y.; Oritani, T. *J. Chem. Soc., Perkin Trans. 1 2001*, 452-456.

2.7. Carbonyl Ylides

Introduction

Carbonyl ylide 1,3-dipole (a dipole of allyl type with an oxygen atom as middle center) has received much less attention in comparison with the other dipoles such as azomethine ylides, nitrilimines, nitrile oxides, azides, or nitrones. Usually carbonyl ylides generated from thermal electrocyclic ring opening of epoxides are reactive intermediate which readily undergo [3+2] cycloaddition with π-bonds to afford tetrahydrofuran ring or oxa-cyclic systems.

2.7.1. Intermolecular Reactions

Microwave irradiation of a series of *p*-substitued epoxides such as *770* yields via the electrocyclic ring opening [1] transient carbonyl ylide *771* which undergoes 1,3-dipolar cycloaddition with various alkenes. The reaction of epoxide *770* with methyl fumarate and phenylmaleimide in refluxing toluene or xylene afforded mixture of regioisomeres *774* and *775*, or *776* and *777* (Scheme 157). Benhaoua [2] has shown that microwave irradiation is beneficial for 1,3-dipolar cycloaddition of carbonyl ylide *771* to obtain better yields in shorter reaction times. The use of solvent is necessary to prevent the decomposition of epoxide either under microwave irradiation or by classical heating. Under solvent free conditions, heating of *770* with alkenes causes decomposition of epoxide, leading to the *para*-substituted aldehydes, which result from hydrolysis of carbonyl ylide. Interestingly, when solvent free reaction of *770* was carried out with aromatic aldehydes at 120°C and lower microwave power (60W), epoxide *770* reacts readily in 1,3-dipolar manner without decomposition. [3]

A number of 1,3-dioxolanes and 1,3-oxazolidines have been synthesized by [3+2] cycloaddition reactions between carbonyl ylides generated from epoxides and aromatic aldehydes or imines in solvent free conditions (Scheme 158). As reported by Mongin, [3] in contrast to the use of aldehydes as dipolarophiles, the reactions performed with imines proceed diastereoselectively. Cycloaddition reactions carried out between 2,2-dicyano-3-(4-substituted)phenyloxiranes *778* and benzaldehydes *779* afforded 2,5-diphenyl-1,3-dioxolane-4,4-dicarbonitriles *780a,b*. To obtain the full conversion, thermal reaction was carried out for 35-68 h, while microwave reactions were complete in 30-40 min. In these reactions, the *cis* products *780a* predominate over the *trans 780b*, whereas the change of thermal to microwave conditions slightly diminishes the *a*/*b* ratio. On the other hand, reactions carried out between *778* and imine *781* furnished oxazolidine *783*, the *cis* diastereoisomer was formed with ratio >90/10. Notable shortening of reaction times from days to 25-60 min in microwave conditions was achieved, and for the most examples small decrease in yield was observed. [4] Epoxide *778* in the 1,3-dipolar cycloaddition reaction with ketones (glyoxylates and isatin) afforded in the case of *782* corresponding 1,3-dioxolanes *784*. [5] The *cis* product *784a* dominated in thermal reactions, but in microwave conditions the preferred isomer was *trans-784b*.

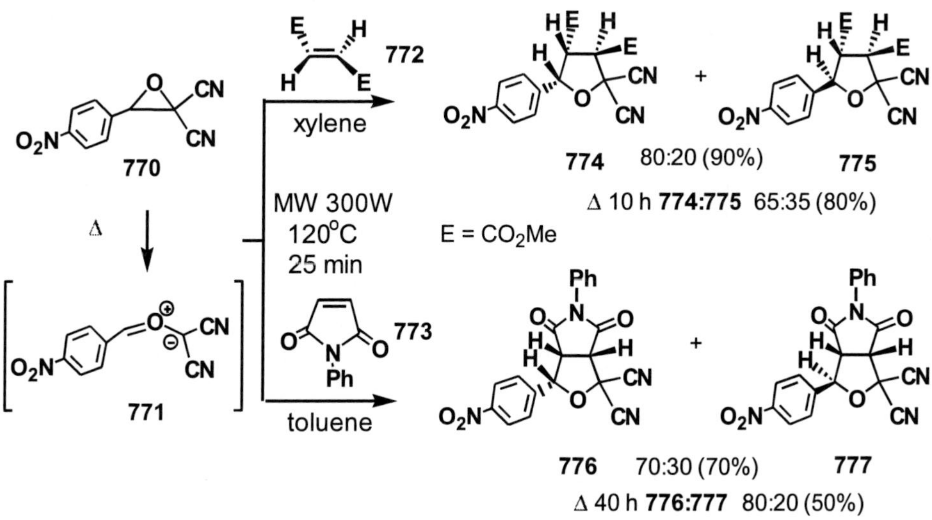

Scheme 157.

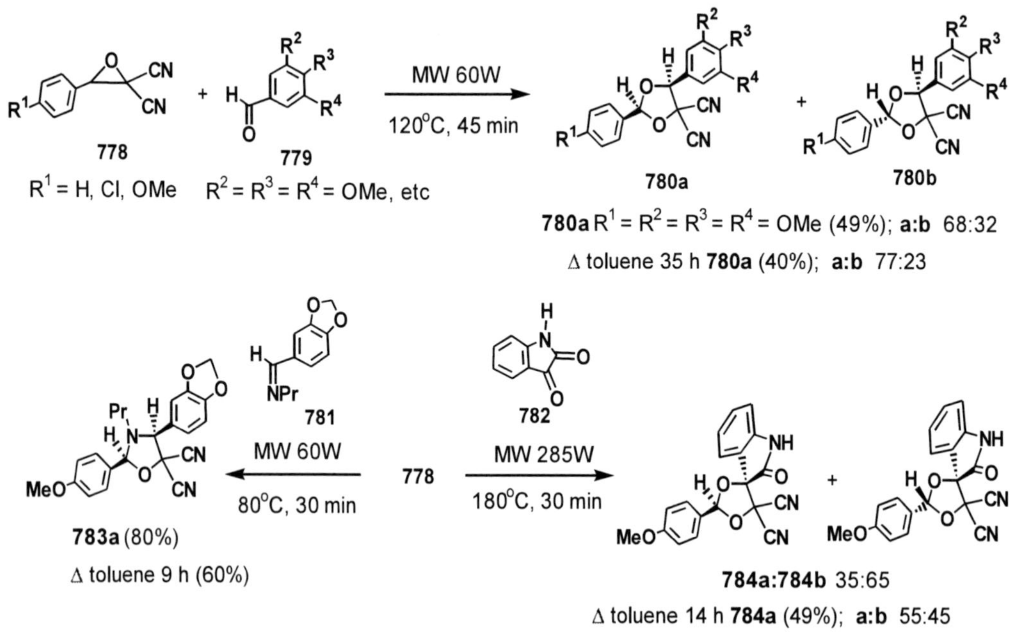

Scheme 158.

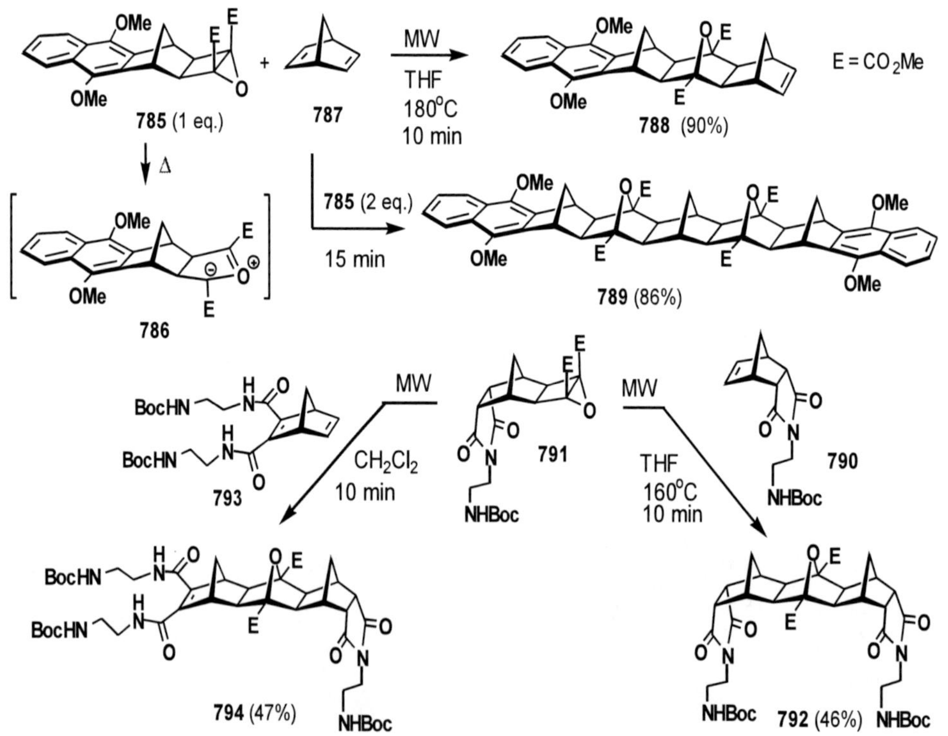

Scheme 159.

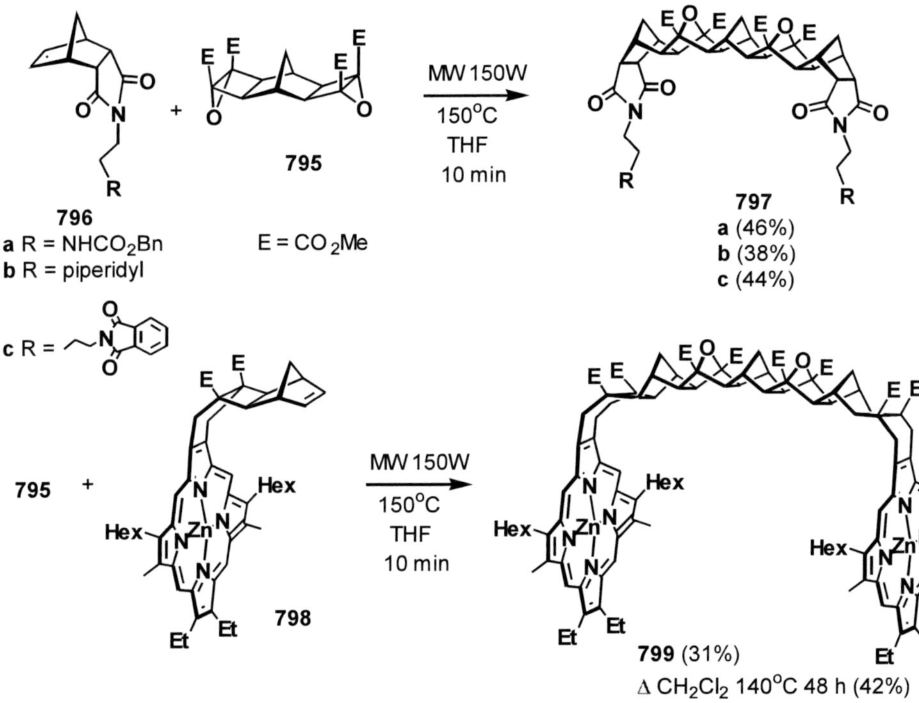

Scheme 160.

Pfeffer [6] used microwave-accelerated 1,3-dipolar cycloaddition for the formation of fused [*n*]polynorbornanes. Cyclobutane epoxides were used for *in situ* generation of ester-stabilized intermediate 1,3-carbonyl dipoles *786*, which react with a variety of norbornene π-bonds as dipolarophiles to form new 7-oxabicyclo[2.2.1]bridge at the ring junction (Scheme 159). [7] Optimal reaction conditions were determined to be 10 and 15 min at 180 °C for *788* and *789*, respectively, using THF as solvent. The advantage of microwave conditions over conventional heating is the apparent reduction in reaction time. In addition, the microwave assisted synthesis of NBoc protected *792* and *794* was far cleaner comparing to results obtained by traditional thermal methods.

The double 1,3-dipolar cycloaddition of carbonyl ylides is the useful synthetic technique to prepare variety of functionalized [*n*]polynorbornanes. [8] In a final reaction step, two norbornene blocks *796* possessing variety of functionality were joined by the double 1,3-dipolar cycloaddition coupling reaction with bis-epoxide *795* by microwave heating to afford [*COCOC*]-[5]polynorbornanes *797a-c* (Scheme 160). In this process symmetrical products were obtained with exclusive π-facial selectivity and stereoselectivity. Superiority of microwave heating over conventional conditions was demonstrated in the synthesis of bisporphyrin *799*: conventional reaction was carried out by heating in a sealed glass tube for long period of time, which was reduced in microwave reactor to merely 30 minutes.

2.7.2. Tandem [4+2]/[3+2] Cycloaddition Reactions

Tandem [4+2]/[3+2] cycloaddition reactions of 1,3,4-oxadiazoles were successfully applied in the synthesis of [n]polynorbornane molecules as well as natural products. [9] The traditional conditions for oxadiazole coupling of norbornenes require strong heating for prolonged periods of time and are not conductive to the preparation of thermally sensitive materials, or the use of thermally sensitive substrates. High pressure facilitated coupling (at 1.4 GPa and room temperature) is advantageous in these cases, but reactions are limited to the use of special high pressure equipment. Only certain functional groups are able to withstand the harsh conditions that are used in oxadiazole reactions: nitrogen protection using phthalimido, ester or benzyl, and carboxylic acid functionality. [10] An improvement in terms of using significantly shorter reaction times was achieved by microwave heating. In reaction of *C*-bridged norbornene *800* with *801*, a cycloadduct *803* was obtained in 15% yield when two reactants were heated at 150°C, for 2 hours, or at 170°C, for 30 minutes (Scheme 161). [11] On the other hand, microwave heating of alkene *800* with 1,3,4-oxadiazole *802*, without the presence of solvent gave almost complete conversion to cycloadduct *804*. Enhanced yield is ascribed to the fact that the reaction is much more concentrated (no solvent). In addition, oxadiazole *802* is solid, while *801* is liquid and is partially in the gas phase in reaction conditions, which further decreases concentration. In this tandem [4+2]/[3+2] cycloaddition reaction oxadiazole reacts as diene partner in an inverse electron demand Diels-Alder reaction. Dinitrogen loss from the Diels-Alder intermediate *805* (which generates highly reactive 1,3-dipole *806*) is a key element in coupling of second equivalent of norbornene. Both [4+2] and subsequent [3+2] cycloaddition reactions are highly π facially selective affording *endo*-isomer exclusively, leading to single product.

The π stereoselectivity observed in conventional heating is preserved when a range of 7-oxanorbornene substrates was used as dienophiles/dipolarophiles in microwave conditions. Microwave reaction of *807* with *801* produced mixture of two O^3-[3]-polynorbornanes *808* and *809*, favoring the *exo,endo-* adduct *809* in approximately 2:1 ratio. These results are in accordance with stereospecificity attained in reaction under classical and high pressure conditions (Scheme 162). A side-reaction was observed for 7-oxabenzonorbornene substrates, which includes thermally induced retro-Diels-Alder fragmentation of intermediate 1,3-dipole *810* to the parent isobenzofuran *811*. Its [4+2] reaction with alkene *807* affords cycloadducts *813* and *814*. This side-reaction was not observed in high pressure reactions conducted at room temperature (0.8 GPa, RT, 3 days, CH_2Cl_2).

2.7.3. Natural Product Synthesis via 1,3-dipolar Reaction of Carbonyl Ylides

Intramolecular 1,3-dipolar reactions of carbonyl ylides were employed in natural product synthesis, as a very effective way to achieve structures with considerable complexity. For example, intramolecular oxadiazole coupling reactions have been used by Boger [12] in synthesis of vindoline alkaloids (Scheme 163). For this purpose, oxadiazole and two alkene components were in-built in the substrate *815*. In the thermal reaction obtained yield was 87%, while microwave conditions gave 70% in shorter time. Remarkable achievement in the course of these tandem [4+2]/[3+2] cycloadditions is that three new rings were constructed with formation of four new C-C bonds and set all six stereocenters about the central six-member ring in a single step obtaining a single diastereoisomer. Classical reaction mechanism is postulated, involving [4+2] addition of oxadiazole in the first step, and formation of intermediate Diels-Alder adduct *816*. At this stage, two stereocenters are defined. Facile loss of dinitrogen from *816* allows the formation for reactive 1,3-dipole intermediate *817* and subsequent 1,3-dipolar addition to the indole moiety to give the final product *818*.

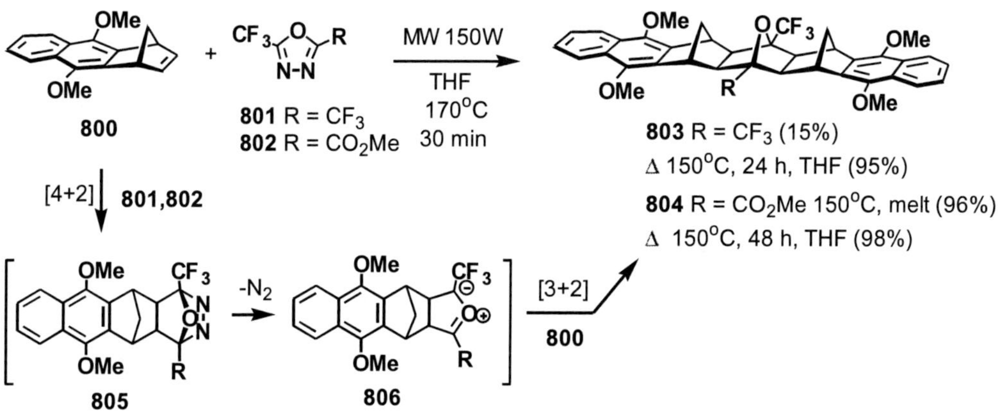

Scheme 161.

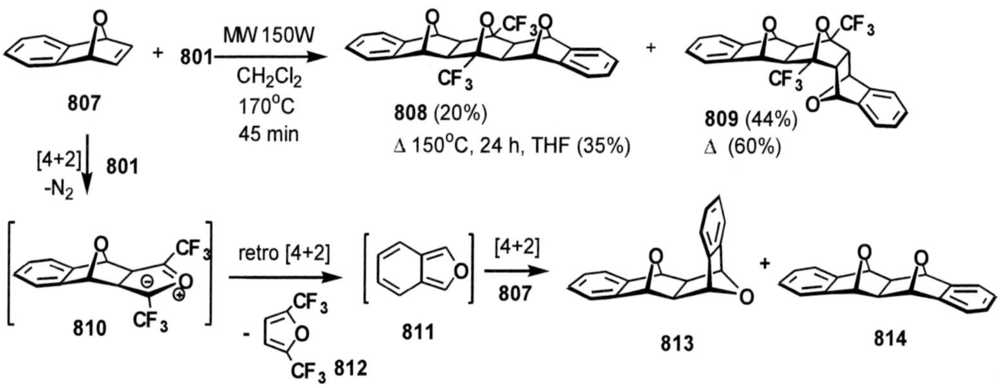

Scheme 162.

In Padwa's intramolecular [3+2] cycloaddition reaction of 1,3-dipoles across heteroaromatic π-systems, the alkaloid derivatives related to vindoline were obtained. [13] Reactive 1,3-dipole *820* was generated *in situ* from the Rh(II)-catalyzed reaction of diazo imides *819* containing tethered heteroaromatic rings, through a rhodium carbenoid species which underwent ready cyclization onto the neighboring imido carbonyl to form dipole *820*. This reactive species undergoes facile intramolecular [3+2] cycloaddition across the 2,3-π-bond to provide novel pentaheterocyclic compound *822* in good yield in a stereocontrolled fashion (Scheme 164).

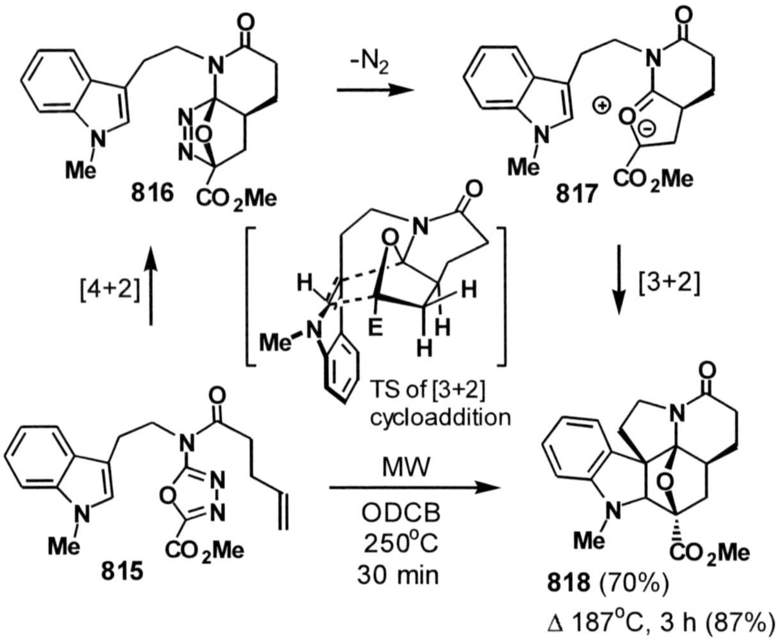

Scheme 163.

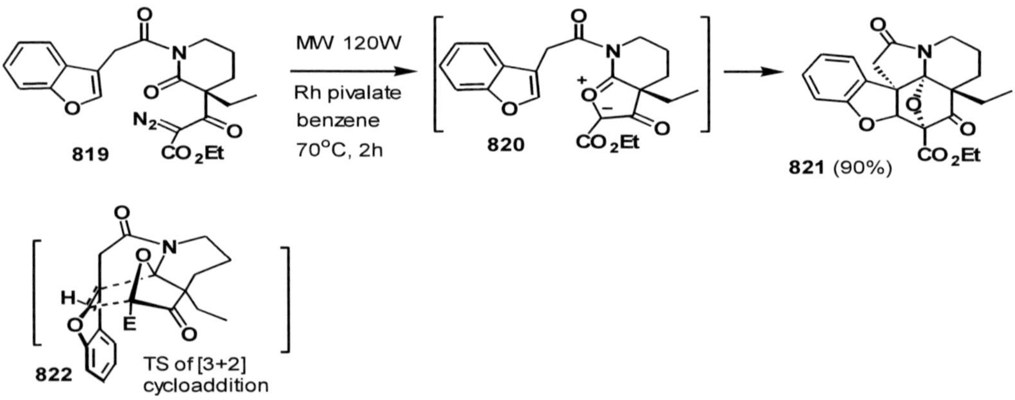

Scheme 164.

References

[1] Huisgen, R. *Angew. Chem. Int. Ed. 1977*, *16*, 572-584.

[2] Bentabed, G.; Derdour, A.; Benhaoua, H. *Synth. Commun. 2003*, *33*, 1861-1866.

[3] Bentabed-Ababsa, G.; Derdour, A.; Roisnel, T.; Sáez, J. A.; Pérez, P.; Chamorro, E.; Domingo, L. R.; Mongin, F. *J. Org. Chem. 2009*, *74*, 2120-2133.

[4] Bentabed, G.; Derdour, A.; Benhaoua, H. *Synth. Commun. 2007*, *37*, 2935-2948.

[5] Bentabed-Ababsa, G.; Derdour, A.; Roisnel, T.; Sáez, J. A.; Domingo, L. R.; Mongin, F. *Org. Biomol. Chem. 2008*, *6*, 3144-3157.

[6] Foitzik, R. C.; Lowe, A. J.; Pfeffer, F. M. *Tetrahedron Lett. 2009*, *50*, 2583-2584.

[7] Warrener, R. N.; Margetić, D.; Foley, P. J.; Butler, D. N.; Winling, A.; Beales, K. A.; Russell, R. A. *Tetrahedron 2001*, *57*, 571-582.
[8] Margetić, D.; Trošelj, P.; Briš, A. manuscript in preparation.
[9] Margetić, D.; Trošelj, P.; Johnston, M. R. *Mini-Reviews in Org. Chem. 2010*, in press.
[10] Shang, M.; Warrener, R. N.; Butler, D. N.; Margetić, D. *9th International Electronic Conference on Synthetic Organic Chemistry (ECSOC-9)*, 30th November *2005*, Seijas, J, A.; Tato, M. P. V. (Eds.).
[11] Margetić, D.; Trošelj, P.; Đilović, I.; Matković-Čalogović, D. *J. Heterocycl. Chem. 2011*, submitted.
[12] Wilkie, G. D.; Elliott, G. I.; Blagg, B. S. J.; Wolkenberg, S. E., Soenen, D. R.; Miller, M. M.; Pollack, S.; Boger, D. L. *J. Am. Chem. Soc. 2002*, *124*, 11292-11294.
[13] Mejía-Oneto, J. M.; Padwa, A. *Org. Lett. 2004*, *6*, 3241-3244.

2.8. Dipolar Other

Introduction

Besides the 1,3-dipoles described in previous sections, small number of other 1,3-dipoles was subjected to cycloadditions in microwave conditions, obtaining variety of heterocyclic systems.

2.8.1. Münchnones

Synthesis of 3-carboxylate pyrroles was carried out by Grassi [1] *via* 1,3-dipolar cycloaddition of 1,3-oxazolium-5-oxides (münchnones) with acetylenic dipolarophiles. One-pot, single step microwave-assisted synthetic methodology between münchnone *824* and α-acetoxyacrylic esters *825* offers the advantage over the classical procedure, which involves the preparation of *824* in the first step from aminoacids and its reaction in the second synthetic step (Scheme 165). Mesoionic derivative *824* was prepared *in situ* by the cyclodehydration from the *N*-substituted amino acid (*N*-methyl-*N*-benzoyl-alanine) *823* with acetic anhydride in dioxane, in the presence of olefins *825* and the mixture was irradiated at 100°C. In these reaction conditions, initially formed 7-aza-2-oxabicyclo[2.2.1]heptan-3-one cycloadduct *826* looses CO_2 and acetic acid to afford mixture of pyrroles *827a,b* and *828a,b*, where products *827* are major isomers.

Similar synthetic methodology was used by Yli-Kauhaluoma [2] in the preparation of pyrroles (Scheme 166). Herein, a series of pyrroles was synthesized from *N*-acetylated amino acids *829* with acetic anhydride as the dehydrating agent. Two alkynes, DMAD and methyl propiolate, were used in a solvent-free, microwave-assisted 1,3-dipolar cycloaddition, and a small library of pyridylpyrroles such as *832* (two regioisomers) was obtained. The advantage of this procedure is the simplicity and generation of 1,3-dipole *in situ*.

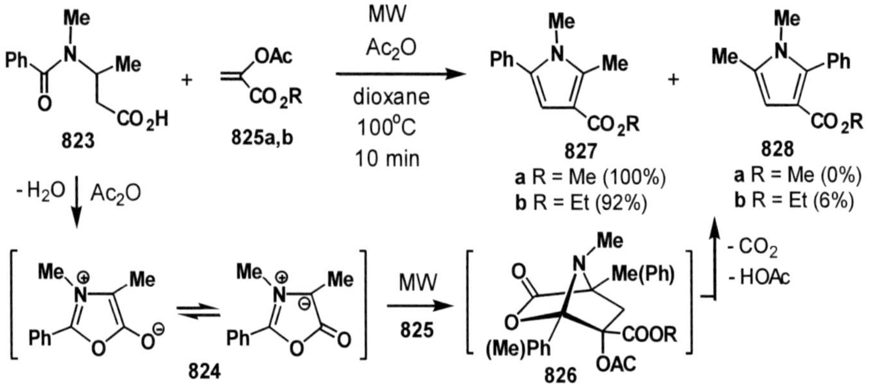

Scheme 165.

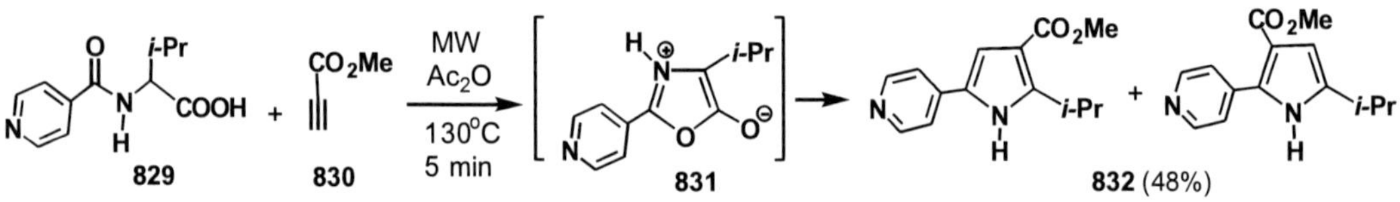

Scheme 166.

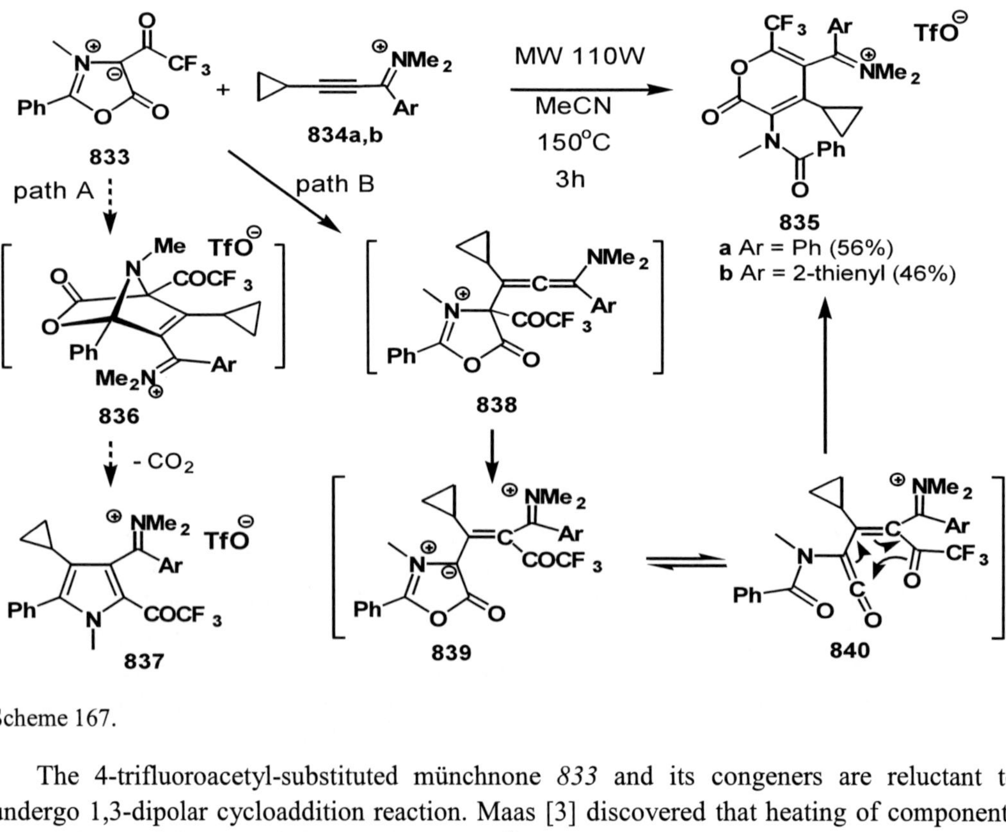

Scheme 167.

The 4-trifluoroacetyl-substituted münchnone *833* and its congeners are reluctant to undergo 1,3-dipolar cycloaddition reaction. Maas [3] discovered that heating of components *833* and *834a,b* in acetonitrile solution at 150°C under microwave irradiation furnished (6-

oxo-2-trifluoromethyl-6H-pyranyl)benzylidene iminium salts *835a,b* (Scheme 167). Instead of 1,3-dipolar cycloaddition reaction leading to intermediate *836* and pyrrole *837* (path A), Michael addition to *834a,b* takes place (path B), followed by the shift of the trifluoroacetyl group to furnish münchnones *839*. Elevated temperature triggers the valence tautomerization of *839* to form an amidoketenes *840* which then undergo a six-electron electrocyclic ring closure to yield pyranones *835*.

2.8.2. Sydnones

Kamble [4] subjected *trans* 4-bromo-3-[*p*-(3'-aryl- acryl-1'-oyl)] phenylsydnones *841a-d* upon classical heating in the presence of acetic anhydride at 135°C or to microwave irradiation. Sydnone ring underwent 1,3-dipolar cycloaddition along with the elimination of bromine as acetyl bromide to afford exclusively *trans* 5-methyl-3-[*p*-(3'-aryl -acryl-1'-oyl)-phenyl]-3*H*-2-oxo-Δ^4-1,3,4-oxadiazoles *840a-d*. The comparison of the synthesis by microwave, with that by conventional method, revealed that the reaction progressed very fast with excellent yield in the former. It was postulated that microwave irradiation facilitates polarization of the molecule under irradiation causing rapid reaction to occur. This is consistent with the reaction mechanism, which involves a polar transition state (Scheme 168). Products *844g-i* were obtained in a similar reaction from sydnones *843d-i*, combining two reaction steps in a single-pot reaction. [5] After treatment of sydnones *843d-i* with bromine in acetic acid, the mixture was subjected to microwave heating until the evolution of carbon dioxide has ceased. Thermally conducted reactions take longer to complete, in lower yield.

The mechanism of this conversion has been explained in terms of 1,3-dipolar cycloaddition of acid anhydride to the intermediate 3-aryl-4-bromosydnone *846* (Scheme 169). [6] Diazabicyclo[2.2.1] product *847* subsequently undergoes several transformations, starting with elimination of CO_2 and acetate to form cationic intermediate *848*. Acetate addition to *848* generates oxadiazoline *849*, followed by bromine and acetyl elimination to afford *850*.

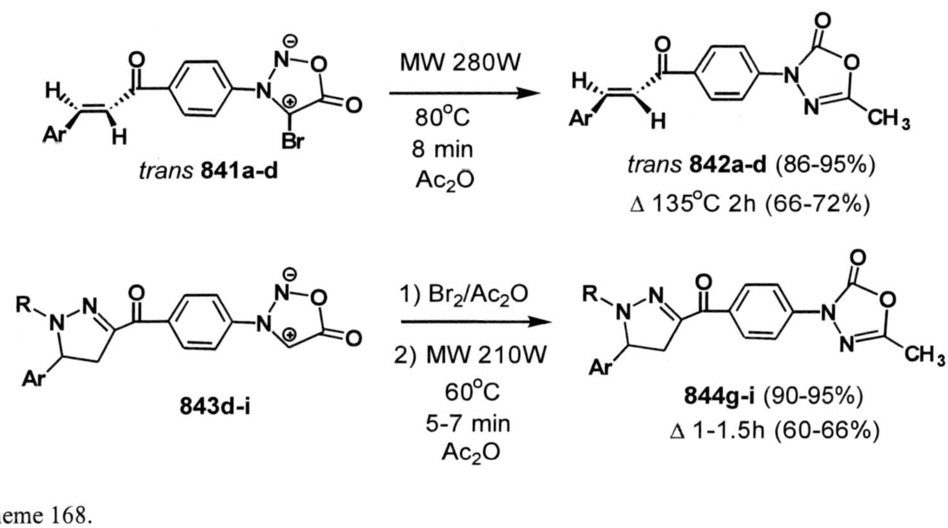

Scheme 168.

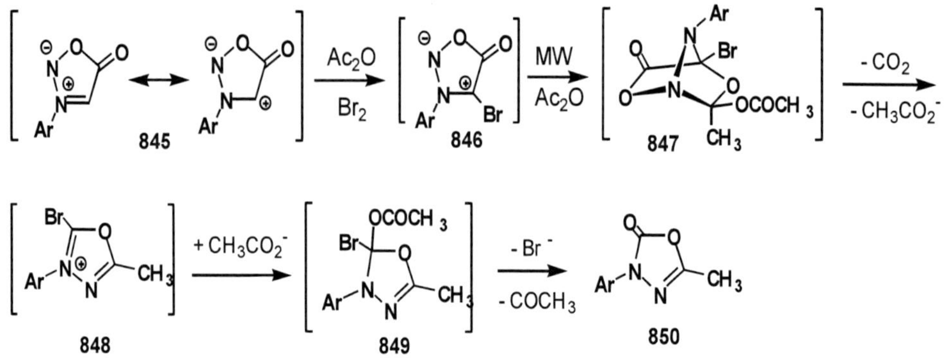

Scheme 169.

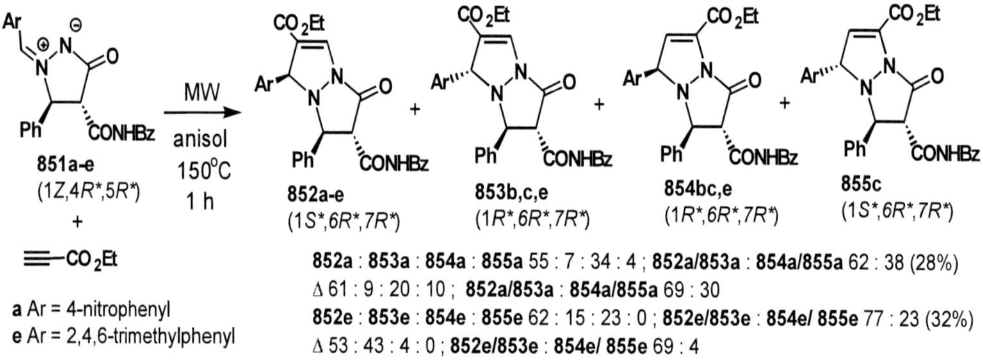

Scheme 170.

2.8.3. Azomethine Imines

Cycloadditions of chiral azomethine imine dipoles *851a-e* (i.e. (1*Z*,4*R**,5*R**)-4-benzamido-3-oxo-5-phenylpyrazolidin-1-ium-2-ides) to ethyl propiolate were carried out by Svete [7, 8] under microwave irradiation in anisole at 150°C. All reactions were non-selective and furnished the corresponding mixtures of the isomeric cycloadducts *852-855a-e*. The major 2-COOEt regioisomer was in all reactions *852/853* and the major (1*S**6*R**,7*R**)-diastereomer *852* with *syn*-oriented aryl groups at positions 1 and 7 (Scheme 170). Furthermore, microwave-assisted cycloadditions of 1,3-dipoles *851d* and *851e* with two *ortho*-substituents followed the same stereocontrol as reactions of their *ortho*-unsubstituted analogs *851a-c*: (1*R**,6*R**,7*R**)-isomer *853d* and (1*S**,6*R**,7*R**)-isomer *852e* were obtained as the major isomers. Thermally conducted reactions showed some small changes in the relative distribution of isomeric cycloadducts. Again, ethyl (1*S**,6*R**,7*R**)-6-benzamido-1-(4-nitrophenyl)-5-oxo-7-phenyl-1,5,6,7-tetrahydro pyrazolo[1,2-*a*]pyrazole-2-carboxylates *852a-e* were the major isomers.

Cossío [9] synthetised bipyrazoles by 1,3-dipolar cycloaddition of hydrazones. 4-Pyrazolyl hydrazones *856a* or *856b* reacted *via* azomethine imine tautomer with electron-poor dipolarophiles, under microwave irradiation within 10-45 min affording the [4,3']-bipyrazoles (Scheme 176). Under microwaves in solvent-free conditions, 1,3-dipolar cycloadditions were dramatically accelerated achieving good yields of the aromatic bipyrazoles in a few minutes. The effect of microwave irradiation is not exclusively reaction acceleration: the hydrazone *856a* and ethyl propiolate *857* did not react by classical heating in an oil bath in comparable reaction conditions (time and temperature). When ethyl propiolate was employed as dipolarophile, two isomeric cycloadducts *858* and *859* resulted from the two possible approaches of the reagents in the transition state. However, regioisomer *858* is major product. In the cycloaddition with dimethyl fumarate shorter reaction times (15 min), allowed the intermediate products *861* and *862* be isolated. The intermediates *861* and *862* spontaneously aromatized *in situ* to *863* by aerial oxidation.

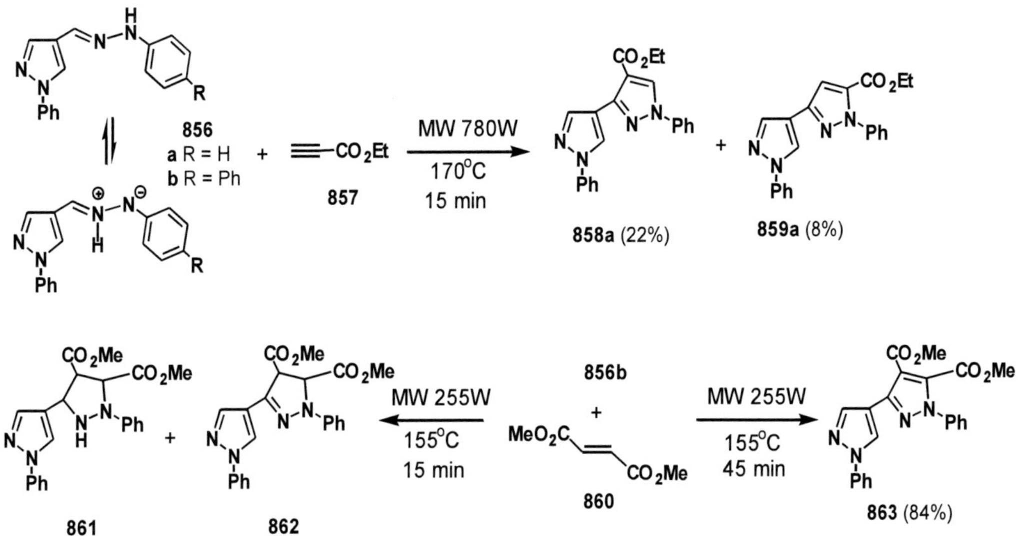

Scheme 171.

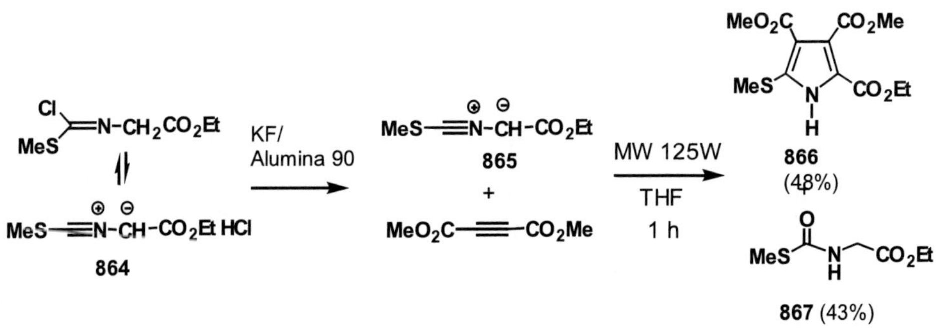

Scheme 172.

2.8.4. Nitrile Ylides

One of the first papers employing microwave irradiation to dipolar cycloadditions was published by Morel [10] in 1992. Imino chlorosulfides *864* were transformed to nitrile ylides *865* by treatment with base, followed by the cycloaddition reaction in solid-liquid media, led to pyrroles. The cycloaddition of *864* with DMAD over KF/Alumina90 support as base led to formation of cycloadduct *866* and also the methyl thioimidate *867*, which is hydrolysis product of dipole *865* (Scheme 172). When dimethyl fumarate was used as dipolarophile, only *867* was isolated (92%).

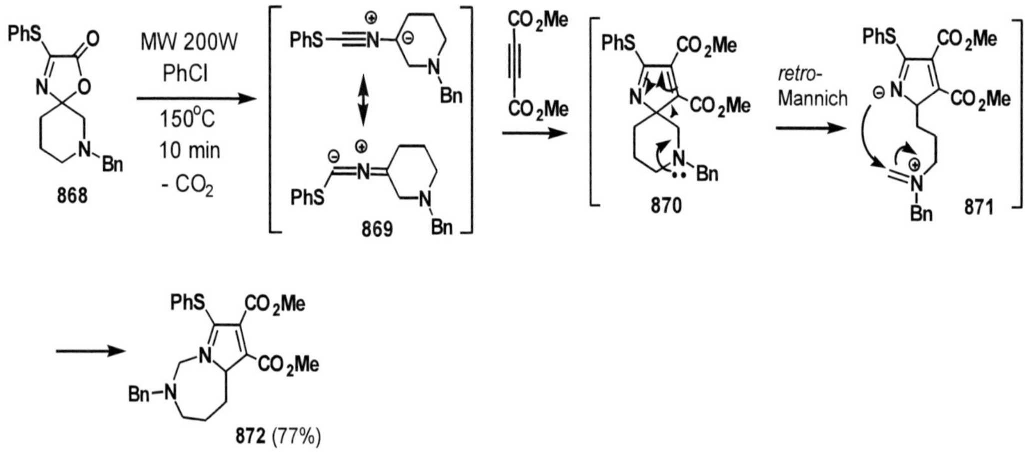

Scheme 173.

Wipf [11] has discovered novel cascade process for ready access to unusual pyrrolo[1,3]-diazepine derivatives, by the reaction of the oxazolinone *868* with electron-deficient alkynes. Heating of *868* in chlorobenzene at 150°C in the microwave reactor released carbon dioxide to generate the intermediate thio-substituted nitrile ylide *869*, which was spontaneously trapped by the powerful dipolarophile dimethyl acetylenedicarboxylate in a 1,3-dipolar cycloaddition reaction (Scheme 173). Rather than the expected spirocycle *870*, the 2,3,4,5-tetrahydro-1*H*-pyrrolo[1,2-*c*][1,3]diazepine *872* was isolated as the end product of a spontaneous retro-Mannich-aminal formation pathway which likely includes the zwitterionic intermediate *871* as yet another step along a largely unprecedented cascade process.

2.8.5. Nitrile Sulfide

Nitrile sulfides could be generated from the oxathiazolone precursors by thermal elimination of CO_2 (Scheme 174). Paton [12] has shown that the classical thermolysis of *873* at reflux failed to trap the pyranosyl nitrile sulfides *874* in synthetically useful yields, instead glucuronic acid nitrile *875* was obtained as a major product. Failure is due to the thermal stability of the oxathiazolone precursors, to the short lifetime of the nitrile sulfides, and also to the need to use high-boiling solvents and long reaction times. The very low cycloadduct yield indicates that the decomposition rate of *874* is greater than the rate of trapping by the dipolarophiles employed. These problems were overcomed using microwave irradiation in place of conventional heating. Thus, irradiation of a solution of xylopyranos-5-*C*-yl oxathiazolone *873* in toluene for 55 min at 200°C with ethyl cyanoformate gave the 3-xylopyranos-5-*C*-yl-1,2,4-thiadiazole *876* in good yield (63%).

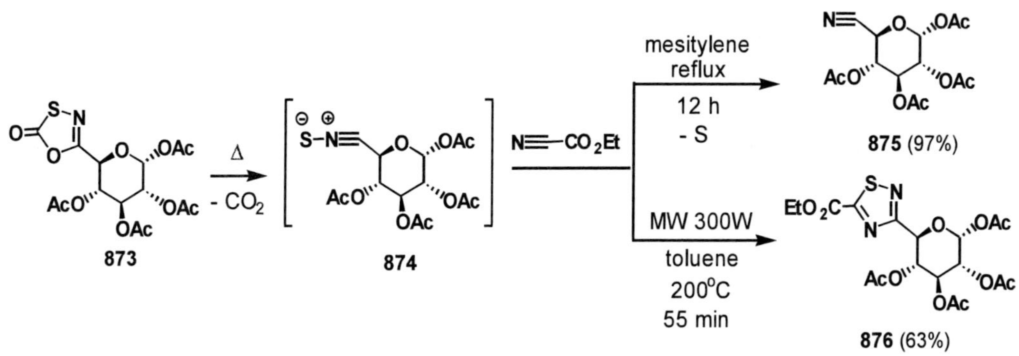

Scheme 174.

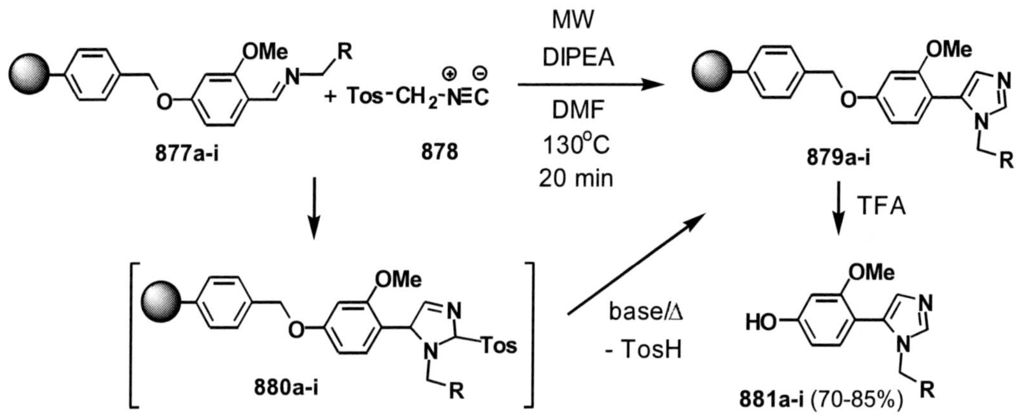

Scheme 175.

2.8.6. Isocyanides

Solid-phase, microwave synthesis of 1,5-disubstituted imidazoles from *p*-tosylmethyl isocyanide (TOSMIC) *878* and aldimines was devised by Yli-Kauhaluoma. [13] The 1,3-dipolar cycloaddition reaction takes place to 3-methoxy-4-hydroxybenzimine *877a-i* which are polymer-bounded on the 4-formyl-3-methoxyphenoxymethyl polystyrene (Ameba resin). In a typical reaction, the polymer-bound imine *877* was swollen in DMF and treated with TOSMIC in the presence of *N,N*-diisopropylethylamine, then irradiated in microwave reactor at 130°C for 20 min (Scheme 175). Mechanism of the one-pot preparation of imidazoles *59a-i* involves in the first step a 1,3-dipolar reaction between tosylmethyl isocyanide and an imine *877a-i* under basic conditions to afford imidazolines *880a-i*. In the second reaction step, imidazoles *879a-i* were obtained upon elimination of the toluenesulfonyl group under basic conditions at high temperatures. Subsequent cleavage from resin in acidic conditions afforded imidazoles *881a-i* in high yields.

References

[1] Grassi, F.; Foti, F.; Risitano, F.; Zona, D. *Tetrahedron Lett. 2005*, *46*, 1061-1062.

[2] Harju, K.; Manevski, N.; Jari Yli-Kauhaluoma, J. *Tetrahedron 2009*, *65*, 9702-9706.

[3] Gerster, H.; Maas, G. *Z. Naturforsch. 2008*, *63b*, 384-394.

[4] Kamble, R. R.; Sudha, B. S. *Phosph. Sulf. and Silic. 2008*, *183,* 1691-1709.

[5] Kamble, R. R.; Sudha, B. S.; Bhadregowda, D. G. *J. Serb. Chem. Soc. 2008*, *73*, 131-138.

[6] Mallur, S. G.; Badami, B. V. *Il Farmaco 2000*, *55*, 65-67.

[7] Pezdirc, L.; Stanovnik, B.; Svete. J. *Z. Naturforsch. 2008*, *63b*, 375-383.

[8] Svete, J. Chapter 5, in *Stereochemistry Research Trends*, Horvat, M. A.; Golob, J. H., eds., Nova Science, Hauppauge NY, *2008*, 1-55.

[9] Arrieta, A.; Carrillo, J. R.; Cossío, F. P.; Díaz-Ortiz, A.; Gómez-Escalonilla, M. J.; de la Hoz, A.; Langa, F.; Moreno, A. *Tetrahedron 1998*, *54*, 13167-13180.

[10] Berrée, F.; Marchand, E.; Morel, G. *Tetrahedron Lett. 1992*, *33*, 6155-6158.

[11] Liang, M.; Saiz, C.; Pizzo, C.; Wipf, P. *Tetrahedron Lett. 2009*, *50*, 6810-6813.

[12] McMillan, K. G.; Tackett, M. N.; Dawson, A.; Fordyce, E.; Paton, M. R. *Carbohydr. Res. 2006*, *341*, 41-48.

[13] Samanta, S. K.; Kylänlahti, I.; Yli-Kauhaluoma, J. *Bioorg. Med. Chem. Lett. 2005*, *15*, 3717-3719.

Chapter 3

OTHER CYCLOADDITIONS

INTRODUCTION

While the main body of synthetic work was carried by the employment of either [4+2] Diels-Alder or 1,3-dipolar cycloadditions, a small number of other cycloadditions carried out under microwave conditions could be found in the literature. Amongst them, the formation of β-lactams from imines and chloroacetyl chloride by an [2+2] reaction is synthetically most explored.

There is a number of other condensation reactions carried out under microwave conditions which yielded cyclic products, which could not be regarded pericyclic cycloadditions. Amongst them, one may found: oxa-6π electrocyclisation, [1, 2] aza-6π-electrocyclisation, [3] cyclobutane ring opening and closing of intermediate diradicals after rearrangement, [4] synthesis of indoles by sequential aryl amination/Heck coupling, [5] amine condensation with ester to afford ketopiperazine, [6] the ring closing olefin methathesis with Grubbs type II catalyst, [7, 8, 9] reaction of metal carbenes and ureas to give uracils, [10] intramolecular Claisen reaction, [11] Niementowski condensation, [12, 13] three component condensation of aldehyde, amine and isocyanide, [14] condensation of β-formylenamide with cyanomethylenes, [15] Pechmann condensation of substituted phenols and β-ketoesters to coumarins, [16] Friedländer condensation of acetophenones with aminoacetophenones, [17] and condensation of amino esters with imidoyl chloride to pyrimidin-4(3*H*)-ones. [18]

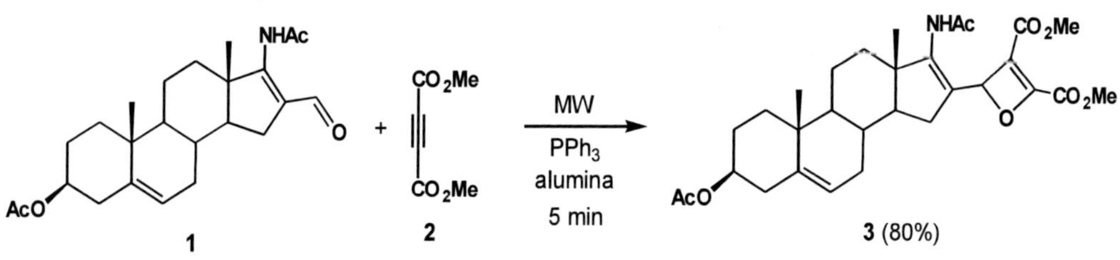

Scheme 1.

3.1. [2+2] Cycloadditions

Boruah [19] developed a convenient synthesis of oxetene ring via [2+2] cycloaddition reaction under microwave irradiation. This synthesis features solvent free, solid phase conditions for reaction of steroidal β-formyl enamides and alkynes (Scheme 1). In a typical reaction, a mixture of 3β-acetoxy-16-formyl-17-acetamido-androst-5,16-diene *1*, triphenylphosphine and dimethyl acetylene dicarboxylate *2* was irradiated in a microwave oven in the presence of PPh_3 to afford steroidal oxetene *3* in 80% yield. The conventional thermal heating of formyl enamides with alkynes in the presence of PPh_3 in protic solvent under prolong heating (72 h) led to very poor yields of oxetenes (15-23%).

It was shown by Dandia [20] that the formation of β-lactams from imines and chloroacetyl chloride could be carried effectively in solvent-free conditions. Solvent free synthesis of β-lactam fused benzothiazepine derivatives *6* was carried out on potassium carbonate surface in few minutes without using any strong base (Scheme 2). Potassium carbonate is a mild base, inexpensive and water-soluble which eliminates the use of organic solvents required for washing the adducts from the solid surface. The reaction was also carried out conventionally in basic medium (triethylamine, dry benzene), but in these conditions the product was obtained in lower yield (20%) after long reaction time (90 h). Mechanistically, ketene which was generated *in situ* from chloroacetyl chloride *5* with the aid of potassium carbonate base undergoes a facile [2+2] cycloaddition to imine bond of benzothiazepine *4* forming β-lactam ring regioselectively. When basic alumina/potassium carbonate was used as solid support, [21] the [2+2] cycloaddition reaction occurred successfully, thus eliminating the requirement of volatile solvent and hazardous base triethylamine. Hence, 3-indolylimine *7* with chloracetyl chloride affords β-lactam *8* in high yield and regioselectivity.

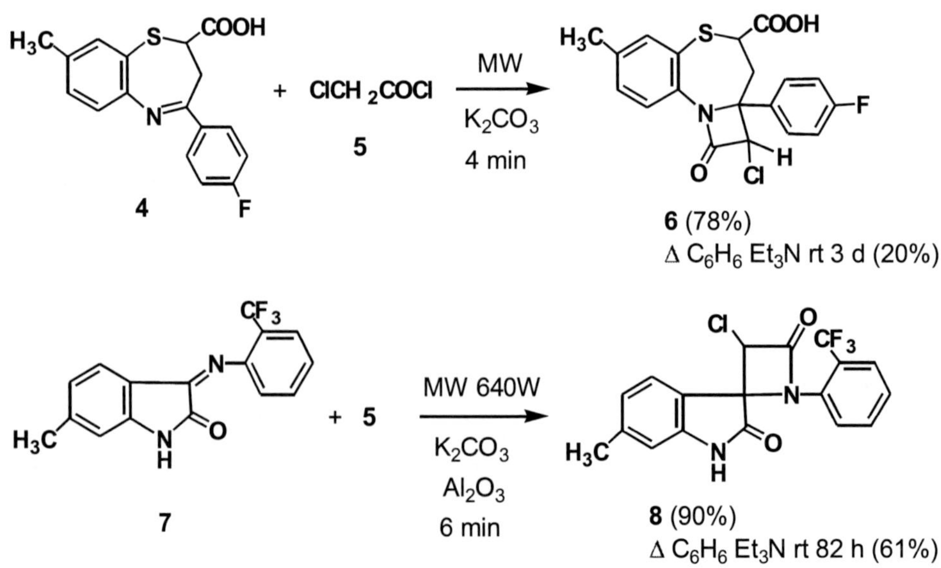

Scheme 2.

Basic alumina was also used as a support and base for the formation of β-lactams in reaction of imines with ethyl-α-mercapto/α-cyanoacetate *10a,b* (Scheme 3). As shown by Kidwai, [22] synthesis under microwave conditions showed clear advantage over conventional heating or microwave reaction carried out in solution.

For microwave assisted synthesis of α-vinyl-β-lactams, reaction conditions used in classical heating reactions were modified: triethylamine as a tertiary base was replaced with the higher boiling *N*-methylmorpholine (bp 116°C). In addition, benzene which absorbs microwave energy poorly, was replaced with chlorobenzene, which has a substantial dipole and a high boiling point (133°C). Improved yields (70-75%) were obtained at elevated temperatures by Manhas. [23] Imine *13* in the reaction with acid chloride *12* produced a mixture of *cis* and *trans* β-lactams *14* and *15* in a ratio depending on the temperature of the reaction mixture (Scheme 4). Higher temperature favored the formation of the *trans* β-lactam, which is thermodynamically more stable. Interestingly, *cis* β-lactam *18* was obtained exclusively from acid chloride *16* and imine *17* irrespective of the temperature of the reaction. The problem of *cis/trans* isomerism disappears when an imine *19* is used for the preparation of the vinyl-β-lactam *20*. However, an isomer *21* was also produced in which the double bond in the side chain is conjugated with the β-lactam carbonyl group (Scheme 4).

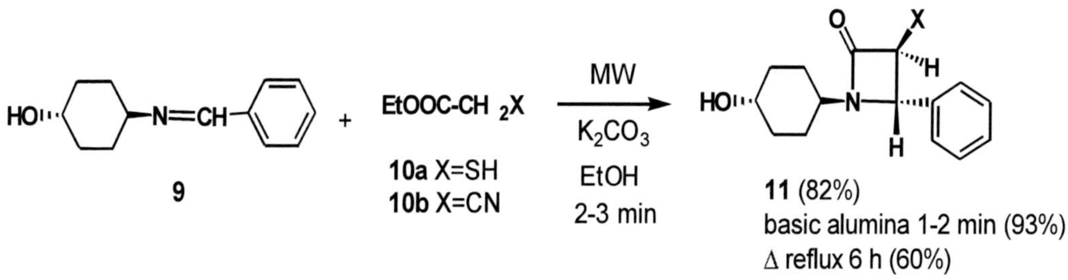

Scheme 3.

Similar rearrangement of double bond was observed by Bose [24] in the [2+2] cycloaddition reactions caried out in DMF, at temperatures 120-140°C, which cointained some NMM. Crotonyl chloride reaction with imine *22* led to the formation of β-lactams *24* and *25*/*26* in the proportion of 1:3 (Scheme 5). Identical products were obtained when *22* was heated classically under reflux with *23* and triethylamine in dichloromethane solution for 2 h.

Staudinger formation of β-lactams from ketenes could be initiated by either photochemical or microwave irradiation, as shown by Xu. [25] In the case of photochemical reaction, side products were also observed, favoring the use of microwave conditions. Hence, bicyclic *β*-lactam *29* was obtained from the reaction of a cyclic imine, substituted dibenzo[*b,f*][1,4]-oxazepine *27*, and *R*-diazoketone *28* stereospecifically under both microwave and photoirradiation conditions (Scheme 6). In the photochemical reaction (high-pressure mercury lamp under a nitrogen atmosphere at -20 °C for 1 h), byproduct *30* of [2+2+2] cyclization was also obtained. However, no similar byproduct was found under microwave conditions, or with alicyclic imines.

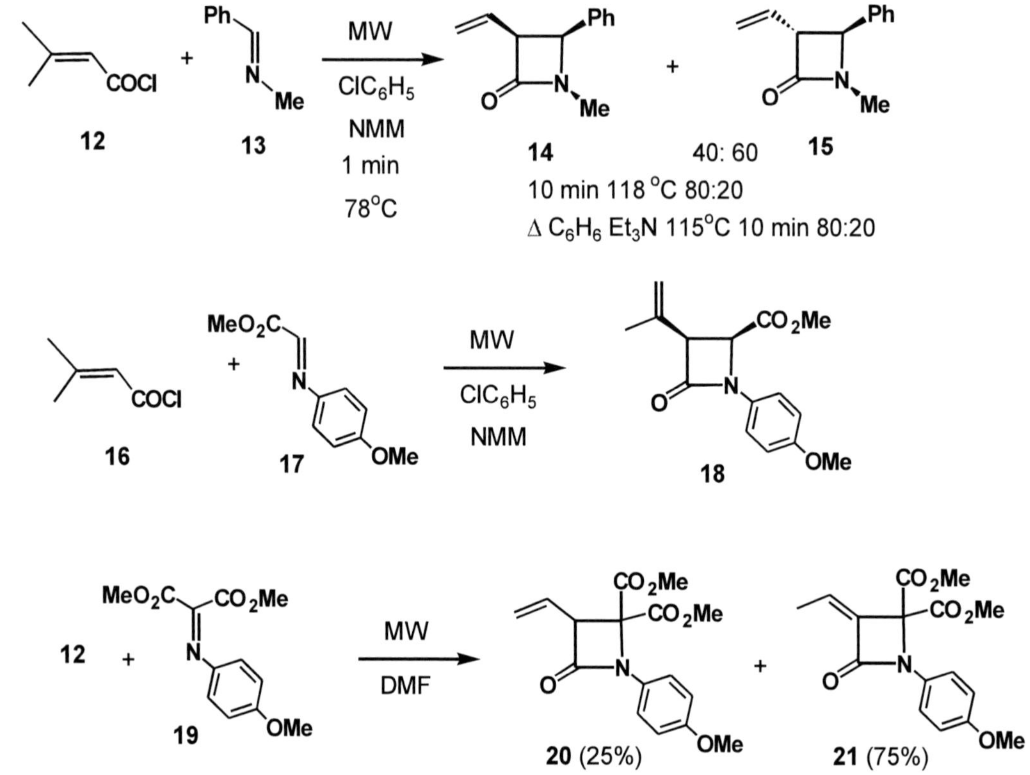

Scheme 4.

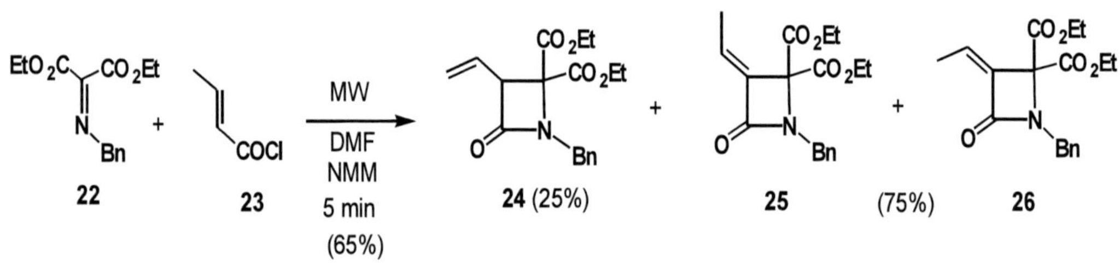

Scheme 5.

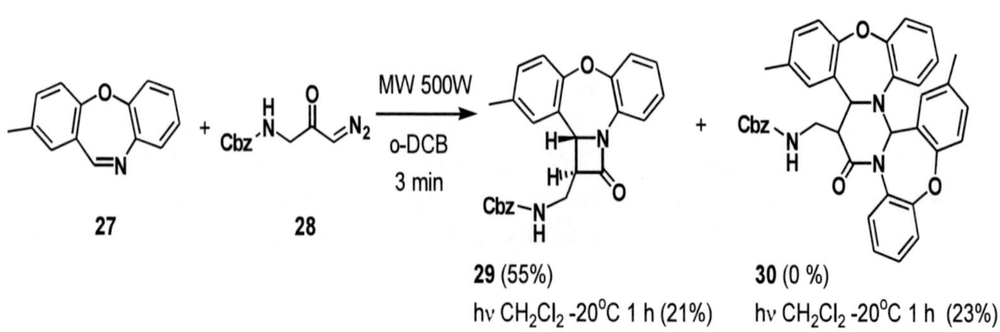

Scheme 6.

The unfavorable use of the high-boiling *o*-dichlorobenzene, which is hard to remove after the Staudinger reaction, could be avoided by performing the reaction in 1,2-dimethoxyethane (b.p. 85 °C). Podlech [26] used modified conditions for reaction of imine *31* and diazoketone *32*, to obtain a similar yield and selectivity of β-lactam *33* (Scheme 7).

An efficient and convenient procedure has been developed for the ruthenium catalyzed [2π+2π] cycloadditions of dimethylacetylene dicarboxylate with norbornenes by Margetić [27]. Reaction is significantly accelerated in microwave conditions, while the commonly used benzene solvent was replaced by environmentally benign tetrahydrofuran. Thermal reaction of norbornenes such as *34* with dimethylacetylene dicarboxylate in the presence of ruthenium catalyst, $RuH_2CO(PPh_3)_3$ (0.1 equiv), takes place specifically at the norbornene *exo-* face to obtain cyclobutene diester *36*. In the case of norbornadiene *37*, a complex product mixture was obtained, consisting mainly of three components, *38*, *39* and *40*, where bishomo [2π+2π+2π] cycloaddition product *30* dominates (Scheme 8).

Another publication on Mitsudo reaction appeared, almost simultaneously in which Pfeffer [28] uses different reaction conditions. Microwave heating in DMF for 2 min at 100°C were found to be the best conditions for variety of norbornene substrates such as *41* (Scheme 9). When these reaction conditions were employed to reaction of *37*, double [2+2] cycloadduct *39* could be obtained in high yield (76%), without the formation of *40*.

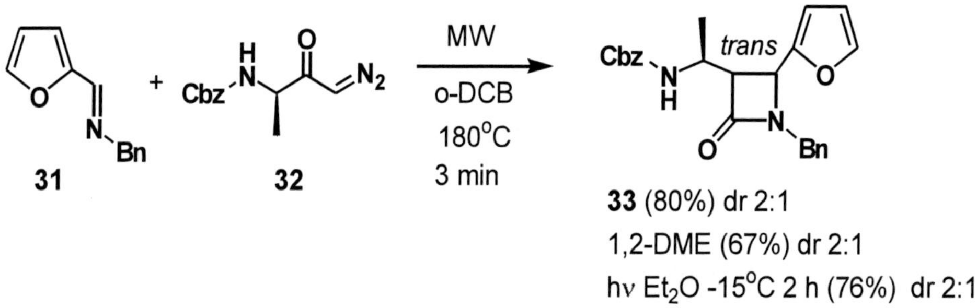

Scheme 7.

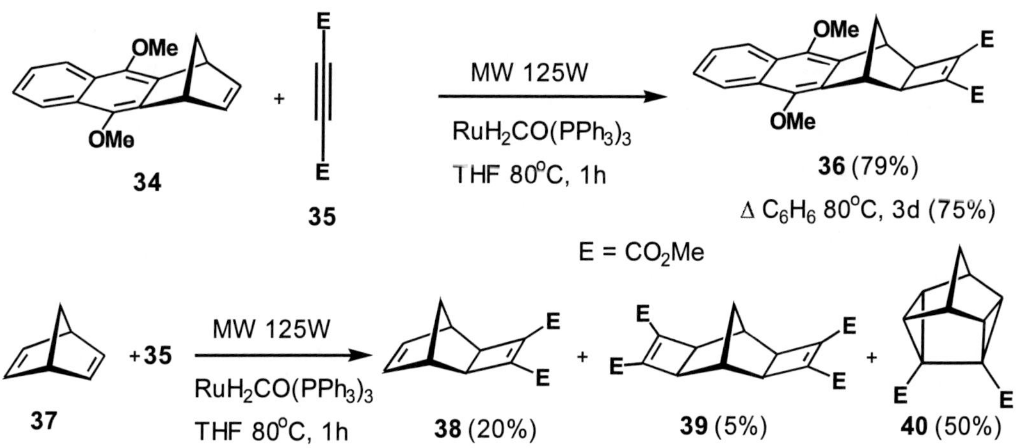

Scheme 8.

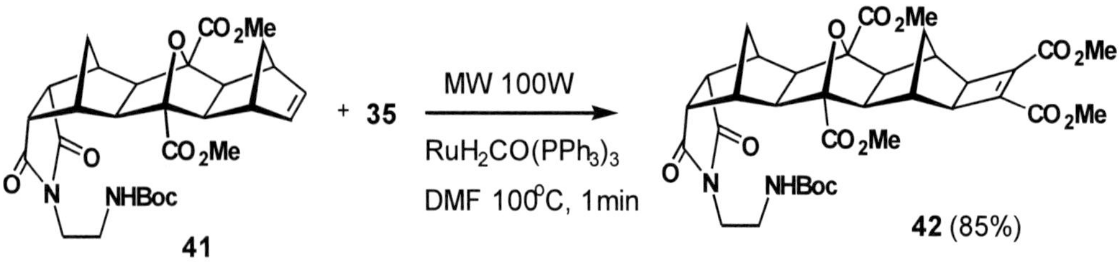

Scheme 9.

3.1.2. Intramolecular [2+2] Cycloadditions

Microwave irradiation of alkynyl allenes affords an intramolecular [2+2] cycloaddition reaction, which provides an efficient route to bicyclomethylenecyclobutenes. Brummond [29] reported that the reaction occurs with complete regioselectivity for the distal double bond of the allene when ionic liquid was added to solution. Required higher temperatures (>200 °C) were obtained by doping the toluene with an ionic liquid, 1-ethyl-3-methylimidazolium hexafluorophosphate to increase the ability of the solution to absorb microwave energy. [30] This is an alternative to use of rather expensive ionic liquids as the solvent. Bicyclo[4.2.0]octadienes and bicyclo[5.2.0]nonadienes have been prepared in high yield (Scheme 10). Hence, microwave heating of *43* in 3 M solution of 1-ethyl-3-methylimidazolium hexafluorophosphate in toluene at 250°C for 15 min affords bicyclo[4.2.0]octadienes *44*. A wide range of allenynes possessing substituents and functionalities was succesfully used and a variety of hetero- and carbocyclic substrates was prepared in such conditions. More recent experimental and computational study favors stepwise mechanism for this reaction. [31]

Tandem [3,3]-sigmatropic rearrangement / intramolecular[2+2] cycloaddition reaction could be thermally-induced by microwave irradiation. Brummond [32] used this approach in synthesis of carbocyclic spirooxindoles. Heating of propargylic acetate *45* to 225 °C in 1,2-dichlorobenzene in the microwave gave the spirooxindole *46* in 60% yield (Scheme 11). This densely functionalized alkylidene cyclobutene is postulated to arise *via* the [2+2] cycloaddition reaction between the alkyne and the allenyl acetate in intermediate *47*, which in turn arises from the thermal [3,3]-sigmatropic rearrangement of substrate *45*. In some instances, *N*-methylpyrrolidinone (NMP) was used as a solvent so the reaction mixture could be heated to higher temperatures (250°C vs. 225°C in 1,2-dichlorobenzene). A further advantage of NMP over 1,2-dichorobenzene was the ease of isolation of the products since the solvent could be removed with aqueous washings instead of chromatography.

Oh [33] reported that intramolecular [2+2] cycloaddition reaction of allenynes under microwave irradiation gives high yields of bicyclo[m,2,0]alkadienes. These reactions are shorter and under lower temperature than classical heating with the $PdCl_2(PPh_3)_2$ catalyst (Scheme 12). Most interestingly is the advantage of microwave heating in some substrates, for which they claimed that the transformation of 1,8-allenyne carboxylates such as *48* into bicyclo[5,2,0]nonadiene *49* could not occur under the stated conditions without microwave irradiation.

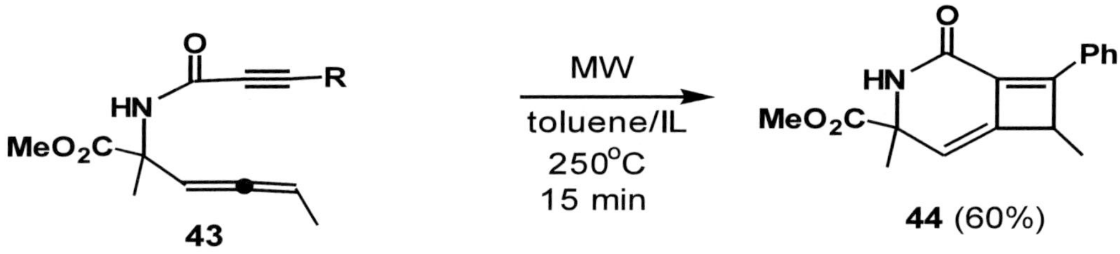

Scheme 10.

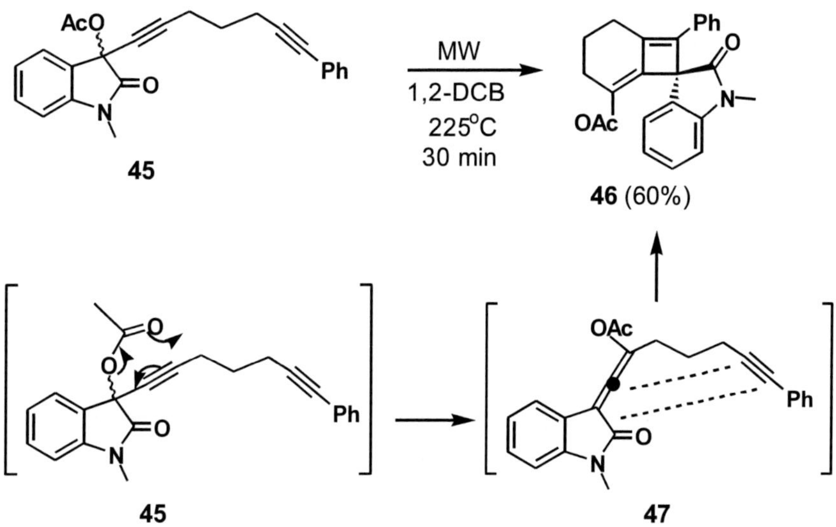

Scheme 11.

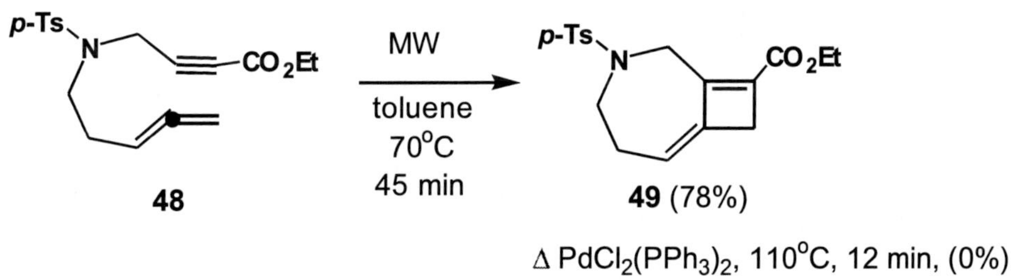

Scheme 12.

Cascade reaction involving formal intramolecular [2+2] cycloaddition was used for total synthesis of artochamins H, I and J. Their unique bicyclo[3.2.0]heptane carbon frameworks was obtained by Nicolaou [34] from an appropriately functionalized stilbene substrate *50* (Scheme 13). When *50* was subjected to microwave heating in the presence of catalytic amounts of Ph_3PO, the tetracyclic core *55* of the artochamins H-J was formed. Mechanism for this remarkable transformation that generates two new rings and three stereogenic centers (d.r. 5:1) starts with two rapid Claisen rearrangements, which are followed by collapse of the two Boc groups leading to *54*. The stilbene is then converted into *55* through a formal [2+2] cycloaddition reaction. Two further mechanistic alternatives for the microwave-promoted reactions were considered. The first, and least probable, involves a [$\pi2_s+\pi2_a$] cycloaddition. The second involves a stepwise diradical mechanism in which formation of the five membered ring precedes diradical recombination. Experiments with (*E*)- and (*Z*)-*50* showed that the formal cycloaddition reaction is not stereospecific with respect to the alkene geometry, indicating that a stepwise radical mechanism, in which bond rotation occurs to some extent before recombination of the diradical, or the redox mechanism are likely.

Microwave methodology allows a straightforward entry to the unique 5-6-4 ring system of sterpurene metabolites. Ovaska [35] demonstrated that intramolecular, microwave mediated [2+2] cycloaddition reactions of allenes and alkynes (1-allenyl-2-propargyl-substituted cyclopentanols *56a-f*) led to formation of cycloadducts *57a-f* (Scheme 14). To avoid acid catalyzed reactions, microwave vials were washed with base (NaOH), and reaction conducted in rarely used solvent phenetole (phenyl ethyl ether). The reaction was shown to be rather general and independent of the type of substituent on the triple bond. Only those allenic systems having the OH functionality and the propargylic moieties in a *cis* orientation were found to be reactive.

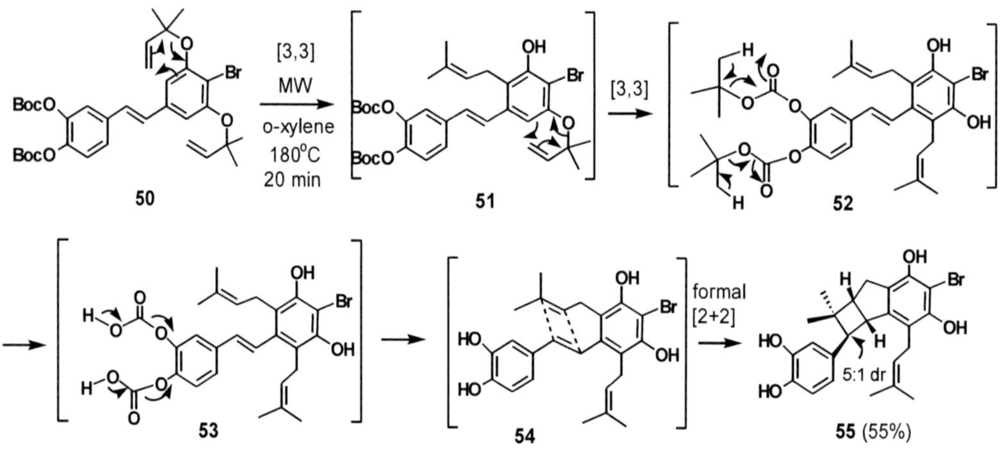

Scheme 13.

3.1.3. [2+2] Retrocyclisation

Thermally less stable products of [2+2] cycloadditions could undergo retroelectrocyclisation. For instance, Stanovnik [36] showed that regiospecific [2+2] cycloaddition of electron-poor acetylenes to (*Z*)-2-acylamino-3-dimethylamino propenoates represents an efficient synthesis of highly functionalised buta-1,3-dienes. When *58* and acetylenes *59* were heated in acetonitrile under microwave irradiation, the corresponding (1*E*,3*E*)-1-acylamino-4-(dimethylamino)buta-1,3-dienes *60a-i* were isolated as the single isomers (Scheme 15). For the cycloaddition of acetylenes *59* to *58* two major pathways were envisioned, a concerted and a stepwise one. In the case of a concerted cycloaddition the reaction proceeds as a [2_s+2_a] cycloaddition resulting in cyclobutene intermediate *62'*, which undergoes a conrotatory retroelectrocyclisation to afford (1*Z*,3*E*)-isomer *64* and/or (1*E*,3*Z*)-isomer *65*. In the two step mechanism, the initially formed intermediate *61* is transformed into *61'*, which gives sterically favored cyclobutene intermediate *62*. Mechanism is then followed by a conrotatory retro-electrocyclisation to produce (1*E*,3*E*)-isomer *60* and/or (1*Z*,3*Z*)-isomer *63*.

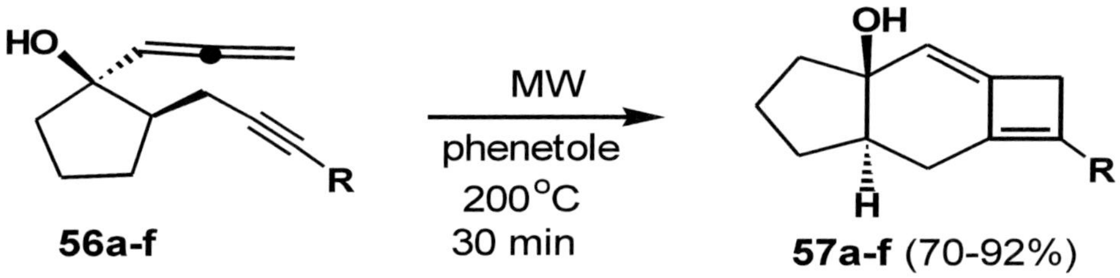

Scheme 14.

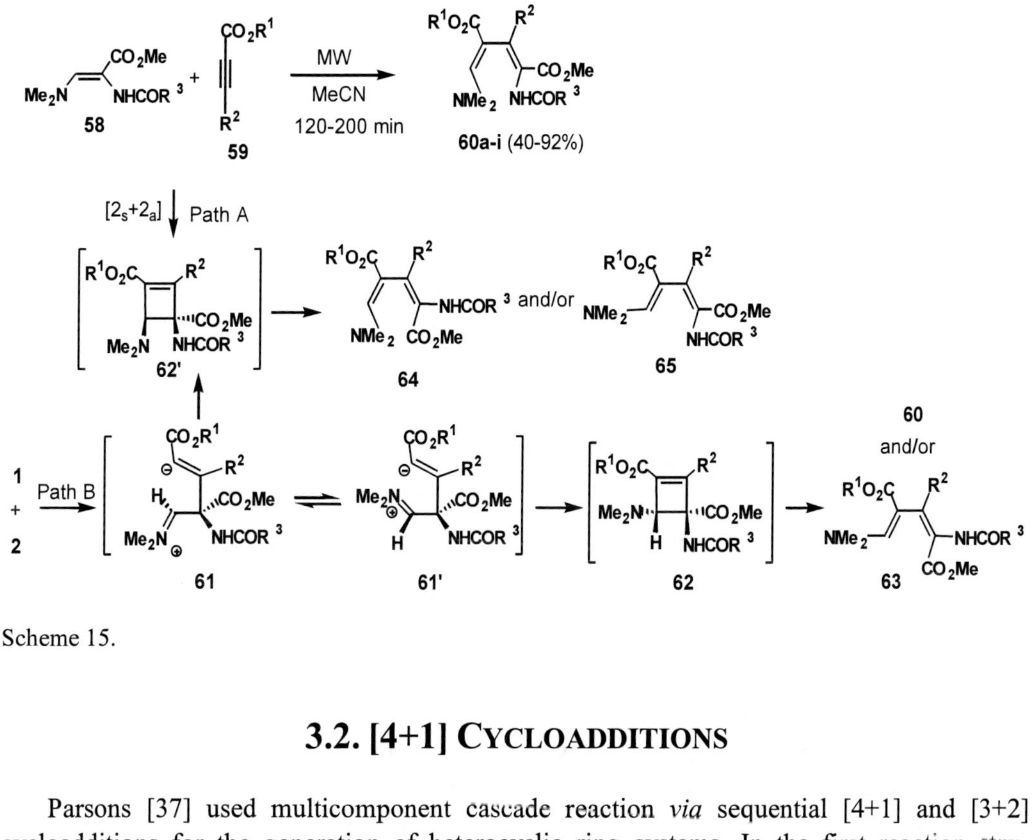

Scheme 15.

3.2. [4+1] Cycloadditions

Parsons [37] used multicomponent cascade reaction *via* sequential [4+1] and [3+2] cycloadditions for the generation of heterocyclic ring systems. In the first reaction step, epoxide reacts with trimethylsilyl cyanide to afford isocyanide *71*, which reacts with nitroalkene *68* in [4+1] fashion to afford nitrile oxide *72*. Then *N*-(isoxazolidinylidene)alkylamine *71* reacts in a [3+2] dipolar cycloaddition with methyl acrylate to form bicyclic intermediate *73*. Molecular rearrangement of bicycle *73* affords the final product *70* (Sheme 16).

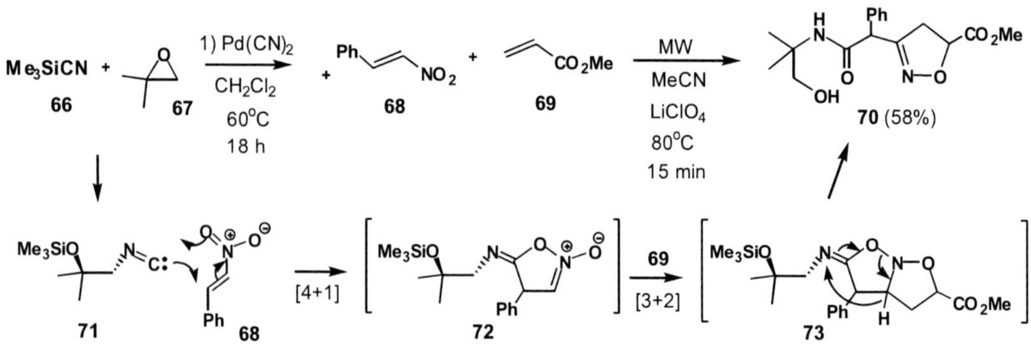

Scheme 16.

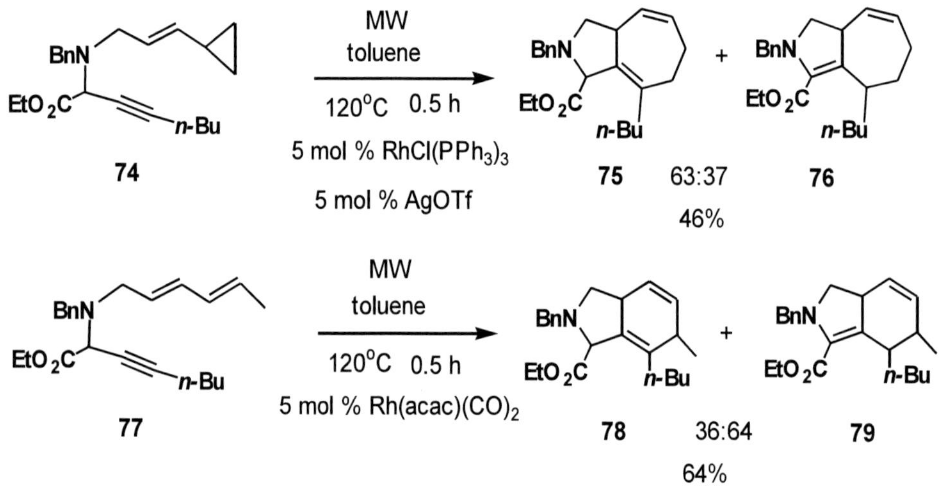

Scheme 17.

3.3. [5+2] Cyclisation

When the glycinate-tethered dienyne and cyclopropylenyne were subjected to rhodium-catalyzed cycloadditions under microwave irradiation by Yamamoto [38] to give the intramolecular [5+2] cycloadducts in moderate yields. When cyclopropylenyne *74* was treated with 5 mol% Rh(acac)$(CO)_2$ catalyst under the microwave irradiation, a seven-membered ring products *75* and *76* were furnished in 33% combined yield (with the ratio of 55:45). Slight improvement in the yield and the selectivity (46%, 63:37) was achieved when the cationic catalytic system, 5 mol% $RhCl(PPh_3)_3$/AgOTf, was applied to the same reaction (Scheme 17). Activity of catalysts was reversed when dienyne *77* was subjected to rhodium-catalyzed cycloadditions. Using $RhCl(PPh_3)_3$/AgOTf, catalyst, product *79* arising from the intramolecular [4+2] cycloaddition was obtained in a low yield. Employing Rh(acac)$(CO)_2$ catalyst under the same reaction conditions, the product yields were improved and cyclohexadiene *78* and its olefin isomer *79* were obtained favoring the later.

When the related enyne *80* was submitted to iridium catalyst (first generation Grubbs catalyst), a domino cycloisomerization/Diels-Alder reaction/dehydrogenative aromatization sequence takes place (Scheme 18). Heating of *80* in the presence of the $[IrCl(cod)]_2$ catalyst, and *N*-phenylmaleimide *81* in the microwave reactor obtained adduct *83*, along with pyrroles *82* and *84*, while only *82* and *84* were obtained under conventional heating conditions. Mechanism of iridium-catalyzed cycloisomerization of *80* in the first step led to formation of exocyclic diene *85*, which undergoes Diels-Alder reaction to *83*. The detailed mechanism is unclear, *84* was presumably formed by the dehydrogenative aromatization of *83* via C-H activation on the 3,4-dehydroproline ring with iridium species.

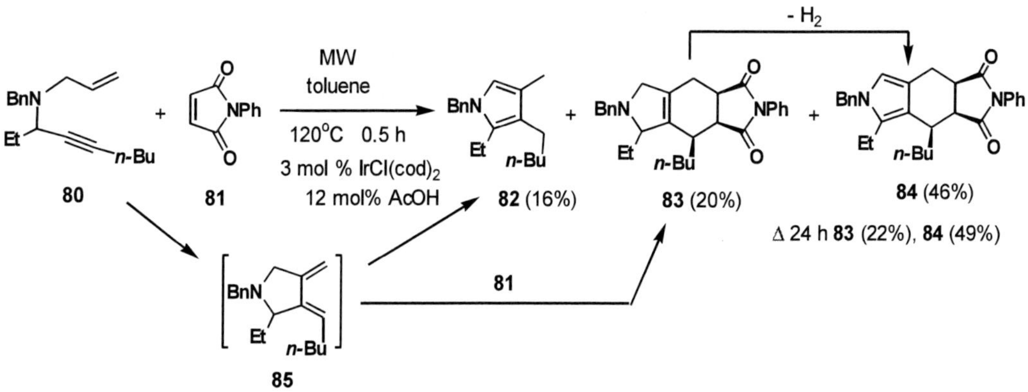

Scheme 18.

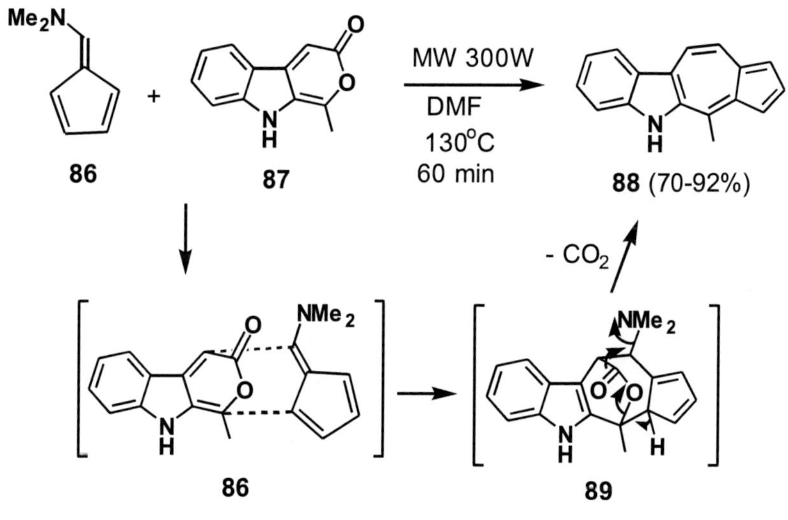

Scheme 19.

3.4. [6+4] Cycloadditions

Hong [39] studied a microwave-enhanced [6+4] cycloaddition reactions between 6-aminofulvene and pyrones followed by CO_2 extrusion providing azuleneindole products which display interesting antineoplastic activity. In this reaction, fulvene *86* acts as dienophile (6π-component), while pyrone *87* is diene (4π-component) (Scheme 19).

Reactions performed at higher microwave power and for longer irradiation times gave lower yields of the azuleneindoles along with decomposition products. The molar ratio of two reagents plays an important role, lower yields were observed when reactant concentrations were higher or lower than 1M, such that 5-fold excess of *86* afforded the best yield. When rarely used DMSO was employed as a microwave solvent lower yields were obtained.

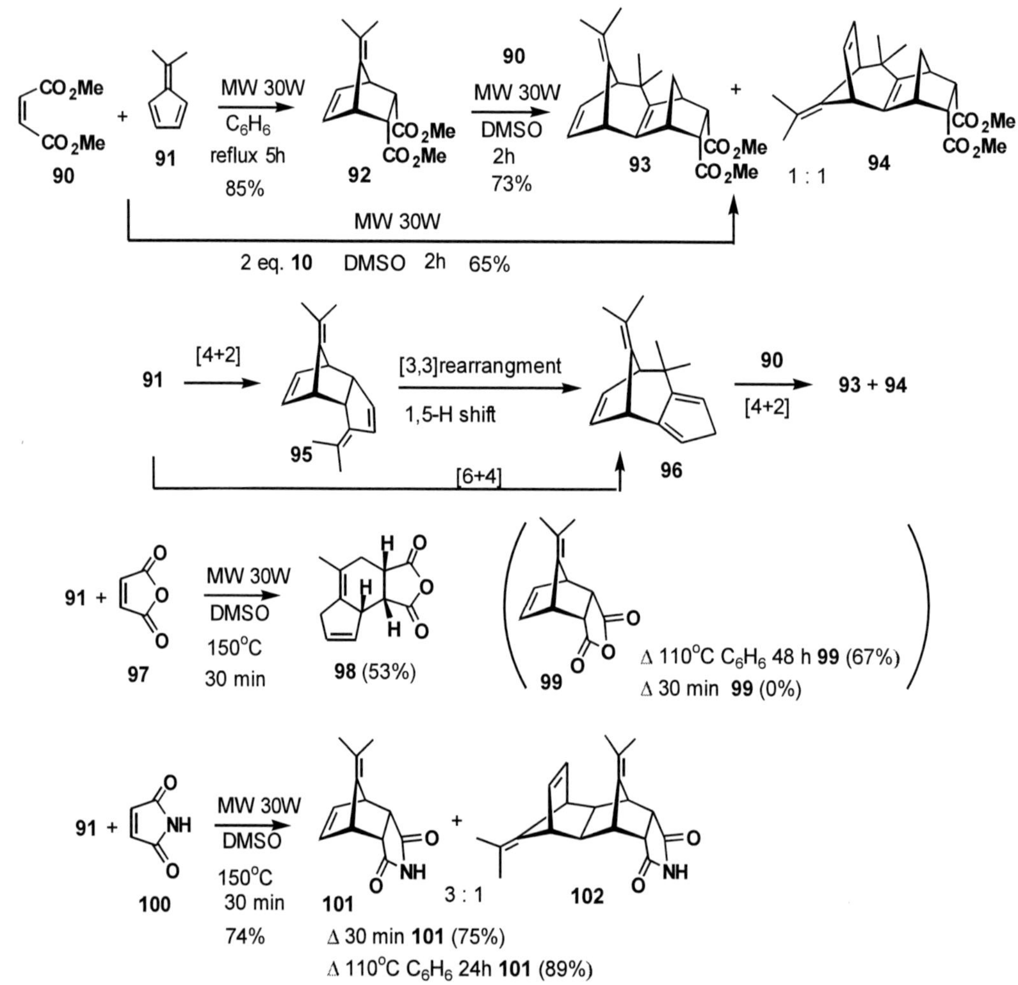

Scheme 20.

For certain cycloadditions the use of microwave irradiation instead of conventional heating leads to the formation of structurally unrelated products, hence microwave irradiation alters the reaction pathway, as demonstrated by Hong. [40] Under conventional heating, 6,6-dimethylfulvene reacts with dimethyl maleate to afford the Diels-Alder adduct *92* in 84% yield. When a DMSO solution of 6,6-dimethylfulvene *90* and dimethyl maleate was subjected to microwave irradiation for 2 h, the tandem [6+4]/[4+2] cycloaddition products *93* and *94* were obtained in 30% (Scheme 20). The combined yield of *93* and *94* increased to 65% when 2 equiv of 6,6-dimethylfulvene were used. A plausible mechanism for the formation of *93* involves the competitive and reversible formation of Diels-Alder adduct *92* and facile dimerization of fulvene *via* [6+4]-cycloaddition to give the dimer *96*. Subsequently, [4+2]-cycloaddition of *96* with dimethyl maleate occurs to generate adducts *93* and *94*. Alternatively, dimerization of fulvene via [4+2]-cycloaddition affords adduct *95*, which can be transformed to dimer *96* by a [3,3]-sigmatropic rearrangement and a 1,5-hydrogen shift. Different mechanistic pathways are followed in reaction of *91* with maleic anhydride under microwave conditions. The [4+2]-cycloadduct *98*, an analogue of the natural product alcyopterosin was obtained. Herein maleic anhydride adds across the C-1 methyl and C-6 atom of the fulvene to afford a hydrindane system. This reaction under conventional heating gave the Diels-Alder adduct *99* in 67% yield. A possible mechanism for the formation of *98* involves microwave-induced isomerization of *91* to 2-isopropenyl-cyclopenta-1,3-diene followed by trapping with maleic anhydride via [4+2]-cycloaddition. On the other hand, microwave reaction of *91* with maleimide yielded *101* and *102* in a 3:1 ratio. In the presence of 2 equiv. of *91*, the yield of *102* increased to 80%, while reaction under conventional heating gave *101* as a single adduct.

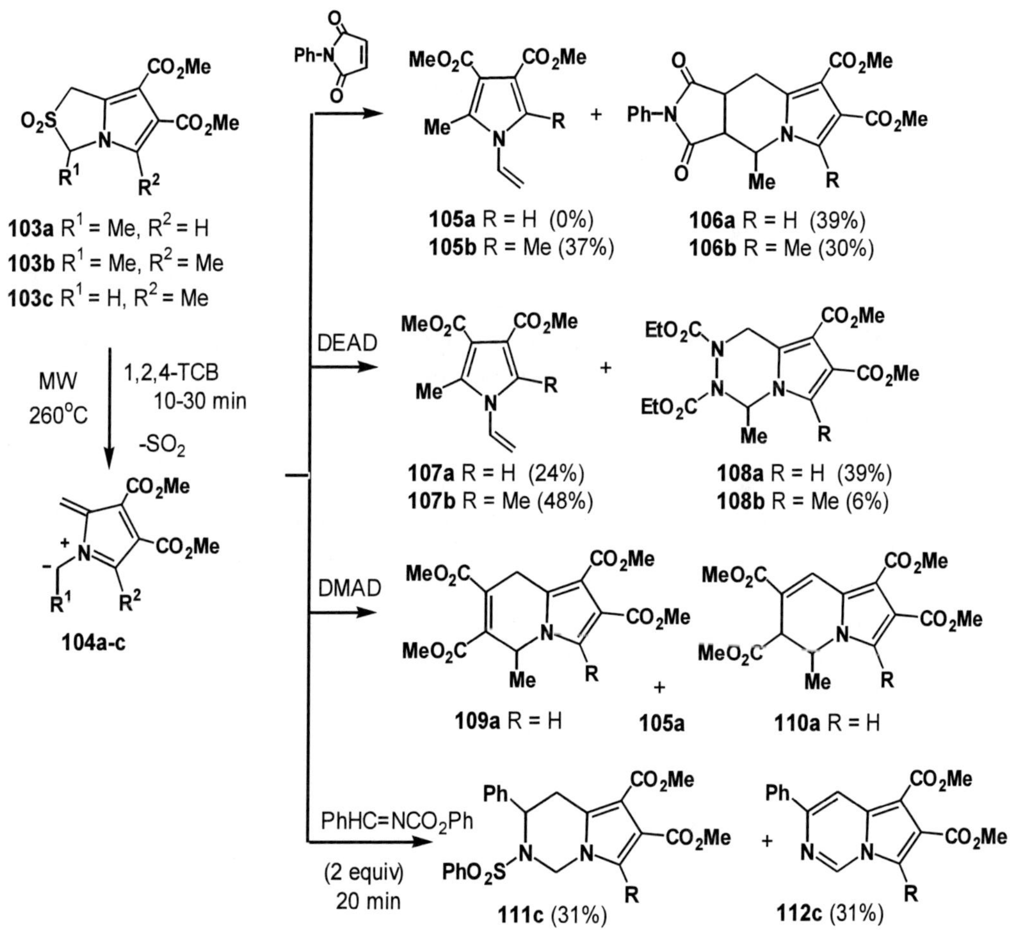

Scheme 21.

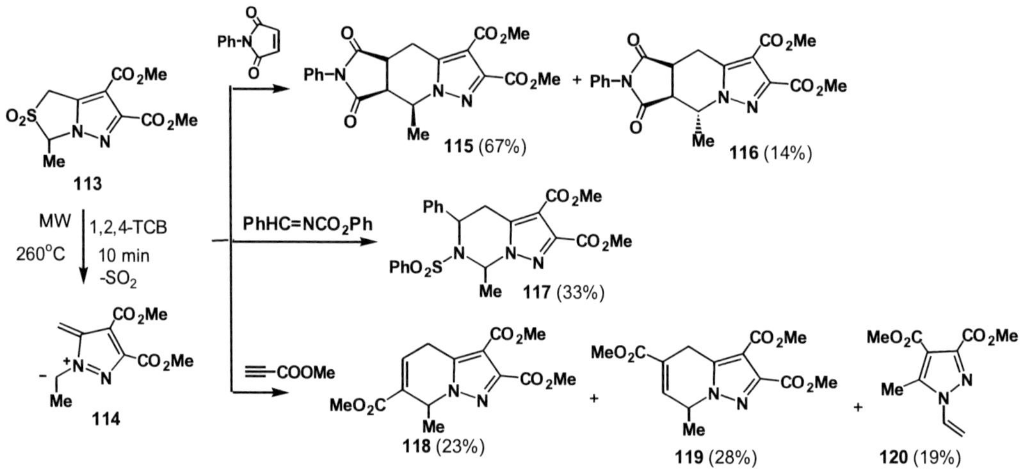

Scheme 22.

3.5. [8+2] Cycloadditions

Pinho e Melo [41] generated azafulvenium methide *104* by the SO_2 extrusion from 3-phenyl-2,2-dioxo-1H,3H-pyrrolo-[1,2-*c*]thiazole sulfones *103* under microwave irradiation in 1,2,4-trichlorobenzene (Scheme 21). The resulting 1,7-dipoles *104* undergo [8π+2π] cycloadditions in the presence of dienophiles. With diethyl diazene-1,2-dicarboxylate cycloadduct *108a* was formed, together with *N*-vinylpyrrole *105a*. The heating of sulfone *103a* in the presence of *N*-phenylmaleimide gave 5,6,7,8-tetrahydroindolizine derivative *106a*. Reaction of *104a* in the presence of DMAD afforded a mixture of 3,5-dimethylindolizine-1,2,6,7-tetracarboxylate and dihydroindolizine-1,2,6,7-tetracarboxylates (*109a* and *110a*) and *N*-vinylpyrrole *105a*. Azafulvenium methide *104b* reacted with DEAD and *N*-phenylmaleimide giving 4,6-dimethyl-1,2,3,4-tetrahydro-pyrrolo[1,2-d][1,2,4]triazine-2,3,7,8-tetracarboxylate *108b* and hexahydro-pyrrolo[3,4-f]indolizine *106b*, respectively. In a similar manner, *104c* reacted with DEAD, *N*-phenylmaleimide but also with the electron-rich dipolarophile bis(trimethylsilyl)acetylene. In the presence of *N*-benzylidene-benzenesulfonamide, the diazafulvenium methide *104c* gave the 1,2,3,4-tetrahydro-5H-pyrrolo[1,2-c]pyrimidine-5,6-dicarboxylate *111c* and the aromatized derivative *112c*. Longer microwave irradiation (40 min) yielded dimethyl 7-methyl-3-phenyl-5H-pyrrolo[1,2-c]pyrimidine-5,6-dicarboxylate *112c* as the sole product.

High reactivity also possesses diazafulvenium methide *114* generated from sulfone *113* under microwave irradiation. In the presence of dipolarophiles *114* participates in [8π+2π] cycloaddition with a range of electron-deficient dipolarophiles giving pyrazolo[1,5-a]pyridine derivatives *115-120* (Scheme 22).

3.6. [2+2+2] CYCLOTRIMERIZATION

Pla-Quintana [42] investigated intramolecular [2+2+2] cycloaddition reactions of cyanodiynes *121a-e* with Wilkinson's catalyst $RhCl(PPh_3)_3$ (10 mol %) catalyst in toluene at 90°C under microwave irradiation (Scheme 23). Good yields of pyridine derivatives *121d* and *121e* were obtained with short reaction times (classical heating 3-8 h, lower yields). Even cyanodiyne *121c*, which had not give product under conventional heating, could be cycloisomerized under microwave heating when reacted in a toluene/chlorobenzene (3:1) mixture. To increase efficiency of reaction, the higher microwave absorbing solvent chlorobenzene was added. When $[Rh(cod)_2]BF_4$/binap (5 mol %) was used as the catalytic system, heating (dichloroethane, 75°C, 30 min) produced *122c* in 67% yield. Reaction works well also when nitrogens were replaced by oxygen and carbon atoms (69-89%). In the case of substrates with pyridyl substituents, DMSO was used as solvent, while for 2-pyridine derivative *121e* mixture of DMF/H_2O (1:1) was employed. Double intramoelcular [2+2+2] cycloaddition process takes place for substrate *123*, yielding cycloadduct *124* in 88 % yield.

Deiters [43] demonstrated that the synergistic application of microwave irradiation and a polymeric solid support to intramolecular cobalt catalyzed [2+2+2] cycloadditions affords a significant increase in the yield, extremely shortened reaction times, and the elimination of catalyst activating additives, excessive heating, or irradiation with light. The cyclotrimerization of trityl-protected dipropargylamine *125* immobilized on a polystyrene resin and benzonitrile was performed in toluene with 10 mol% $CpCo(CO)_2$ under microwave irradiation. The protecting group was removed with TFA, to obtain *127* in excellent 92% yield (Scheme 24). Without microwave irradiation, only 9% product formation was observed after 24 h at 110°C. Reaction conditions make the [2+2+2] cyclotrimerization applicable to a variety of substrates and attachment points. For instance immobilized alkynyl nitrile *128* afforded the fused pyridine *130* in complete regioselectivity, excellent yield, and high purity (>90%). The cyclotrimerization of alkynes and isocyanates is an effective way of generating pyridones, while carbodiimides in cyclotrimerization reactions provide a synthetic entry to 2-iminopyridines. Conversions are highly efficient comparing classical thermal conditions (110 °C, 24 h), where virtually no formation of *132* was detected, or generally show low yield of *133*.

Intramolecular rhodium and cobalt catalyzed [2+2+2] cycloaddition methodology allows efficient approach toward rings system of 6-oxa-allocolchicinoids. Prokop [44] used the intramolecular [2+2+2] cycloaddition (Reppe-Vollhardt alkyne cyclotrimerization) of triyne precursors of type *rac-134a*, and *rac-134b*, in microwave conditions with $CpCo(CO)_2$ catalyst to afford tetracyclic structures of the 6-oxa-allocolchicinoids *rac-135a*, and *rac-135b*, in 85 and 65% yield, respectively (Scheme 25). Under conventional thermal conditions only disappointing results were obtained.

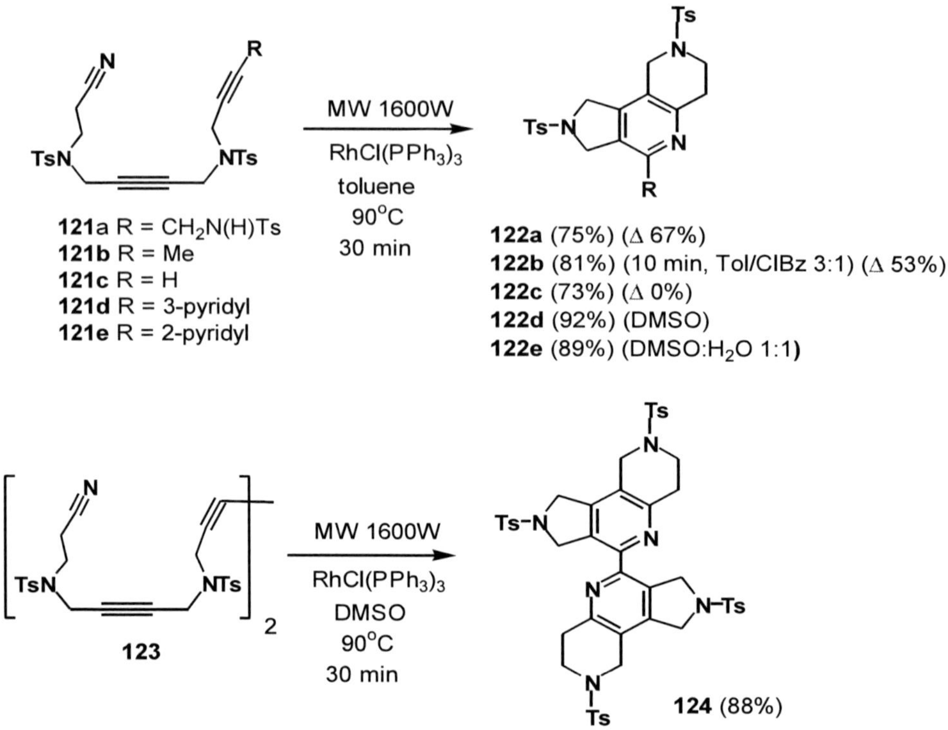

Scheme 23.

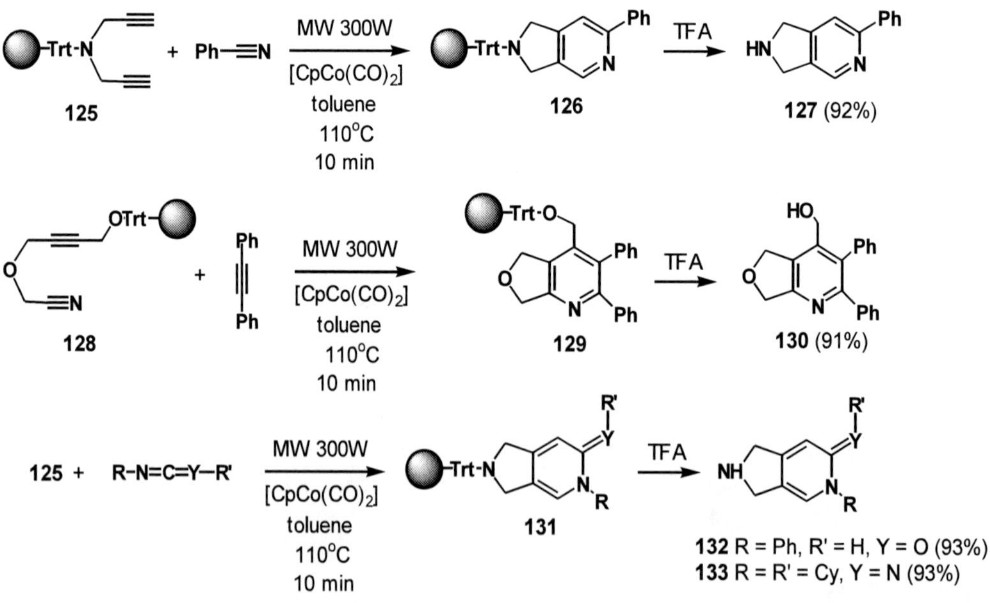

Scheme 24.

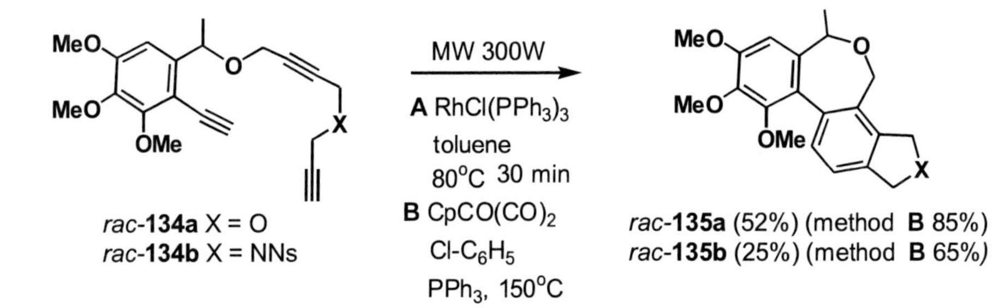

Scheme 25.

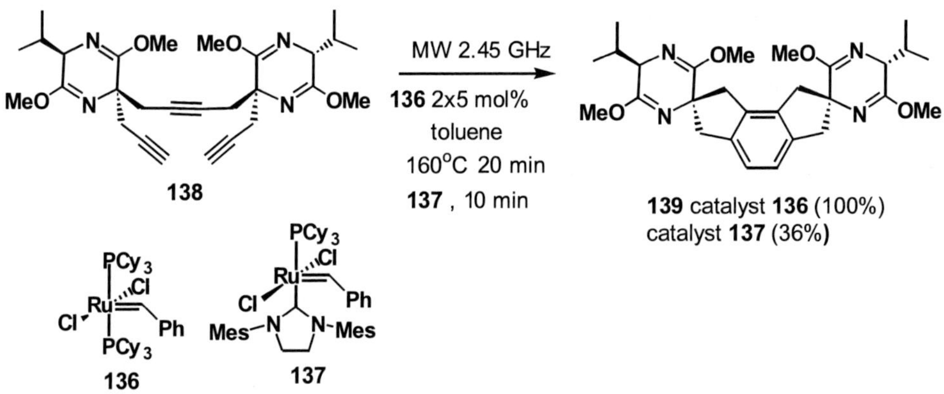

Scheme 26.

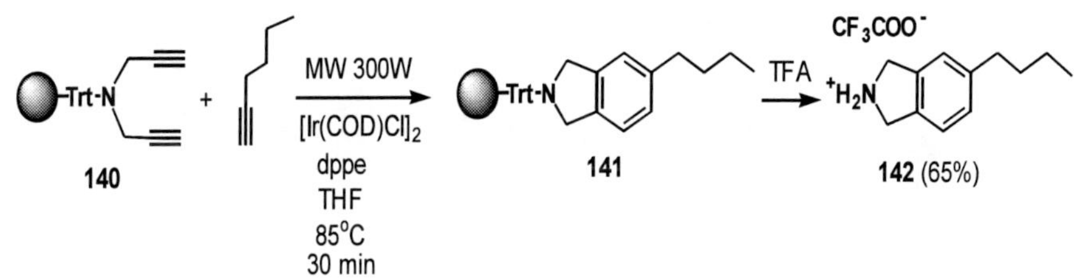

Scheme 27.

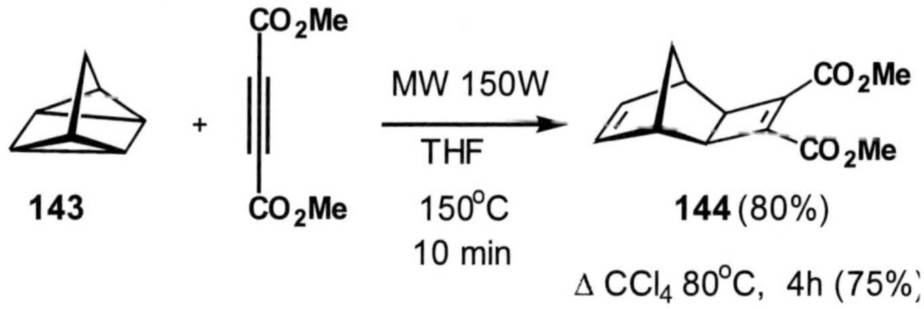

Scheme 28.

Microwave assisted intramolecular [2+2+2] cycloaddition with the standard Grubbs' catalyst *136* as well as with the more recent and more active ruthenium catalyst system *137* were used in conjuction with triyne substrate *138*. Undheim [45] reported that under conventional conditions this reaction requires heating above 80°C and reloading with catalyst.

Five mol% of catalyst gave 60% conversion to the product *139* after 5 h heating. Almost complete conversion was achieved when the same amount of catalyst was added a second time, and the heating continued for a total of 14 h. Under microwave conditions, 100% conversion to product *139* was observed after 20 min (Scheme 26). Surprisingly, the [2+2+2] cycloaddition reaction with the second generation Grubbs' catalyst *137* under microwave heating was less satisfactory, and 36% conversion was achieved after 10 min.

The [2+2+2] cycloaddition catalyzed by iridium catalyst could be also carried out on solid support, as demonstrated by Martinez. [46] The [2+2+2] cycloaddition of dipropargylamine *140* (bound on trityl chloride resin) with 1-hexyne afforded cycloadduct *142* (Scheme 27). Reaction was catalyzed by the $[Ir(COD)Cl]_2$/dppe system (10 mol%), followed by the TFA cleavage of cycloadduct *141* and subsequent catch and release purification of TFA salt. It was found that the addition of diphenylphosphiono ethane (dppe) ligand affords the best yield of isoindoline cycloadduct *142* in 65% yield. While the use of other bidentate ligands such as dppp and dppb or monodentate ligand, PPh_3 was less effective. The microwave [2+2+2] cycloaddition of resin-bound dipropargylamine was used to obtain variety of isoindolines in 50-70% yields.

Rare are examples of [2+2+2] cycloadditions without the use of catalysts in microwave conditions, with the single example given by Margetić. [47] Quadricyclane *143* participates in [2+2+2] cycloaddition reaction with dimethylacetylene dicarboxylate under microwave irradiation to obtain tricyclic product *144* in yields comparable to these obtained by classical heating in CCl_4 (Scheme 28). Notable is the change from highly toxic CCl_4 to environmentally more friendly solvent THF.

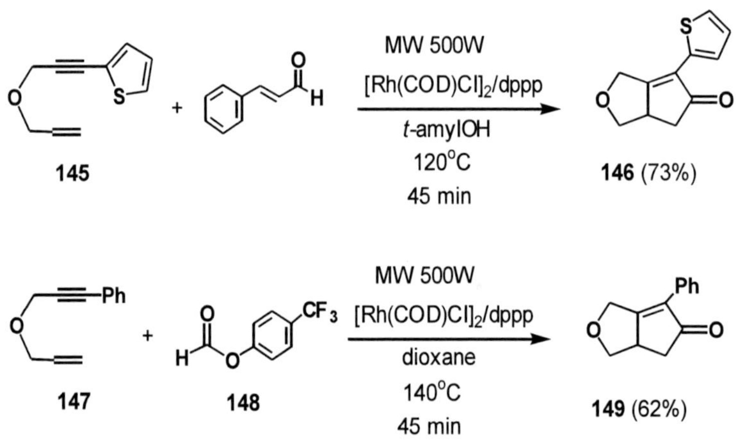

Scheme 29.

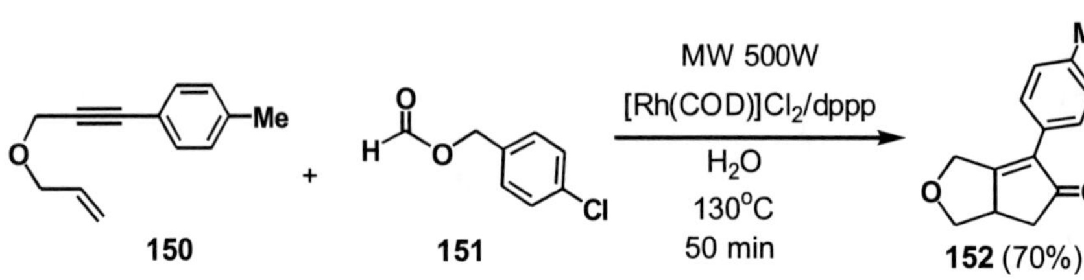

Scheme 30.

3.7. [2+2+1] Cyclisation - Pauson-Khand Reaction

Kwong [48] carried out microwave-assisted dual catalysis of various aromatic oxygen and nitrogen-tethered 1,6-enynes *145* for the Pauson-Khand-type cyclization employing decarbonylation of cinnamaldehyde with (carbonyl)(dppp)-rhodium(I) chloride catalyst (3 mol%). In this catalytic system, for instance, the intramolecular [2+2+1] cycloadduct *146* was obtained in 73% yield after microwave irradiation for 45 min at 120°C in *tert*-amyl alcohol (Scheme 29). Reaction performed under conventional heating at 120 °C for 36 h yielded only traces of product. First enantioselective version of microwave-assisted cascade transformation was achieved by use of chiral diphosphane ligands in place of dppp. (*S*)-BisbenzodioxanPhos provided the best results in terms of product yield and enantioselectivity. In the case of *146*, 58% yield, and 90% ee was obtained by addition of $[Rh(COD)Cl]_2$ (3 mol-%), (*S*)-BisbenzodioxanPhos ligand (6.6 mol %), and microwave heating at 100 °C for 45 min. When microwave-assisted cascade decarbonylation-Pauson-Khand-type reaction was carried out with the use of formate *148* as a CO surrogate cycloadduct *149* was formed in 62% yield in 45 min. Under conventional oil-bath heating, 3 days were required for completion of the reaction with 43% isolated yield.

Further development of the catalytic cascade systems by Chan [49] enabled formate esters as the CO surrogate and the use of water as the solvent in [2+2+1] carbonylative cyclization. The first aqueous Pauson-Khand-type reaction utilizes 4-chlorobenzyl formate ester *151* as the condensed CO source under microwave-assisted conditions with $[Rh(COD)Cl]_2$ (3 mol%) catalyst. This interesting methodology effectively accelerates the carbonylative reaction to provide the corresponding cycloadducts such as *152* in a short period of time and can be achieved for asymmetric version with suitable choice of chiral phosphine ligands (Scheme 30). Excellent product enantioselectivities (up to 98%) were obtained with axially chiral diphosphine ligands such as bisbenzodioxanPhos (*SYNPHOS*) and the modified MeO-BIPHEP (Ar=3,5-di-*t*-Bu-4-OMe-C_6H_2).

As indicated by the results of Groth, [50] toluene solvent seemed to work best for Pauson-Khand [2+2+1] cycloadditions of norbornene *153* to phenylacetylene *154* under microwave irradiation with $Co_2(CO)_8$ catalyst (Scheme 31). The cyclohexylamine additive is thought to replace carbon monoxide as a ligand and promotes additional ligand liberation by stabilizing coordinatively unsaturated complex. The optimal amine/catalyst ratio was determined to be 6:1, where 81% yield was obtained. Optimized conditions were 20 mol% $Co_2(CO)_8$ and 1.2 equivalents of cyclohexylamine by heating the reaction mixture by microwave irradiation for 5 min. Other alkynes were also used: 3-methyl-but-1-yne *156* reacted with norbornene to afford *157* in 71% yield, along with the [2+2+2] cyclotrimerisation side product *158*. When diethyl acetylenedicarboxylate was used a 48% yield was obtained. Intramolecular cyclizations of the enyne *159* also provided the Pauson-Khand product *160* although in lower yield.

When $Co_2(CO)_8$ catalyst was employed by Helaja [51] in Pauson-Khand cyclisation for estrone E-ring extension, it was found that reaction yields depend on the mode of addition of cobalt octacarbonyl/alkyne complex. In all cases, alkynyl-cobalt complex was prepared first by stirring in DME. In the single mode addition equimolar amount of alkynyl-cobalt complex and 16-Ene-3-methoxyestra-1,3,5(10)-triene *161* was heated with microwave for 150 min to obtain two [2+2+1] regioisomeric products *163a* and *163b* in 24% yield and 1:1.3 ratio

(Scheme 32). The gradual addition (addition 4x0.25 equiv) increases yield to 33%, while gradual addition of excess of alkynyl-cobalt complex (6x0.25 equiv, total 360 min) afforded products in 60% yield. Suspension of charcoal in the reaction mixture prior to first microwave irradiation did not influence yields or ratio of regioisomers. Charcoal was used as an easy procedure for removal of undesired organocobalt species, but also as promoter for transfer of microwave energy to reacting system.

Almost identical conditions were employed by Evans. [52] The intermolecular Pauson-Khand reactions between the preformed trimethylsilylacetylene dicobalthexacarbonyl complex *165* (0.2 equiv) and norbornadiene *166* proceeded efficiently under microwave assisted conditions. Use of 1,2-dichloroethane as the solvent gave excellent yields of the corresponding cyclopentenone *167* after only 5 minutes of microwave irradiation (Scheme 33). Microwave technology could be efficiently applied for the extremely rapid synthesis of various cyclopentenones, starting with acetylenes, both intermolecular and intramolecular reactions such with substrate *168*.

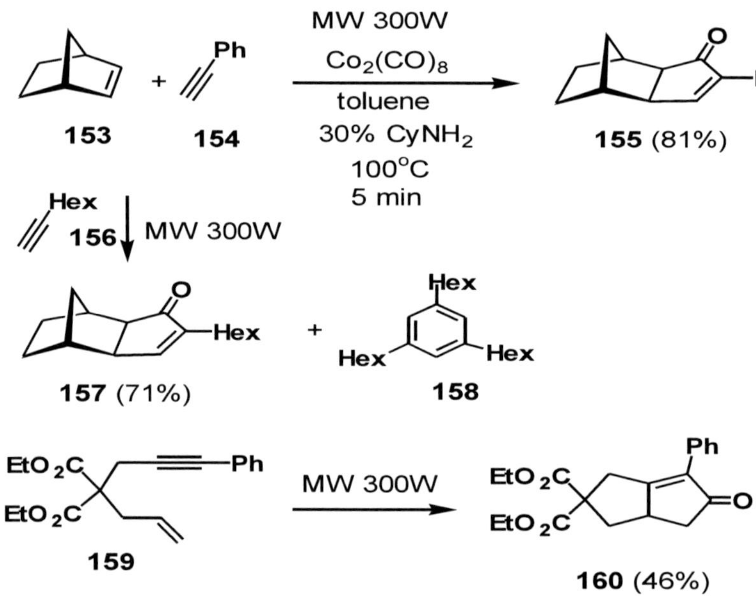

Scheme 31.

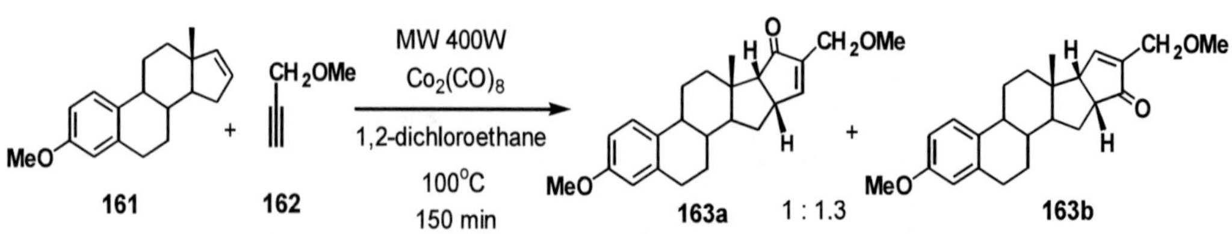

(24 %) single addition (1 equiv)
(33 %) 4x0.25 equiv (4x30 min)
(60 %) 6x0.25 equiv (6x1h)
(28 %) in charcoal 4x0.25 equiv (4x30 min)
(63 %) in charcoal 6x0.25 equiv (6x1h)

Scheme 32.

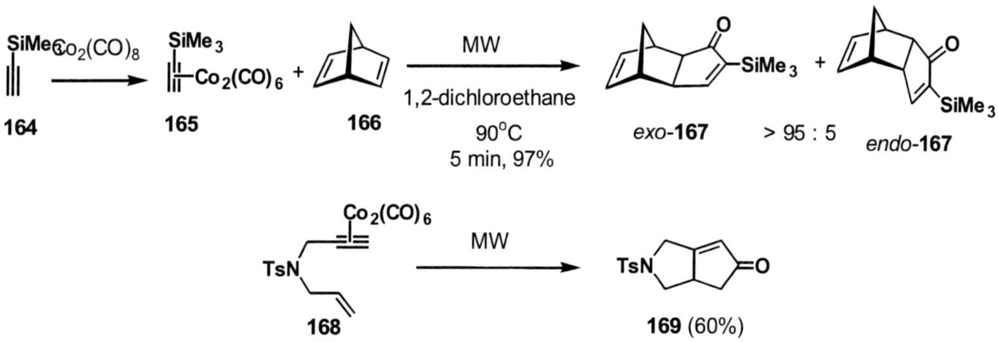

Scheme 33.

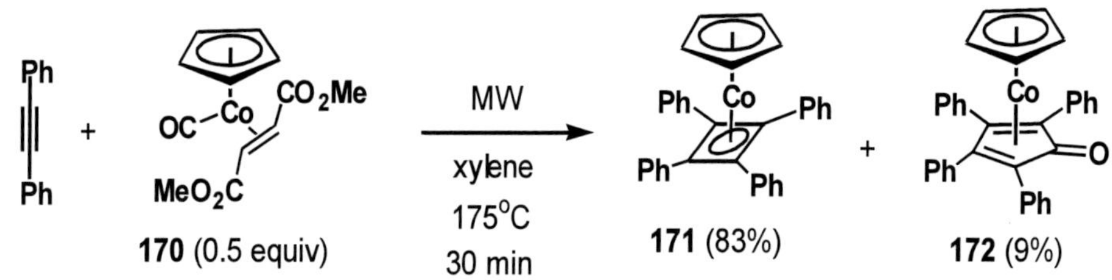

Scheme 34.

The cobalt dimethyl fumarate catalyst *170*, developed by Aubert [53] presents major advantages compared to common catalysts such as [$CpCo(CO)_2$] and [$CpCo(C_2H_4)_2$]. It is air-stable and retains its activity after several months of storage. Using 5 mol% of catalyst *170*, the synthesis of the [CpCo]-complexed cyclobutadiene [2+2] adduct *171* was accomplished under microwave irradiation (Scheme 34). Interestingly, the yield (83%) was much better than that reported with [$CpCo(CO)_2$] (52%), owing to significantly lower contamination by the [2+2+1] adduct *172* (9% vs. 40% with [$CpCo(CO)_2$]).

3.8. [6+3] Cycloadditions

The 1,3-dipolar cycloaddition of adamantylidenefulvene *173* with 2 equiv of nitrile oxide *174* prepared *in situ* in THF solution gave a mixture of several products in total yields of 40-60%. As published by Chung, [54] the reaction took 24 h under reflux condition but was completed within 20 min under microwave irradiation and concurrently with higher yields. The major 1:1 cycloadducts were *175* and *176*, while minor 1:2 cycloadducts are: *177* (15%), *178* (9%), *179* (12%), and *180* (total yields, 50-80%) (Scheme 35). With other aromatic nitrile oxides, only a trace of [6+3] product *180* was observed. The increase of molar ratio between *173* and benzonitrile oxide does not change product distribution, just yield increased slightly at higher concentrations of the nitrile oxide.

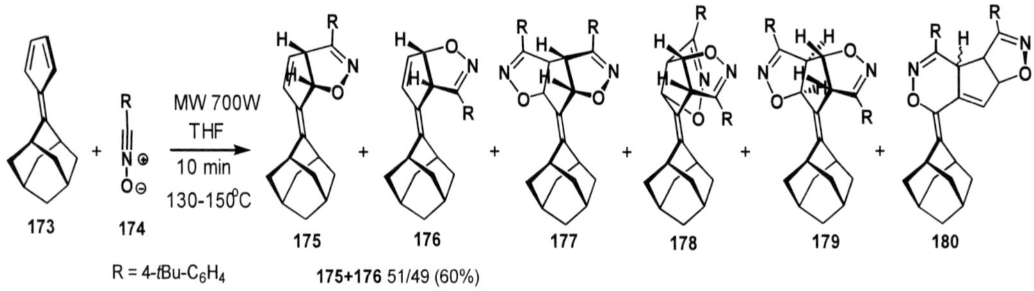

173 174 175 176 177 178 179 180

Scheme 35.

References

[1] Dintzner, M.; Wucka, P.; Lyons, T. W. *J. Chem. Educ. 2006*, *83*, 270-272.

[2] Dintzner, M. R.; Lyons, T. W.; Akroush, M. H.; Wucka, P.; Rzepka, A. T. *Synlett 2005*, 785-788.

[3] Cado, F.; Jacquault, P.; Dozias, M.-J.; Bazureau, J. P. ; Hamelin, J. *J. Chem. Res. (S) 1997*, 176-177.

[4] Northrop, B. H.; O'Malley, D. P.; Zografos, A. L.; Baran, P. S.; Houk, K. N. *Angew. Chem., Int. Ed.* 2006, *45*, 4126-4130.

[5] Shore, G.; Morin, S.; Mallik, D.; Organ, M. G. *Chem. Eur. J. 2008*, *14*, 1351-1356.

[6] Artman III, G. D.; Grubbs, A. W.; Williams, R. M. *J. Am. Chem. Soc. 2007*, *129*, 6336-6342.

[7] Garbacia, S.; Desai, B.; Lavastre, O.; Kappe, C. O. *J. Org. Chem. 2003*, *68*, 9136-9139.

[8] Mayo, K. G.; Nearhoof, E. H.; Kiddle, J. J. *Org. Lett. 2002*, *4*, 1567-1570.

[9] Undheim, K.; Efskind, J. *Tetrahedron 2000*, *56*, 4847-4857.

[10] Spinella, A.; Caruso, T.; Pastore, U.; Ricart, S. J. *Organomet. Chem. 2003*, *68*, 266-268.

[11] Dintzner, M. R.; Lyons, T. W.; Akroush, M, H.; Wucka, P.; Rzepka, A. T. *Synlett 2005*, 785-788.

[12] Alexandre, F.-R.; Berecibar, A., Wrigglesworth, R.; Besson, T. *Tetrahedron 2003*, *59*, 1413-1419.

[13] de Fatima Pereira, M.; Picot, L.; Guillon, J.; Léger, J.-M.; Jarry, C., Thiéry, V.; Besson, T. *Tetrahedron Lett. 2005*, *46*, 3445-3447.

[14] Varma, R. S.; Kumar, D. *Tetrahedron Lett. 1999*, *40*, 7665-7669.

[15] Sharma, U.; Ahmed, S.; Boruah, R. C. *Tetrahedron Lett. 2000*, *41*, 3493-3495.

[16] Frère, S.; Thiéry, V.; Besson, T. *Tetrahedron Lett. 2001*, *42*, 2791-2794.

[17] Song, S. J.; Cho, S. J.; Park, D. K.; Kwon, T. W.; Jenekhe, S. A. *Tetrahedron Lett. 2003*, *44*, 255-257.

[18] Yang, K.; He, X.; Choi, H.; Wang, Z.; Woodmansee, D. H.; Liu, H. *Tetrahedron Lett. 2008*, *49*, 1725-1728.

[19] Longchar, M.; Bora, U.; Boruah, R. C.; Sandhu, J. S. *Synth. Commun. 2002*, *32*, 3611-3616.

[20] Dandia, A.; Singh, R.; Khaturia, S. *J. Fluor. Chem. 2007*, *128*, 524-529.

[21] Dandia, A.; Singh, R.; Sharma, P. *Heteroatom Chem. 2003*, *14*, 468-473.

[22] Kidwai, M.; Venkataramanan, R.; Kohli, S. *Synth. Commun. 2000*, *30*, 989-1002.
[23] Manhas, M. S.; Banik, B. K.; Mathur, A.; Vincent, J. E.; Bose, A. K. *Tetrahedron 2000*, *56*, 5587-5601.
[24] Banik, B. K.; Manhas, M. S.; Robb, E. W.; Bose, A. K. *Heterocycles 1997*, *44*, 405-415.
[25] Liang, Y.; Jiao, L.; Zhang, S.; Xu, J. *J. Org. Chem. 2005*, *70*, 334-337.
[26] Linder, M. R.; Podlech, J. *Org. Lett. 2001*, *3*, 1849-1851.
[27] Margetić, D.; Trošelj, P.; Murata, Y. *Synth. Commun. 2011*, in press.
[28] Johnstone, M. D.; Lowe, A. J.; Henderson, L. C.; Pfeffer, F. M. *Tetrahedron Lett. 2010*, *51*, 5889-5891.
[29] Brummond, K. M.; Chen, D. *Org. Lett. 2005*, *7*, 3473-3475.
[30] Ley, S. V.; Leach, A. G.; Storer, R. I. *J. Chem. Soc. 2001*, 358-361.
[31] Siebert, M. R.; Osbourn, J. M.; Brummond, K. M.; Tantillo, D. J. *J. Am. Chem. Soc. 2010*, *132*, 11952-11966.
[32] Brummond, K. M.; Osbourn, J. M. *Beilst. J. Org. Chem. 2010*, *6(33)*, doi:10.3762/bjoc.6.33
[33] Oh, C. H.; Gupta, A. K.; Park, D. I.; Kim, N. *Chem. Commun. 2005*, 5670-5672.
[34] Nicolaou, K. C.; Lister, T.; Denton, R. M.; Gelin, C. F. *Angew. Chem. Int. Ed. 2007*, *46*, 7501-7505.
[35] Ovaska, T. V.; Kyne, R. E. *Tetrahedron Lett. 2008*, *49*, 376-378.
[36] Uršič, U.; Grošelj, U.; Meden, A.; Svete, J.; Stanovnik, B. *Tetrahedron Lett. 2008*, *49*, 3775-3778.
[37] Fédou, N. M.; Parsons, P. J.; Viseux, E. M. E.; Whittle, A. J. *Org. Lett. 2005*, *7*, 3179-3182.
[38] Yamamoto, Y.; Hayashi, H. *Tetrahedron 2007*, *63*, 10149-10160.
[39] Hong, B.-C.; Jiang, Y.-F.; Kumar, E. S. *Bioorg. Med. Chem. Lett. 2001*, *11*, 1981-1984.
[40] Hong, B.-C.; Shr, Y.-J.; Liao, J.-H. *Org. Lett. 2002*, *4*, 663-666.
[41] Soares, M. I. L.; Pinho e Melo, T. M. V. D. *Tetrahedron Lett. 2008*, *49*, 4889-4893.
[42] Garcia, L.; Pla-Quintana, A.; Roglans, A.; Parella, T. *Eur. J. Org. Chem. 2010*, 3407-3415.
[43] Young, D. D.; Deiters, A. *Angew. Chem. Int. Ed. 2007*, *46*, 5187-5190.
[44] Nicolaus, N.; Strauss, S.; Neudörfl, J.-M.; Prokop, A.; Schmalz, H.-G. *Org. Lett. 2009*, *11*, 341-344.
[45] Efskind, J.; Undheim, K. *Tetrahedron Lett. 2003*, *44*, 2837-2839.
[46] Shanmugasundaram, M.; Aguirre, A. L.; Leyva, M.; Quan, B.; Martinez, L. E. *Tetrahedron Lett. 2007*, *48*, 7698-7701.
[47] Margetić, D.; Briš, A., unpublished results
[48] Lee, H. W.; Lee, L. N.; Chan, A. S. C.; Kwong, F. Y. *Eur. J. Org. Chem. 2008*, 3403-3406.
[49] Lee, H. W.; Kwong, F. Y.; Chan, A. S. C. *Synlett 2008*, 1553-1556.
[50] Fischer, S.; Groth, U.; Jung, M.; Schneider, A. *Synlett 2002*, 2023-2026.
[51] Fager-Jokela, E.; Kaasalainen, E.; Leppänen, K.; Tois, J.; Helaja, J. *Tetrahedron 2008*, *64*, 10381-10387.
[52] Iqbal, M.; Vyse, N.; Dauvergne, J.; Evans, P. *Tetrahedron Lett. 2002*, *43*, 7859-7862.

[53] Geny, A.; Agenet, N.; Iannazzo, L.; Malacria, M.; Aubert, C.; Gandon, V. *Angew. Chem. Int. Ed. 2009*, *48*, 1810-1813.
[54] Chu, J.-H.; Li, W.-S.; Chao, I.; Lee, G.-H.; Chung, W.-S. *Tetrahedron 2006*, *62*, 7380-7389.

Chapter 4

RETROCYCLIZATION REACTIONS

INTRODUCTION

Retro Diels-Alder reactions often require drastic conditions, high temperatures and even sometimes flash vacuum pyrolysis. [1] Microwave irradiation has been used as an alternative to these harsh conditions and significant reductions in reaction times and high yields of the desired products have been observed. Furan and C_{60} cycloadducts were often found to be thermally labile, and several of them was subjected to retro Diels-Alder reaction. However, the employment of microwave irradiation is not always successful to obtain clean retro Diels-Alder products in reasonable yields, since the microwave heating at higher temperatures often leads to further fragmentations. [2] In contrast, thermal 1,3-dipole cycloreversions [3] are much less studied, and examples are limited to the area of fullerenes (retro-Prato reaction).

4.1. RETRO DIELS-ALDER REACTIONS

Shushan [4] has discovered that in the anthracene/maleic anhydride Diels-Alder reaction carried under microwave conditions without the presence of solvent, within only 3 minutes, cycloadduct *3* was afforded in 90% yield, while prolonged heating for 10 minutes reduces reaction yield to 48% (due to retro-Diels-Alder reaction, Scheme 1). Similar results were obtained for the reaction of maleic anhydride and *trans,trans*-1,4-diphenyl-1,3-butadiene *4*. The cycloadduct *5* was obtained within 3 min (85% yield), while in experiment which was carried out for 10 min, yield was only 41%.

Dubac [5] has shown that on graphite support Diels-Alder microwave reactions of anthracene with dimethylacetylene dicarboxylate, diethyl fumarate and maleic anhydride without the use of solvent, afford cycloadducts *8-10* in high yield (Scheme 2). It was observed that these adducts, upon more powerful irradiation, were decomposed by a retro Diels-alder reaction.

Suárez [6] reported that the Diels-Alder reaction of levoglucosenone *11* and the aromatic dienes such as 9-substituted anthracene derivatives *12a-e* yields two cycloadducts *13a-e* and *14a-e* which were identified as the *ortho* and *meta* regioisomers, respectively. In general, the products were obtained in good to excellent yields with high *ortho* selectivity (Scheme 3). Much shorter time was required than conventional reflux in toluene, which was ascribed to microwave irradiation, but also to the increase in concentration of reaction mixture. Cycloaddition of *12a* in various solvents indicated the existence of retro Diels-Alder reaction, in the case when the final temperatures were high, or employing longer reaction times the *ortho/meta* ratio decreases. In the separate experiment, it was shown that upon microwave heating (using longer reaction times, final temperature 165°C), *13a* produces a mixture of *11*, *12a* and *13a*. The *ortho* isomer *13a* undergoes a retro Diels-Alder reaction leading to the formation of the thermodynamically more stable *meta* isomer *14a*.

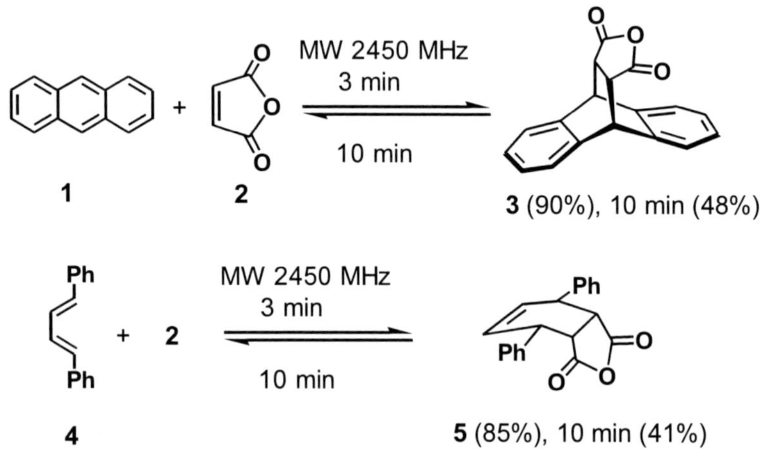

Scheme 1.

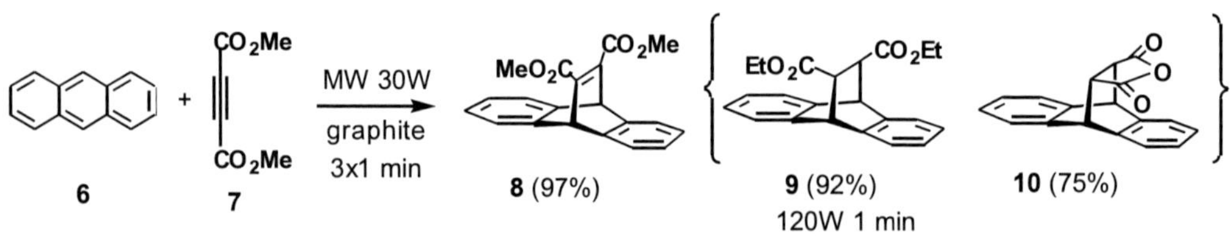

Scheme 2.

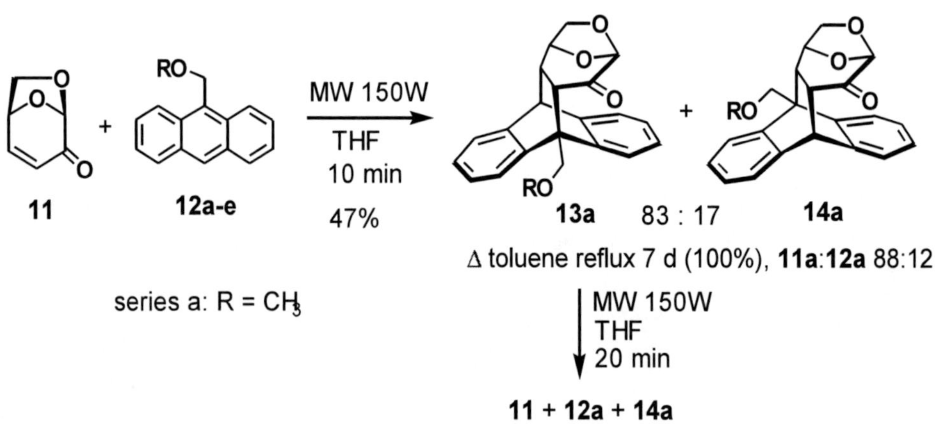

Scheme 3.

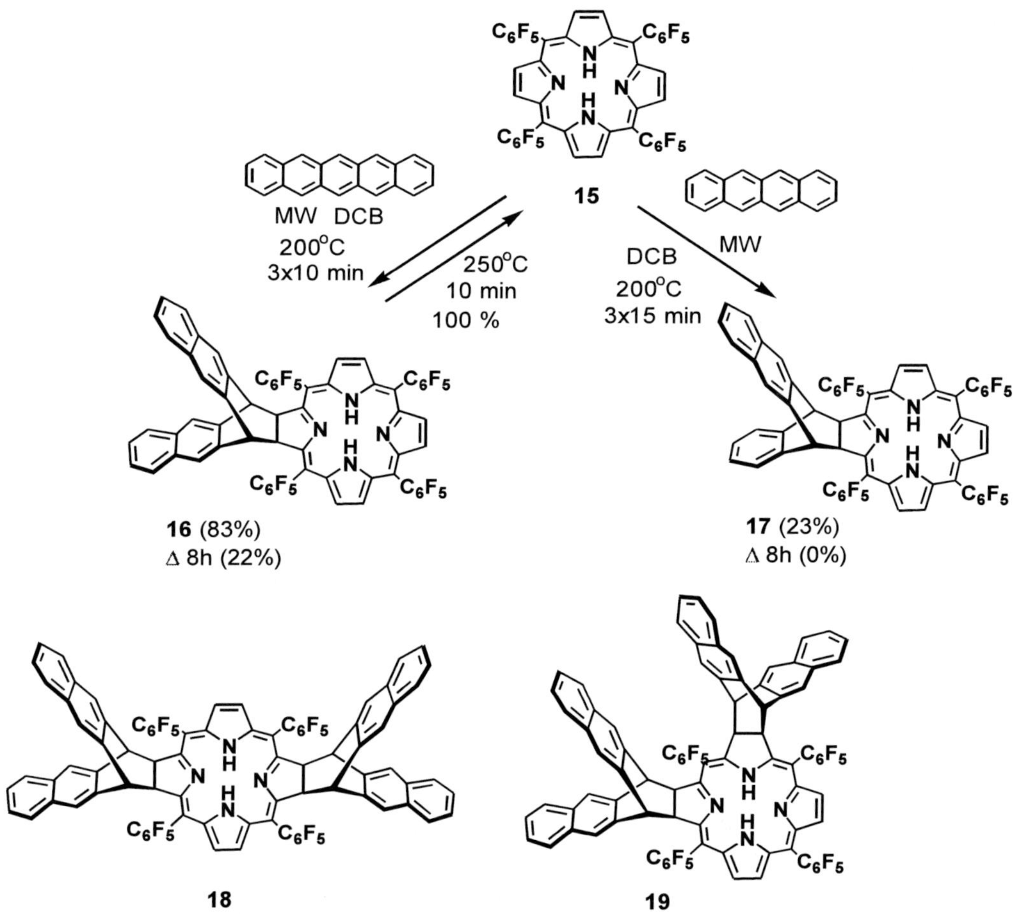

Scheme 4.

Cavaleiro and Kappe [7] have shown that the use of microwave technique results in an impressive improvement in the yield of the Diels-Alder reactions between *meso*-tetrakis (pentafluorophenyl) porphyrin *15* and pentacene or naphthacene (Scheme 4). The highest yield of chlorin *16* was obtained when the reaction was performed by adding the pentacene in small portions (3x1 equiv), followed by irradiation. In this case the chlorin *16* was isolated in 83% yield (bisadducts were also formed). Under classical heating conditions *16* was obtained in very low yield, while with naphthacene no reaction is observed. It was found that the retro-Diels-Alder process decreases the yield at microwave temperatures above 200°C. When *16* was submitted to microwave heating at 200°C for 10 min a small amount of the starting porphyrin *15* was formed, while irradiation at 250°C for 10 min leads to the complete regeneration of *15*. By increase of concentration of reaction mixture from 0.1 and 0.2 M solutions of *15* in DCB, and irradiation the resulting mixtures with microwaves at 200°C for 20 min, *16* was obtained in 66% and 77% yield, respectively. Furthermore, for both dienes, a small amount of bisadducts *18* and *19* was also isolated; this contrasts with the results obtained by conventional heating where no bisaddition was observed.

Bortolussi [8] employed microwave irradiation for retro Diels-Alder synthesis of aminoalcohols. Quantitative yields were obtained for MW-assisted thermolysis of neat dibenzylamino alcohols *21a-d* (Scheme 5, Table 1). Reactions were completed without the presence of solvent, and in significantly shorter reaction times than classical thermolysis in xylene. These conditions were highly advantageous, considering that both classical thermolysis and FVP (leading to decomposition) are poorly productive (or non-operative).

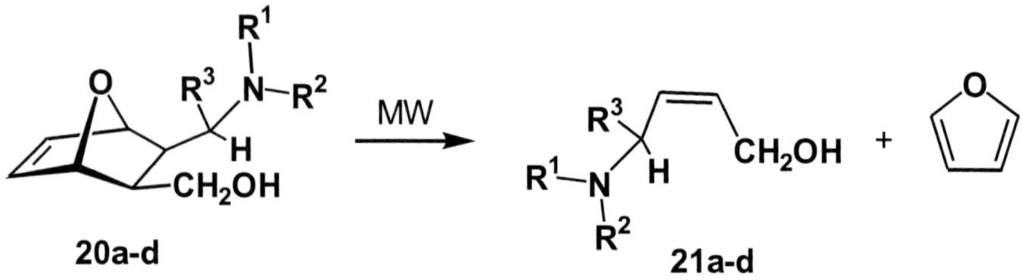

Scheme 5.

Table 1.

				Microwave reaction			Classical heating		
Subst-rate	R^1	R^2	R^3	T (°C)	Time (min)	yield (%)	T (°C)	Time (min)	yield (%)
1a	Bn	Bn	*n*-Bu	120	15	66	140	180	70
				140	10	≥98	140	480	84
1b	Bn	Bn	$(CH_2)_3OBn$	140	10	≥98			
1c	H	Bn	*n*-Bu	120	15	≥98	160	360	74
1d	H	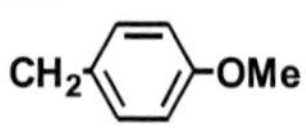	$(CH_2)_3OBn$	120	15	≥98			
1e	H	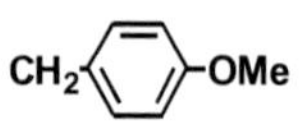	*n*-Bu	100	15	≥65	120	180	50
				120	15	≥98	120	600	84

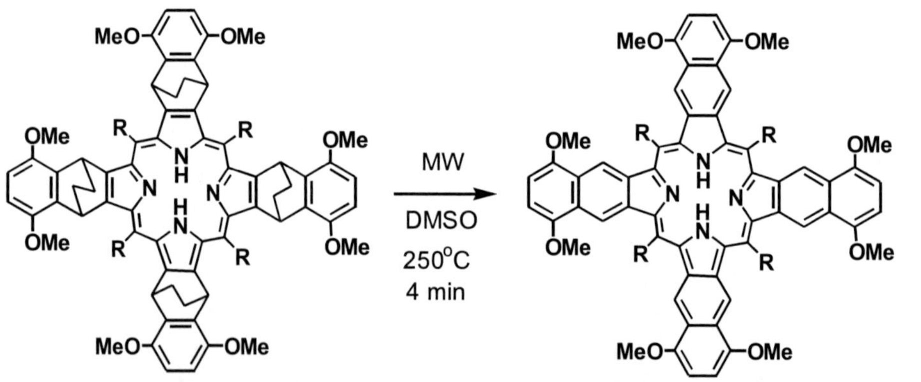

22a R=H
22b R=2,4,6-$(CH_3)_3$-C_6H_2

23 (quant.), Δ 4 min, 250°C (25%)

Scheme 6.

Tietze [9] has shown that elimination of ethylene by retro Diels-Alder process could be facilitated by microwave irradiation. Porphyrins *22a,b* in DMSO were subjected to microwave irradiation for 4 min at 250 °C to give a quantitative yield of porphyrins *23a,b*

(Scheme 6). Under conventional heating (sand bath) for the same period of time and at the same temperature, the annealed porphyrin *23* was obtained in a very low yield (25%).

Solvent free Diels-Alder reactions with microwave irradiation of reactants supported on graphite were dramatically accelerated. The application of these reaction conditions by Garrigues [10] is illustrated in Scheme 7. Firstly, retro-Diels-Alder reaction of cyclopentadiene dimer *24* at high temperature led to the formation of cyclopentadiene, which subsequently reacted with methyl or ethyl maleate *25,26* to afford *exo/endo* substituted norbornenes *27* and *28*. These cycloadditions were in some cases conducted in the presence of Lewis acid, such as reaction of methyl acrylate with isoprene.

Retro Diels-Alder / hetero Diels-Alder sequence of reactions was also carried out in solid phase using cycloadducts attached to resin. Miller [11] showed that the nitroso adduct resins *32a* and *32b*, derived from α-terpinene, were exposed to various dienes in a microwave reactor, then products *35* were cleaved from the resin with TBAF (Scheme 8). While the nitrophenyl adduct *32a* provided quantitative retro DA adducts, the pyridyl derivative *32b* was less reactive under the same conditions.

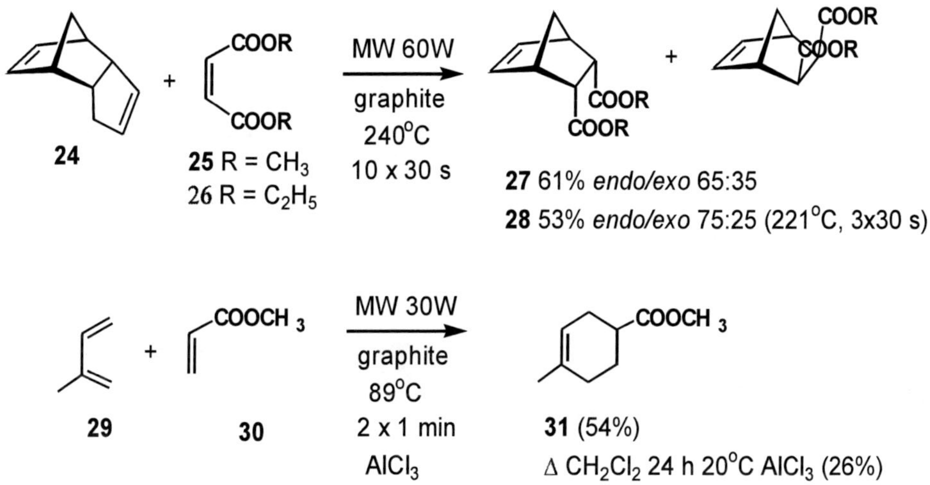

Scheme 7.

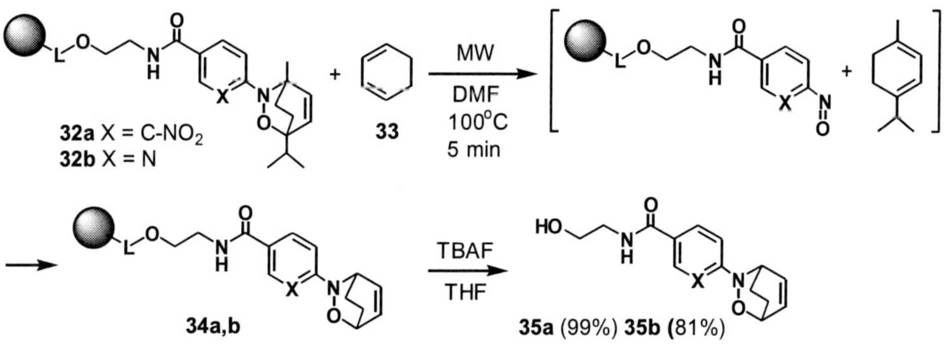

Scheme 8.

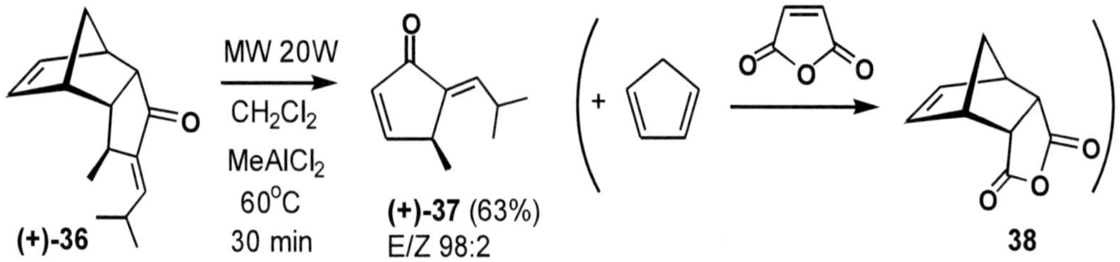

Scheme 9.

Considerable enhancement of the Lewis acid-catalysed retro Diels-Alder reaction rate is possible for the *exo*-cyclic enones such as *36*, when the [4+2]-cycloreversion step was carried out in a microwave reactor, as reported by Eddolls. [12] Previous reports on Lewis acid-catalysed retro Diels-Alder reactions involving *exo*-cyclic enones similar to *36* required reaction times of between 1 and 24 h. For instance, microwave-mediated reaction of *exo*-cyclic enone *36* in employing $MeAlCl_2$ as Lewis acid catalyst and an excess of maleic anhydride as a cyclopentadiene trap produced in short time product *37* (Scheme 9). Continuation of irradiation for longer periods eroded the yields of *37* which were generally very respectable.

Cascade Diels-Alder/retro-Diels-Alder reaction was used in the synthesis of substituted aniline derivatives *43* by Kranjc. [13] Synthesis starts with Diels-Alder reaction of ethyl vinyl ether *40* (acetylene equivalent) with 2*H*-pyran-2-ones *39*, followed by the retro-Diels-Alder elimination of CO_2 from the initially formed cycloadduct *41* and subsequent aromatization (Scheme 10). Microwave heating also promotes elimination of ethanol and aromatization. Neutral organic bases such as DABCO (10 mol%) were most suitable to catalyse reaction of *39* with *40*.

Fülöp [14] has discovered that temperatures required for the cyclopentadiene elimination from cycloadducts by microwave irradiation are above 250°C. Pure diastereoisomers, pyrrolopyrimidine tetracycles (+)-*44a,* (+)-*44b* and (+)-*45b* were subjected to microwave heating, resulting in the enantiomeric pyrroloquinazolinone (-)-*46b* from (+)-*44b* and its counterpart (+)-*46b* from (+)-*45b*, respectively, in yields of over 90% and with ee >99% (Scheme 11).

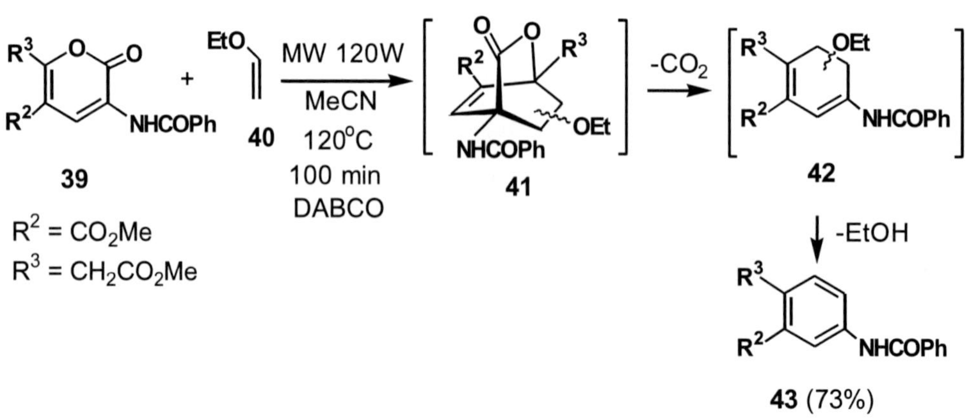

Scheme 10.

The first example of intramolecular and also retro-Diels-Alder reactions conducted in microwave conditions was published by Takano. [15] Racemization of (-)-enantiomer of vincamine indole alkaloid *47* effectively takes place in DMF using a microwave oven, via concurrent Diels-Alder cycloreversion and intramolecular cycloaddition (Scheme 12). Thus, single step reaction by heating of optically active vincadifformine (-)-*47* in domestic microwave oven) obtained racemic vincadifformine (±)-*47* in 92% yield. While the (-)-form of vincadifformine alkaloid is unnecessary product in the synthesis, its (+)-form shows significant biological activity, which could not be obtained in sufficient quantitites from natural sources. Solvent choice was found to be the most important in this process, since the solvents other than DMF yielded much less product. This is also first example of cycloaddition reaction conducted in water in microwave conditions, [16] however in lower yield.

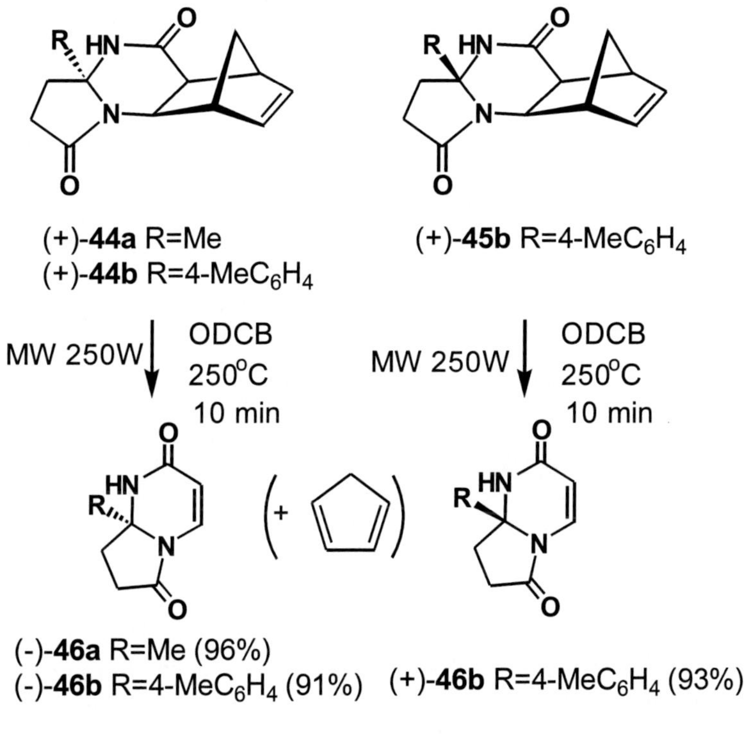

Scheme 11.

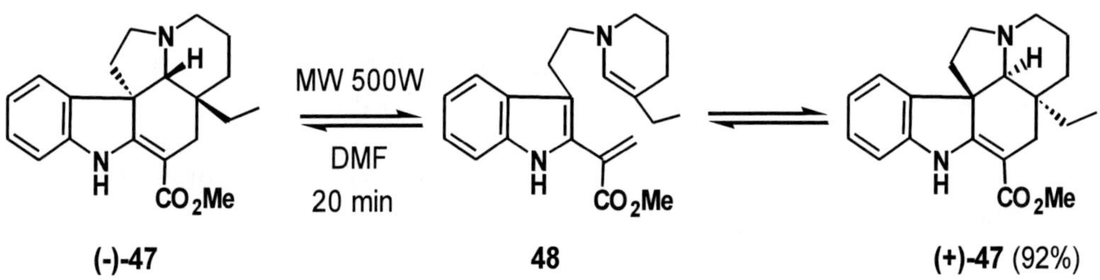

Scheme 12.

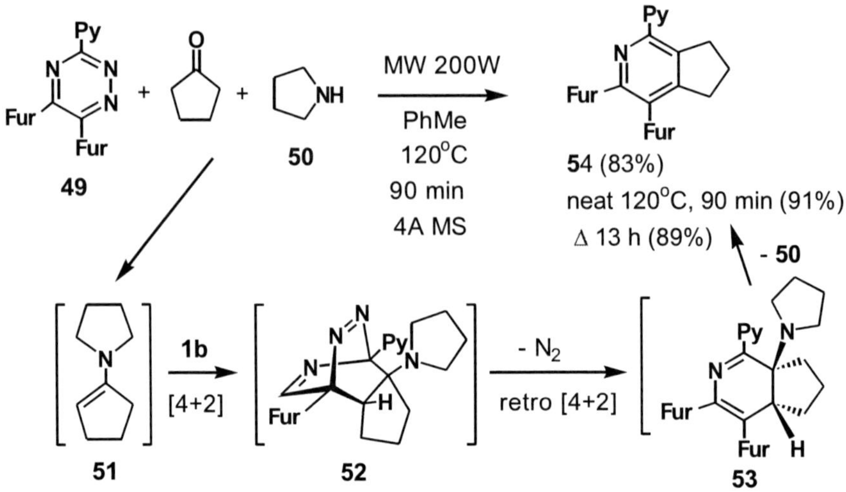

Scheme 13.

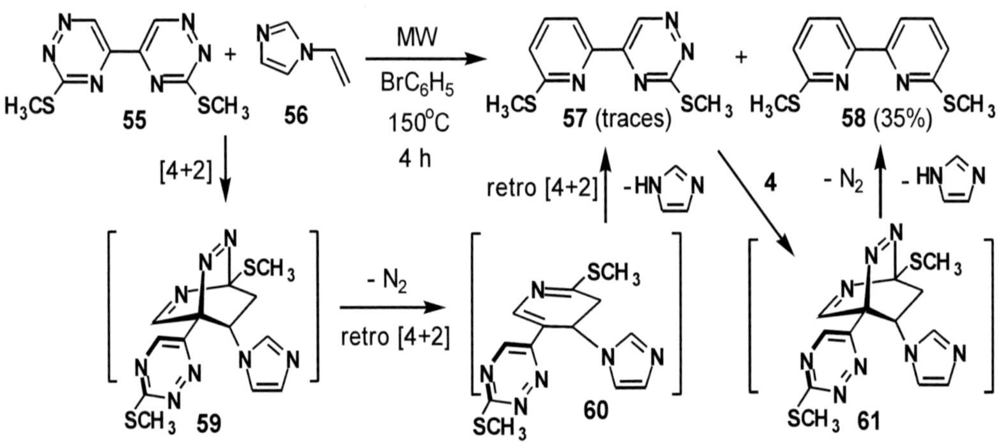

Scheme 14.

The synthesis of heterocyclic and heteroaromatic compounds using heterocyclic azadienes in Diels-Alder reactions, followed by extrusion of part of the resulting adduct in a retro-[4+2]-reaction is another example of retro-Diels-Alder reaction facilitated by microwaves. This methodology was used by Taylor [17] in the synthesis of highly substituted pyridines from 1,2,4-triazines via solvent-free microwave synthesis. When 5,6-difuran-2-yl-3-pyridin-2-yl-[1,2,4]triazine *49* was heated with cyclopentanone and pyrrolidine in microwave reactor, pyridine *54* was obtained in a yield of 83% (in toluene under reflux the yield was 89% after 13 h at reflux). In a solvent-free environment, this reaction is even more effective, and these optimum conditions were employed in the reactions with number of ketones such as cyclopentanone (Scheme 13). Mechanistically, pyrrolidine *50* with cyclopentanone affords a dienophilic precursor *51*, which acts as dienophile in an inverse electron demand hetero Diels-Alder reaction with 1,2,4-triazine *49* across C3/C6 position. Resulting bicyclo[2.2.2] adduct *52* in a facile retro Diels-Alder process looses N_2, to form dihydropyridine intermediate *53*. Subsequent elimination of pyrrolidine and aromatization affords pyridine *54*, which is the problematic elimination step greatly facilitated by microwaves.

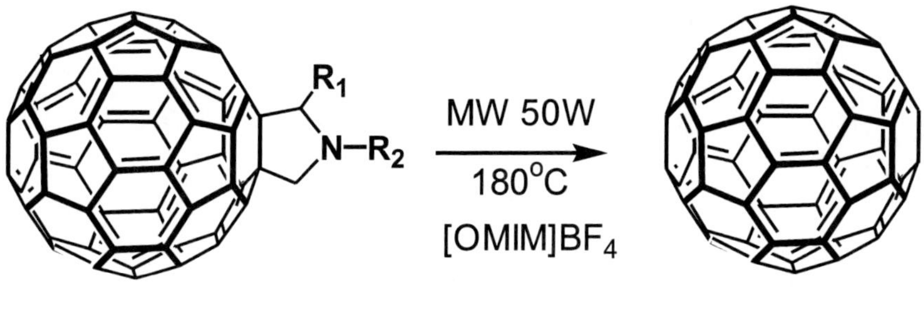

62

a R_1= H , R_2 = CH_3 >99%, 10 min
b R_1= C_6H_{13} , R_2 = CH_3 >99%, 5 min
c R_1= *p*-CH_3O-C_6H_4 , R_2 = CH_3 >99%, 5min
d R_1= 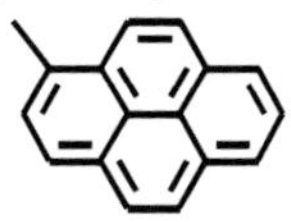 R_2 = CH_3 >45%, 10 min
e R_1= $COOC_2H_5$, R_2 = CH_3 >99%, 5 min
f R_1= 4-pyridyl , R_2 = CH_3 >99%, 5 min
g R_1= H , R_2 = *p*-OH-C_6H_4 >99%, 5 min

Scheme 15.

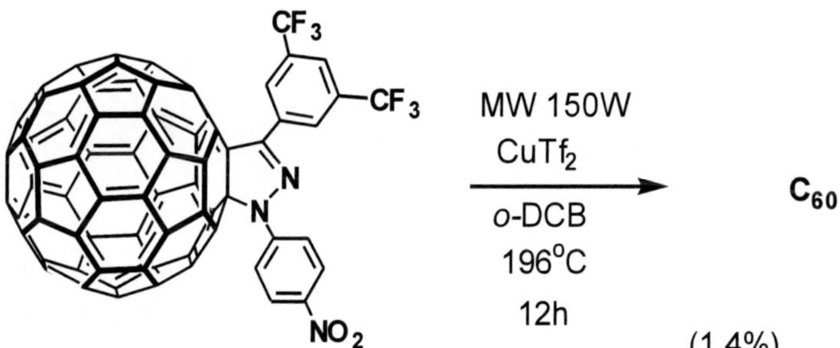

63

(1.4%)
Δ 24 h (1%)
1,2,4-$Cl_3C_6H_3$ 200W, 12h (5%)

Scheme 16.

Hetero Diels-Alder / retro-Diels-Alder sequence is also the key reaction step in cycloadditions of 3,3'-bis(methylsulfanyl)-5,5'-bi-1,2,4-triazine *55*. Branowska [18] reported that *55* undergoes the regiospecific inverse electron demand hetero Diels-Alder reaction with 1-vinylimidazole *56* as dienophile in boiling bromobenzene to give a product *57* together with traces of 6,6'-bis(methylsulfanyl)-2,2'-bipyridine *58* (Scheme 14). The reaction took a different course during microwave irradiation of the reaction mixture. Under these conditions the proportion of bipyridine compound *58* increased considerably to 35% yield when the reaction mixture was irradiated for 4 hrs at 150°C. This process employs tandem Diels-Alder reaction which affords cycloadduct *59*, followed by retro- Diels-Alder reaction (elimination of nitrogen) with formation of *60*, and subsequent elimination of imidazole to restore the aromaticity of pyridine ring. Second equivalent of *56* reacts with *57* in an analogous manner via intermediate *61* to obtain bipyridine *58*.

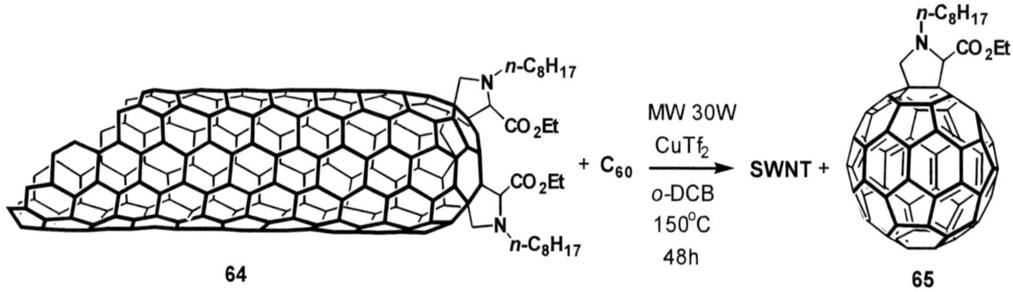

Scheme 17.

4.2. 1,3-DIPOLE CYCLOREVERSIONS

The study of thermal stability of fulleropyrrolidines (pyrrolidino[3,4:1,2,60]fullerenes) by Maggini and Prato [19] has shown that 1,3-dipole cycloreversion (retro-Prato Reaction) of fulleropyrrolidines could be easily achieved in minutes, under microwave irradiation. The interplay of microwave irradiation and ionic liquids offers the unique opportunity of very efficient flash-thermal activation, in conjunction with a strong stabilization of polar/ionic intermediates/reactants. By using 1-methyl-3-*n*-octyl imidazolium tetrafluoroborate ionic liquid as solvent, and by applying microwave irradiation, fulleropyrrolidines *62a-g* undergo quantitative 1,3-dipole cycloreversion to C_{60} (Scheme 15). Amongst the other tested ionic liquids, in 1-methyl-3-n-butyl imidazolium tetrachloroaluminate ([BMIM]$AlCl_4$) quantitative cycloreversion was also achieved.

Langa and Martín [20] carried out assays on the thermal stability of *C*-aryl-*N*-aryl-2-pyrazolino[4,5:1,2][60]fullerenes. Under microwave irradiation as the energy source instead of classical heating, mixture of pyrazoline *63*, copper(II) triflate, maleic anhydride in *o*-DCB showed that the rate of retro-cycloaddition reaction is very similar (1.4%) to that observed by classical heating (1%) (Scheme 16). With the use a higher boiling point solvent, 1,2,4-trichlorobenzene (temperature profile of 204 °C; b.p. 214 °C), even after 12 h of irradiation at 200 W, only 5% of [60]fullerene was formed, indicating that these *C*-aryl-*N*-aryl-2-pyrazolino[60]fullerenes are highly stable fullerene derivatives under thermal conditions.

Finally, Prato [21] showed that microwave-assisted cycloadditions of aziridines to carbon nanotubes are reversible processes. When mixture of functionalized SWNTs *64* was heated with $CuTf_2$ as catalyst in the presence of fullerene in order to trap the corresponding azomethine ylide that results from the thermal retro-cycloaddition, functional groups were lost from *64*, and fullerene monoadduct *65* was obtained (Scheme 17).

REFERENCES

[1] McNab, H. *Aldrichim. Acta* 2004, *37*, 19-26.

[2] Margetić, D.; Butler, D. N.; Warrener, R. N.; Murata, Y. *Tetrahedron* 2011, *67*, 1580-1588.

[3] Filippone, S.; Barroso, M. I.; Martín-Domenech, Á.; Osuna, S.; Solà, M.; Martín, N. *Chem. Eur. J.* 2008, *14*, 5198-5206.

[4] Rongshun, Z.; Pinjie, H.; Shushan, D. *Synth. Commun.* 1994, *24*, 2417-2421.

[5] Garrigues, B.; Laporte, C.; Laurent, R.; Laporterie, A.; Dubac, J. *Liebigs Ann. Chem.* 1996, 739-741.

[6] Sarotti, A. M.; Joullié, M. M.; Spanevello, R. A.; Suárez, A. G. *Org. Lett.* 2006, *8*, 5561-5564.

[7] Silva, A. M. G.; Tomé, A. C.; Maria G. P. M. S. Neves; Cavaleiro, J. A. S.; Kappe, C. O. *Tetrahedron Lett.* 2005, *46*, 4723-4726.

[8] Bortolussi, M.; Bloch, R.; Lopy, A. *J. Chem. Res. (Synop.)* 1998, 34-35.

[9] Elghamry, I.; Tietze, L. F. *Tetrahedron Lett.* 2008, 49, 3972-3975.

[10] Laporte, C.; Oussaid, A.; Garrigues, B. *C. R. Acad. Sci. Paris, Ser. IIc* 2000, *3*, 321-325.

[11] Yang, B.; Lin, W.; Krchnak, V.; Miller, M. J. *Tetrahedron Lett.* 2009, *50*, 5879-5883.

[12] Eddolls, J. P.; Iqbal, M.; Roberts, S. M.; Santoro, M. G. *Tetrahedron* 2004, *60*, 2539-2550.

[13] Kranjc, K.; Kočevar, M. *Synlett* 2008, 2613-2614.

[14] Fülöp, F.; Miklós, F.; Forró, E. *Synlett* 2008, 1687-1689.

[15] Takano, S.; Kijima, A.; Sugihara, T.; Satoh, S.; Ogasawara, K. *Chem. Lett.* 1989, 87-88.

[16] Review on microwave reactions in water: Dallinger, D.; Kappe, C. O. *Chem. Rev.* 2007, *107*, 2563-2591.

[17] Fernández Sainz, Y.; Raw, S. A.; Taylor, R. J. K. *J. Org. Chem.* 2005, *70*, 10086-10095.

[18] Branowska, D. *Molecules* 2005, *10*, 274-278.

[19] Guryanov, I.; Montellano López, A.; Carraro, M.; Da Ros, T.; Scorrano, G.; Maggini, M.; Prato, M.; Bonchio, M. *Chem. Commun.*, 2009, 3940–3942.

[20] Delgado, J. L.; Oswald, F.; Cardinali, F.; Langa, F.; Martín, N. *J. Org. Chem.* 2008, *73*, 3184-3188.

[21] Brunetti, F. G.; Antonia Herrero, M.; dc M. Muñoz, J.; Giordani, S.; Díaz-Ortiz, A.; Filippone, S.; Ruaro, G.; Meneghetti, M.; Prato, M.; Vázquez, E. *J. Am. Chem. Soc.* 2007, *129*, 14580-14581.

Chapter 5

Applications of Microwave Assisted Cycloadditions in Material Sciences

Introduction

A wide variety of cycloadded fullerene materials was synthetized due to the remarkable property of fullerenes to function either as 2π electron-deficient dienophiles or dipolarophiles. [1] Since chemical functionalisation of fullerenes and nanotubes became important in material sciences, [2] a number of novel fullerene cycloadducts was prepared with the aid of microwave heating, in superior yields than prepared by conventional heating. For the application of microwave irradiation in fullerene chemistry, Diels-Alder and mostly 1,3-dipolar cycloadditions were employed. [3] The [4+2] cycloadditions selectively afford the adducts on the fullerene 6,6-ring junctions. In some cases the cycloaddition products undergo facile retro-Diels-Alder reaction as a consequence of their low thermodynamic stability, or in some cases, multiply-substituted cycloadducts were obtained. Chemical functionalization of single-walled nanotubes is research that has attracted considerable attention over the past decade since the remarkable physical, electronic, and chemical properties of carbon nanotubes (NTs) for applications in many fields. In the field of carbon nanotube functionalization, several microwave-assisted reactions have been reported.

5.1. Fullerenes

5.1.1. Diels-Alder Reactions

Wudl [4] published the first report on the use of microwave irradiation for functionalization of C_{60}. Diels-Alder reactions of C_{60} with *o*-quinodimethanes *2a,b* (generated *in situ* from sultines *1a,b*) produced functionalized fullerenes *3a,b* (Scheme 1). When the time increases, reaction yield decreases, due to the formation of multiple adducts. The cycloadduct *3b* obtained from the *p*-dihexyloxy substituted sultine *1b* showed much better solubility than unsubstituted cycloadduct, which may be reason of its formation in better yield.

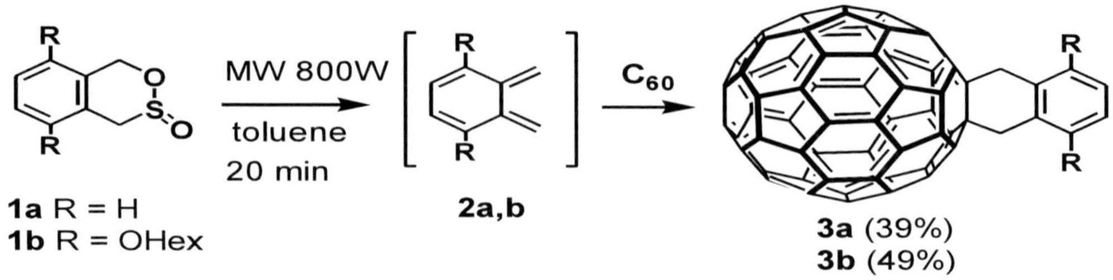

Scheme 1.

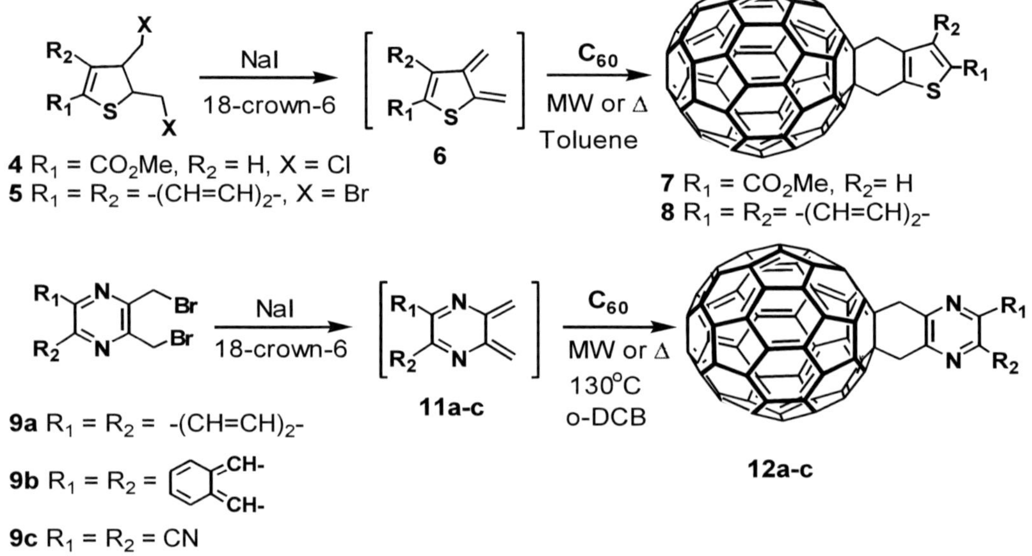

Scheme 2.

Table 1. Comparison of cycloadditions using classical and microwave procedures

Product	microwave[a]		classical heating		
	time (h)	yield (%)	T (°C)	time (h)	yield (%)
7	1.25	12	toluene reflux	72	25
8	1.16 [b]	23	toluene reflux	17	43
11a	0.5	26	ODCB reflux	24	13
11b	0.5	22	ODCB reflux	24	5
11c	0.33	9	ODCB reflux	5	16

[a] power 60 W; [b] power 70 W

Martín [5] synthetised Diels-Alder adducts of fullerene with heteroaromatic analogues of *o*-quinodimethane with the assistance of microwave irradiation (Scheme 2). The C_{60}-pyrazine and thiophene containing adducts were readily prepared, thus providing an expeditious access to [6,6]-closed fullerene derivatives covalently attached to different heterocycles. Thiophene containing derivatives *7-8* were synthetised from *o*-quinodimethane *6* prepared *in situ* from their respective precursors. Hence, treatment of methyl 2,3-bis(chloromethyl)thiophene-5-

carboxylate *4* with tetrabutylammonium bromide in toluene afforded *7* in 15% yield. Instead of TBAB, *4* could be treated with KI/18-crown-6 to afford 16% yield (at 105W, 75 min.). In an analogous manner, reaction of C_{60} with 2,3-bis(bromomethyl)benzo[*b*]-thiophene *5* under microwaves (70W, 70 min) yielded cycloadduct *8* in 23%. Similarly, 2,3-pyrazinoquinodimethane derivatives *10a-c* were generated *in situ* by treatment of 2,3-bis(bromomethyl)pyrazine derivatives *9a-c* with sodium iodide in *o*-dichlorobenzene at 130 °C. Reactive intermediates *10a-c* were easily trapped as the respective Diels-Alder adducts *11a-c* by reaction with [60]fullerene acting as the dienophile (Table 1). Longer reaction times result in lower yields due to the formation of higher adducts.

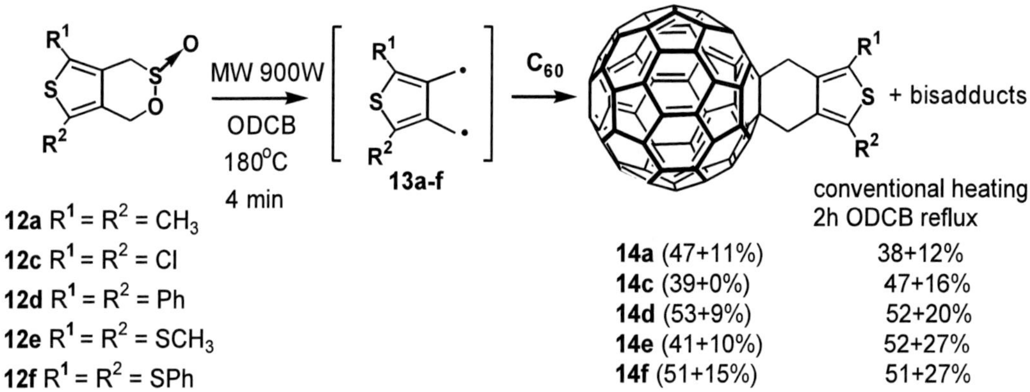

Scheme 3.

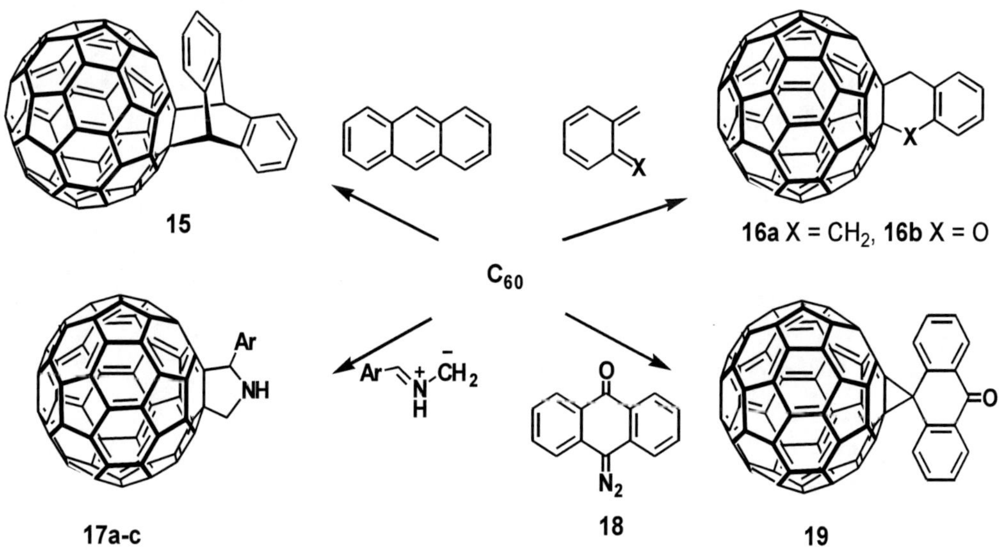

Scheme 4.

Differently oriented [60]fullerene derivatives *14a-f* containing thiophene ring were sythetized by Chung [6] using thienosultines *12a-f* as precursors for the corresponding *ortho*-quinodimethanes (Scheme 3). The reaction was markedly accelerated by microwave irradiation and gave comparable yields to the thermal reaction, but more important is the observation that the ratio of monoadduct *13a-f* to bisadduct was increased when microwave irradiation was applied.

Table 2. Comparison of results obtained in cycloadditions to C_{60} using microwave and classical heating

Product	Microwave irradiation				Conventional heating		
	Solvent	Power (W)	Time (min)	Yield (%)	Solvent/T (°C)	Time (h)	Yield (%)
15	Toluene	800 [a]	15	35	Benzene/reflux	12	25
16a	Benzene	800 [a]	20	49	Benzene/reflux	18	30
16b	ODCB	800 [a]	4	27	ODCB/180	0.5	31
16b	ODCB	400 [a]	8	21			
17a	ODCB	60 [b]	120	37	Toluene/reflux	'several'	35
17b	ODCB	60 [b]	60	30			
17c	ODCB	70 [b]	70	15			
19	ODCB	800 [a]	60	25	ODCB/69-70	16	46

[a] Modified domestic microwave oven perforated at the top with inserted reflux condenser, and the turnable dish replaced by magnetic stirrer; [b] focused microwave reactor

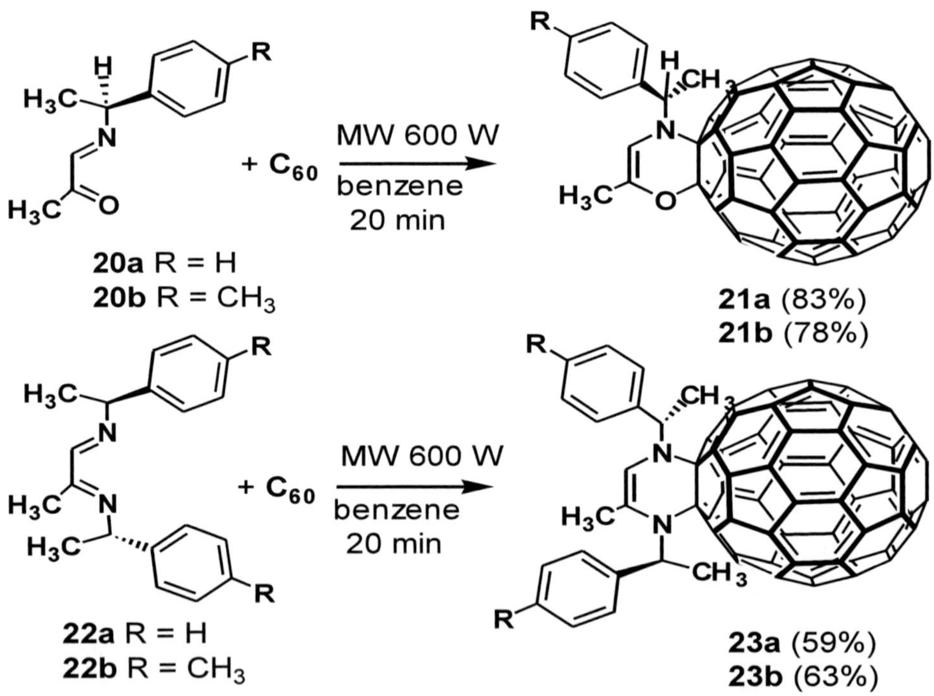

Scheme 5.

Martín and Langa [7] in their study of Diels-Alder and 1,3-dipolar cycloadditions to C_{60} reported that microwave heating is advantageous since in these conditions higher cycloadducts were not formed. Diels-Alder reactions with anthracene, o-quinodimethane (generated *in situ* from sultine, as described in their initial report shown in Scheme 1[4]) gave products *15*, *16a*. Furthermore, *o*-quinone methide (generated *in situ* from o-hydroxybenzyl alcohol), undergoes hetero Diels-Alder reaction affording cycloadduct *16b*, while 1,3-dipolar reaction with azomethine ylides (Prato reaction) forms fulleropyrrolidines *17a-c* and with diazoanthrone *18* produces *19* (Scheme 4, Table 2). [8]

The reaction of C_{60} with anthracene facilitated by microwave irradiation improved the yield of cycloadduct *15* (as compared to the conventional heating: 25%, reflux benzene 12 h), and at the same time avoided the formation of multiple adducts.

Gutiérrez [9] described microwave-assisted synthesis of new adducts from the hetero Diels-Alder [4+2]-cycloaddition reaction of chiral α-oxo imines *20a,b* and α-diimines *22a,b* with fullerene C_{60}, starting from (-)-(*S*)-1-phenylethylamine or (-)-(*S*)-1-(4-methylphenyl)ethylamine and pyruvaldehyde (Scheme 5). Upon reaction of C_{60} with pure α-oxo imines *20a,b* under microwave conditions, cycloaddition products *21a,b* were formed in 83%, and 78% yield (for *21a* and *21b*, respectively). Reactions of α-diimines *22a,b* gave adducts *23a,b* in lower yields. In conventional reactions, these heterodienes require longer heating (72 hours) and obtained yields are considerably lower.

5.1.2. 1,3-dipolar Reactions

The first account on the use of microwave heating in 1,3-dipolar cycloaddition to fullerenes was published by Langa. [10] The C_{60}-donor dyads were constructed by reaction of pyrazolylhydrazones with fullerene under microwave conditions. When hydrazones *24a-c* were reacted with C_{60} in o-dichlorobenzene or trichlorobenzene under microwave irradiation, (after an efficient hydrazone-azomethine imine tautomerization) adduct *25a* was formed in 6% yield (trichlorobenzene, 210°C, 120 min), while only traces of *25b* were detected and *2c* was not formed at all. Alternative route, nitrile amines were generated *in situ* from the corresponding hydrazones *24a-c* and NBS in the presence of Et_3N and reacted with C_{60} under microwave irradiation (Scheme 6). By classical heating used as an alternative source of energy yields are considerably lower (*25a*: 9%, *25b*: 21%, *25c*: 4%), even when longer reaction times (4-8 h) were employed. Similarly, hydrazones *26a-c* were reacted with NBS in benzene, followed by the addition of Et_3N and C_{60} and irradiated in a microwave reactor to afford pyrazolylpyrazolino[60] fullerenes *27a-c* in moderate to good yields (22-44%). Longer reaction times decreased the yield and promote the formation of bisadducts (Scheme 6). [11]

1,3-dipolar cycloadditions of pyrazole nitrile oxides to C_{60} was used by Langa and Martín [12] in the synthesis of electroactive 3'-(*N*-phenylpyrazolyl)isoxazoline[4',5':1,2][60]fullerene dyads. Nitrile oxides *30* were generated *in situ* from corresponding oximes *28* and *29* with *N*-bromosuccinimide in the presence of triethylamine and subsequently reacted with C_{60} in toluene, to obtain adducts *31* and *32* in 22% yield (Scheme 7). Longer reaction times afforded greater amounts of bisadducts. In thermal conditions, similar yields were obtained (14 and 17%, respectively), with significant accelerating effect (10 minutes vs. 24 hours).

Employing these reaction conditions, Langa [13] prepared a library of functionalized isoxazolo[60]fullerenes *35a-h* in one pot from aldoximes under microwave irradiation and C_{60}. Generation of nitrile oxides involved the use of hydroxymoyl chlorides or nitroalkanes as starting materials, then corresponding oxime *33a-h* was treated with NBS or NCS in the presence of Et_3N (Scheme 8). The resulting nitrile oxides *34a-h* were reacted *in situ* with C_{60} under microwave irradiation, to afford adducts *35a-h* in moderate yields (16-36%).

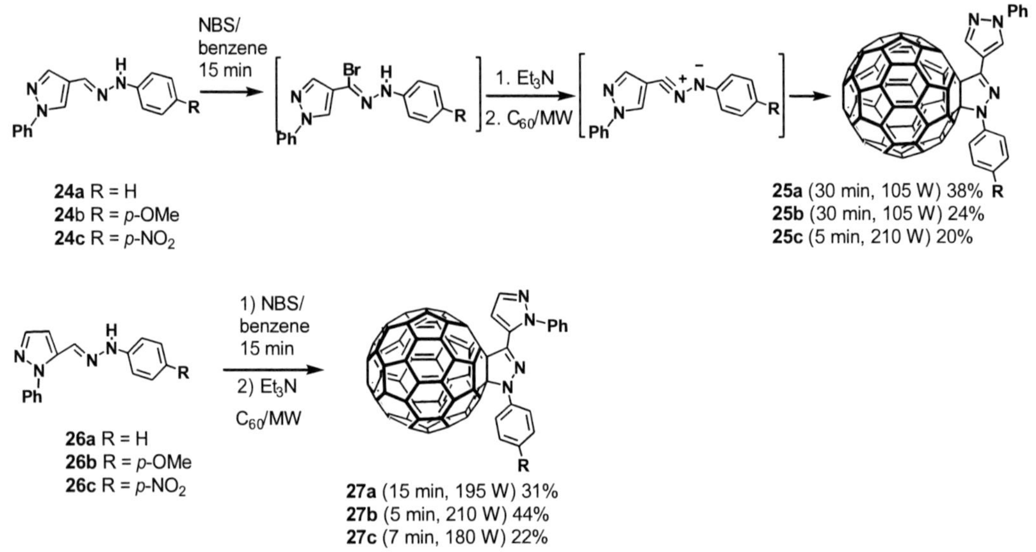

Scheme 6.

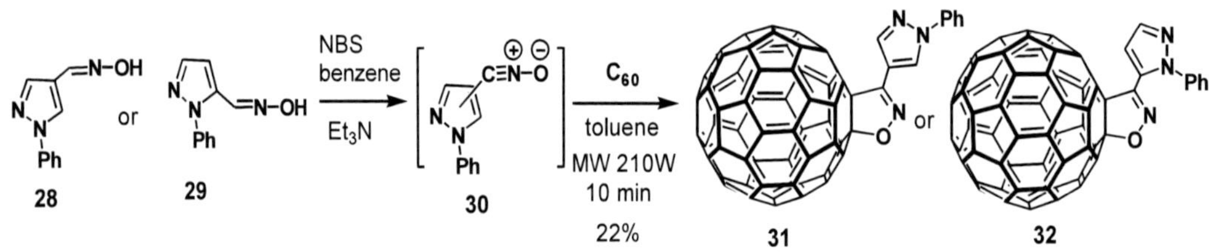

Scheme 7.

This synthetic methodology was transferred by Langa [14] to substrates of increasing complexity. Dendrimer hydrazones *36a,b* were obtained from aldehydes and were reacted with NCS to afford the corresponding dipolar nitrilimine intermediate, which was subsequently reacted *in situ* with C_{60} under microwave irradiation giving dendritic pyrazolylpyrazolino[60] fullerenes *37a-b* in good yields (*37a*: 31%, *37b*: 34%). Long chain hydrocarbons on terminal aromatic groups were introduced to increase solubility of products (Scheme 9).

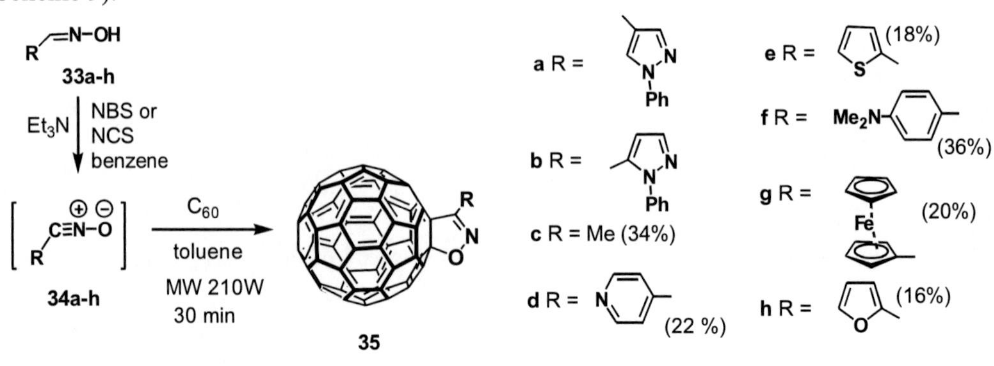

Scheme 8.

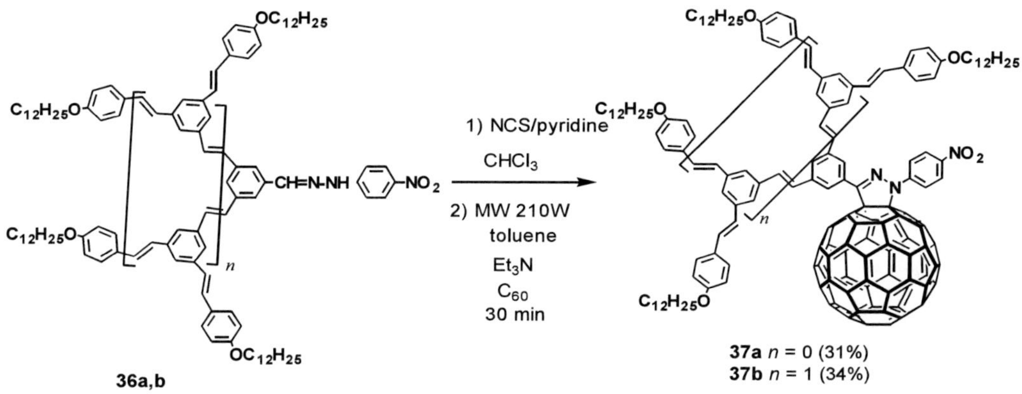

Scheme 9.

In an analogous manner, Langa and Nierengarten [15] have shown that the hydrazones *38* and *39* when reacted with NCS afforded the corresponding nitrile imine intermediates, which undergo 1,3-dipolar cycloaddition reaction *in situ* with C_{60} under microwave irradiation (210W, 40 min) in a focused microwave oven. The corresponding pyrazolino[60]fullerenes *40* and *41* with an oligophenylenevinylene moiety were obtained (Scheme 10).

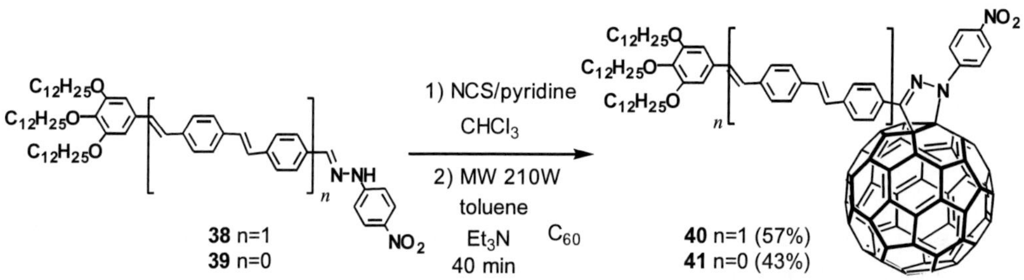

Scheme 10.

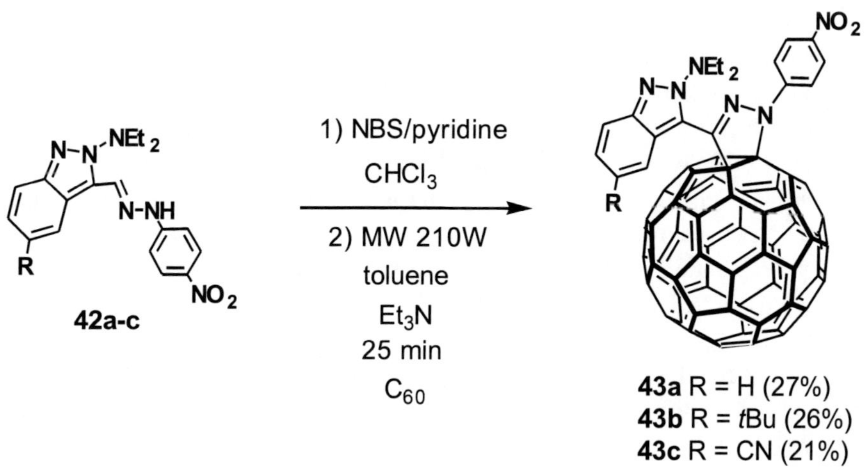

Scheme 11.

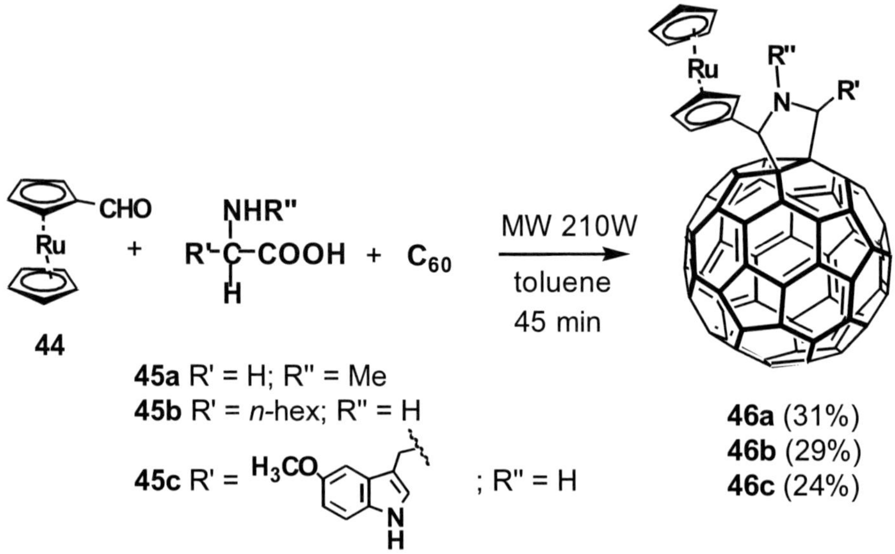

Scheme 12.

In addition, the nitrile imine generated *in situ* from isoindazole hydrazones *42a-c*, readily reacted with C_{60} under microwave irradiation to afford isoindazolylpyrazolino[60] fullerenes *43a-c* in 21-27% isolated yield (Scheme 11). [16]

5.1.3. Prato Reaction

It was also established that the pyrrolidino[60]fullerene systems could be effectively prepared by the Prato reaction in microwave conditions. Functionalization of fullerenes and nanotubes was achieved with azomethine ylides in a 1,3-dipolar cycloaddition. Herein the amino acid reacts with aldehyde when heated at reflux in toluene forming an ylide which reacts with a double bond in a [6,6] ring position in a fullerene in a 1,3-dipolar cycloaddition to yield a pyrrolidinofullerene or pyrrolidino [3,4:1,2][60]fullerenes.

Using the Prato reaction, Langa [17] prepared donor-acceptor dyads *46a-c* possessing the ruthenocene moiety on C_{60} as the electron-donor fragment (Scheme 12). The synthesis of these pyrrolidino[60]fullerene systems was carried out by 1,3-dipolar cycloaddition between ruthenocenecarboxaldehyde *44*, *N*-methylglycine *45a* and C_{60} in toluene under microwave irradiation in a focused microwave oven. The very low solubility of adduct *46a* was overcomed by preparation of adducts *46b* and *46c*, in which an *n*-hexyl and tryptophan moiety were incorporated into the molecule by reaction of *44*, C_{60} and the corresponding amino acid *45b,c*.

Prato reaction could also be carried out in ODCB as a solvent. Zhu [18] prepared structurally related pyrrolidino[60]fullerene systems (2,5-disubstituted C_{60}-pyrrolidine derivatives *49* and *50*) using microwave irradiation and the resulting yields were comparable to those obtained by classical heating (Scheme 13). A notable reduction in the reaction time was the main advantage of the microwave approach. The ratio of *cis-* and *trans-* isomers could be determined to show that in some reactions *cis-/trans-* ratio is improved by microwave conditions (product of *49cb/50cb*).

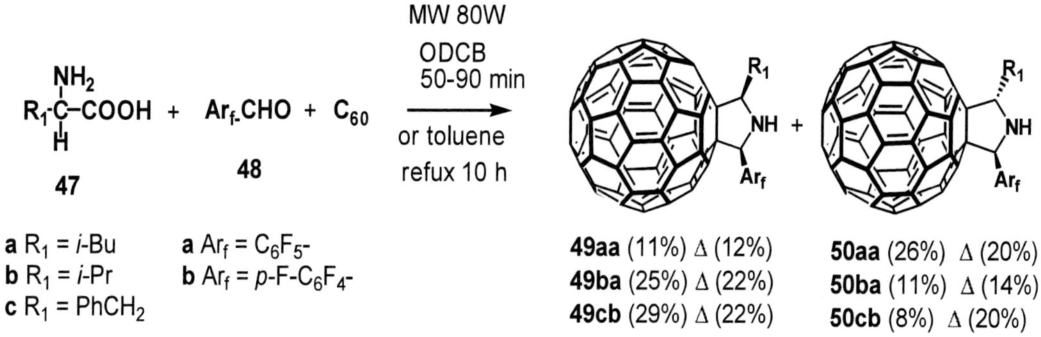

Scheme 13.

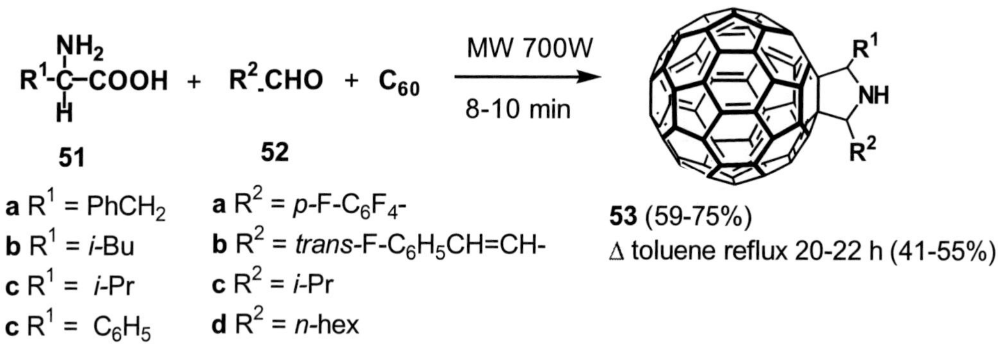

Scheme 14.

Remarkable are the results of Prato reaction carried out without the presence of solvent. A small library of pyrrolidino[60]fullerenes *53* was prepared by Zhu [19] using microwave irradiations in modified domestic microwave oven with a reflux condenser without the use of solvent (Scheme 14). The importance of this procedure lies in the fact that the obtained yields were better than those obtained by classical heating, in much shorter reaction time. Optimal ratio of reactants was found to be *52/51*/C_{60} = 2:4:1. For the microwave reaction of *52a* with *51a*, solid supports were also tested: silica gel and alumina did not promote the reaction (trace of products), while white clay, and montmorillonite clay promoted the reaction (adduct *53aa* was obtained in 59, and 74% yield, respectively). Since the yield of reaction without the solid support gave identical results (71%), the carrier-free conditions were selected, which simplified the workup procedure.

Zeng and co-workers [20] described the synthesis of several donor-acceptor systems based on amino-pyrrolidino[60]fullerene where the C_{60} (electron acceptor) and amino group (electron donor) are covalently bonded with a short linker. Prato reactions of sarcosine *55* and C_{60} were conducted under microwave irradiation with *N*-ethylcarbazole-3-carboxaldehyde, 4-(*N,N*-diphenylamino) benzaldehyde and 4-(di-*p*-tolylamino) benzaldehyde *54a-c* afforded fullerene adducts *56a-c* in reasonable yields (Scheme 15).

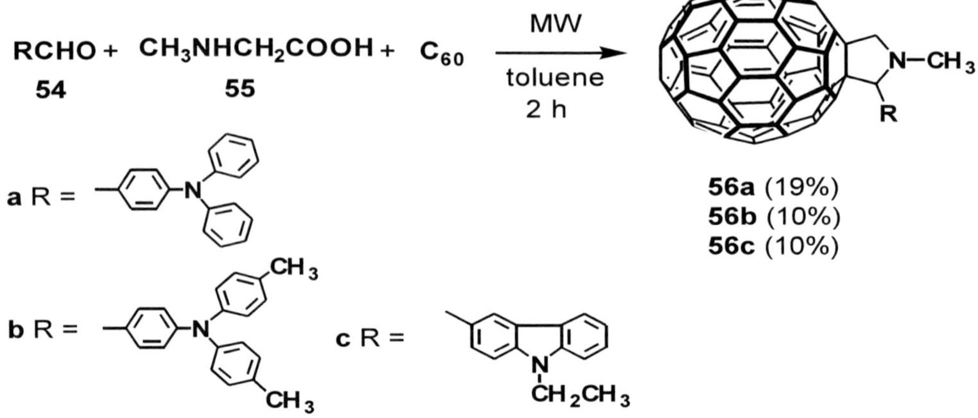

Scheme 15.

Troshin [21] has successfully employed 1,3-dipolar cycloaddition reactions of azomethine ylides for fullerene functionalisation. The reaction of butyric acid, 3-picolylamine and pyridine-3-carbaldehyde under microwave induced heating (in a domestic microwave oven) for 15-25 min with 1,2-DCB and 1,2,4-trichloro benzene as the solvents afforded pyrrolidinofullerene *59* in 70-75% yields (Scheme 16). The above described reaction is the first example of the acid catalyzed [2+3] cycloaddition to fullerene. The mechanism for the generation of azomethine ylides *62* by tautomerization of 3-picolylamine based imines *60* by the acid catalyzed tautomerization initially includes the protonation of imine by AcOH and then the acetate anion abstraction of H^+ from the CH_2 group of the protonated imine.

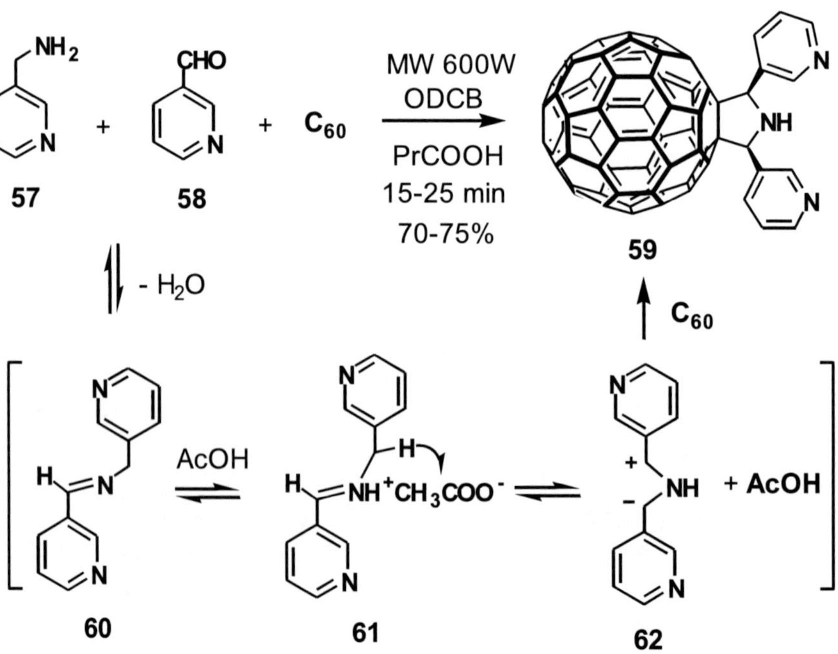

Scheme 16.

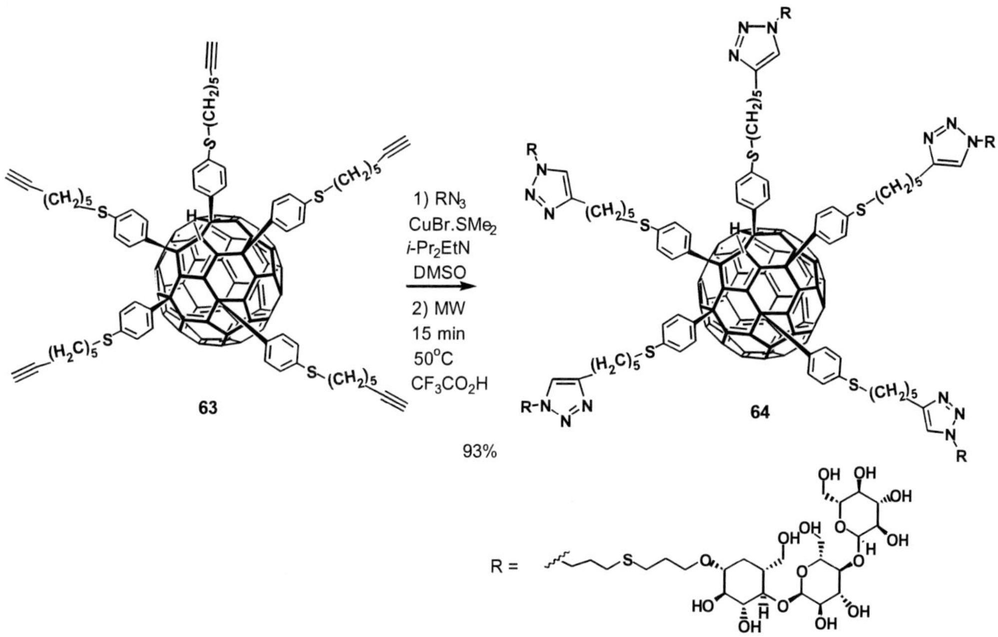

Scheme 17.

Microwave functionalization of fullerene side-chains was carried out by Isobe [22] using 1,3-dipolar cycloaddition of pentameric saccharide azides and alkynylfullerene. Using such a substrate, cycloaddition reaction to the functionality takes place, not direct addition to the fullerene moiety. The synthesis of fullerene-carbohydrate conjugates *64* using a copper-catalyzed [3+2] cycloaddition reaction was facilitated by microwave irradiation (Scheme 17). In such reaction conditions, the union of an azido functionalized sugar (Gb3-trisaccharide) and a pentaalkynyl[60]fullerene *63* is straightforward. Heating of pentaalkynylfullerene *63* with azide-bearing unprotected glucose (maltotriose derivative bearing an azide group at the terminus of a long tether) in the presence of copper bromide (50 mol %) and diisopropylethylamine (5.0 equiv) in DMSO at 40 or 50 °C for 3 days, gave adduct *64* in 91% and 88% yield, respectively, while microwave irradiation dramatically accelerated the reaction to furnish *64* in 93% isolated yield after only 15 min.

The first triazole linked porphyrin-fullerene dyads was synthetised via the Prato reaction of porphyrin aldehydes and fullerene by Fazio and Schuster. [23] For this purpose, the key reaction components were triazole-linked porphyrin aldehydes *69-71*. The connection of the porphyrin alkyne group in *66* and *67* with the azide group of aromatic aldehydes *65* and *66* was achieved using copper iodide and sodium ascorbate in a microwave-assisted dipolar cycloaddition reaction (Scheme 18).

The study of thermal stability of fulleropyrrolidines by Maggini and Prato [24] has shown that 1,3-dipole cycloreversion of fulleropyrrolidines is easily achieved in minutes, under microwave irradiation. The interplay of microwave irradiation and ionic liquids offers the unique opportunity of very efficient flash-thermal activation, in conjunction with a strong stabilization of polar/ionic intermediates/reactants. By using 1-methyl-3-*n*-octyl imidazolium tetrafluoroborate ionic liquid as solvent, and by applying microwave irradiation, fulleropyrrolidines *72a-g* undergo quantitative 1,3-dipole cycloreversion to C_{60} (Scheme 19).

Amongst the other tested ionic liquids, in 1-methyl-3-*n*-butyl imidazolium tetrachloroaluminate ([BMIM]$AlCl_4$) quantitative cycloreversion was also achieved.

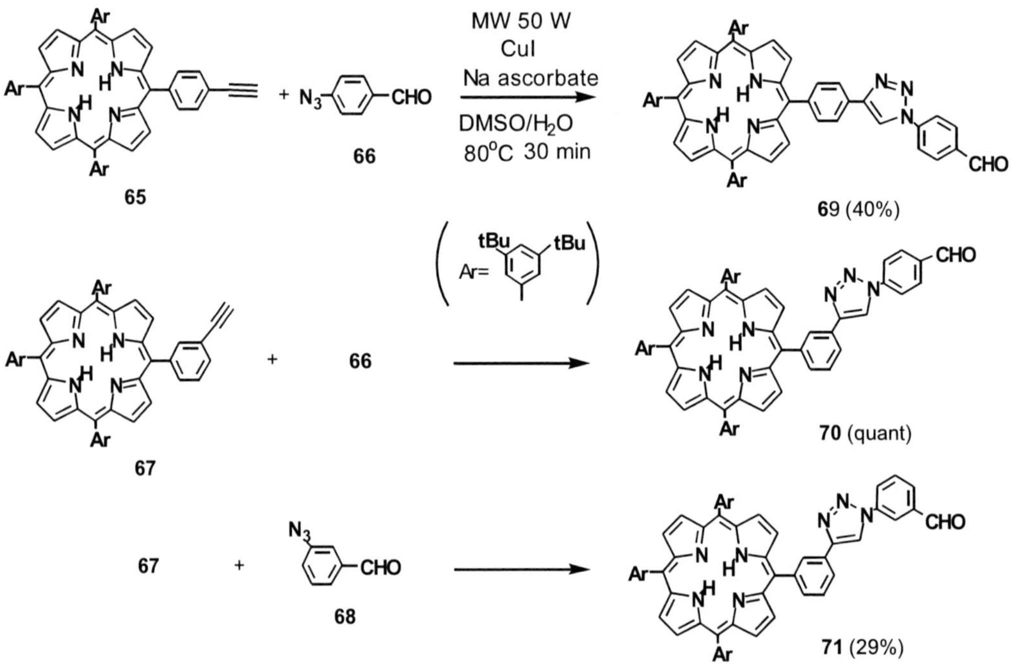

Scheme 18.

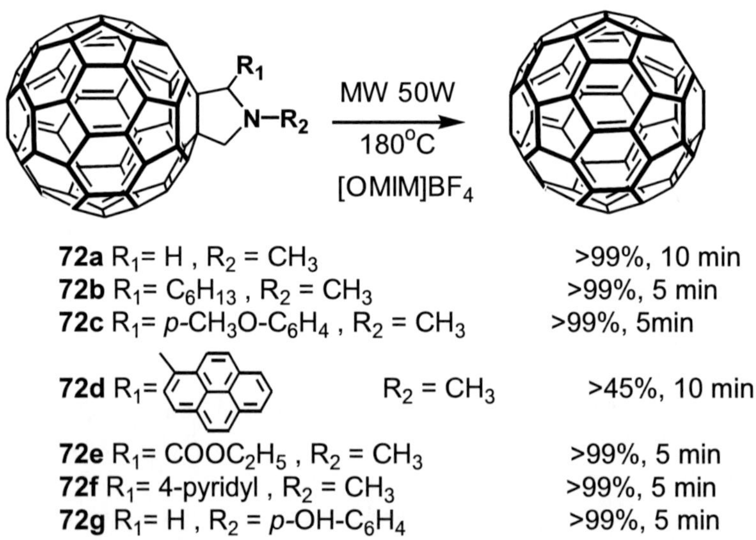

Scheme 19.

Much less work was published on C_{70} cycloadditions in microwave conditions. Langa [25] reported modified regioselectivity in cycloadditions to C_{70} by microwave-assisted synthesis of *N*-methylpyrrolidinofullerenes *73a-c* (Scheme 20). The regioselectivity of the Prato reaction of *N*-methylazomethine ylide to C_{70} was modified by using microwave irradiation as the source of energy. Under microwave irradiation and by choosing the appropriate solvent, the [5,6] isomer *73b* is the major product, a situation that is in contrast to conventional heating where the [1,2] isomer *73a* predominates. Moreover, [7,21] isomer *73c* is not formed under microwave irradiation and with ODCB as solvent. Significant changes of the the product ratios observed in polar solvents such as ODCB and chlorobenzene (which absorb microwaves efficiently) could be related to: a) the presence of the macroscopic "hot spots"; b) the heating rate induced by microwaves, which cannot be achieved by conventional heating; or c) a selective interaction of the electromagnetic field with the transition state that gives rise to compounds *73a*, *73b*, and *73c* if they have different polarities.

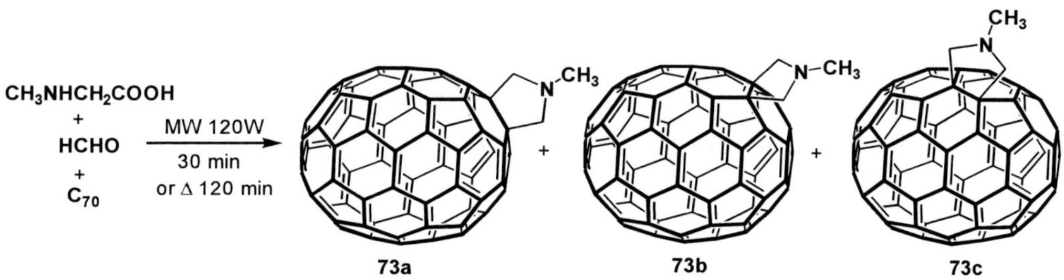

	73a	73b	73c
toluene (32%)	(48%)	(44%)	(8%)
ClC_6H_5 (31%)	(45%)	(52%)	(3%)
o-DCB (39%)	(50%)	(50%)	(0%)
Δ toluene (48%)	(46%)	(41%)	(13%)
Δ ClC_6H_5 (35%)	(50%)	(43%)	(7%)
Δ *o*-DCB (32%)	(46%)	(46%)	(8%)

Scheme 20.

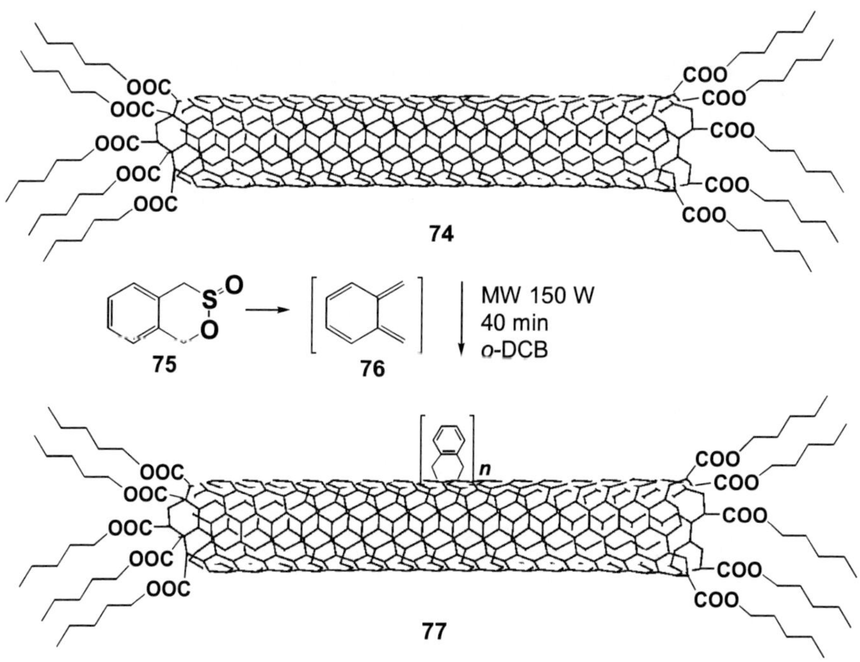

Scheme 21.

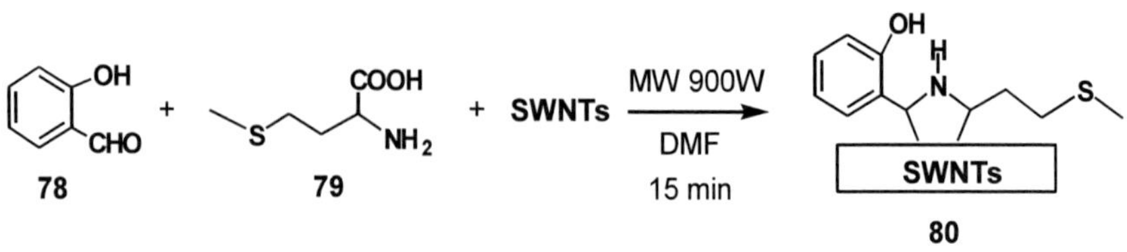

Scheme 22.

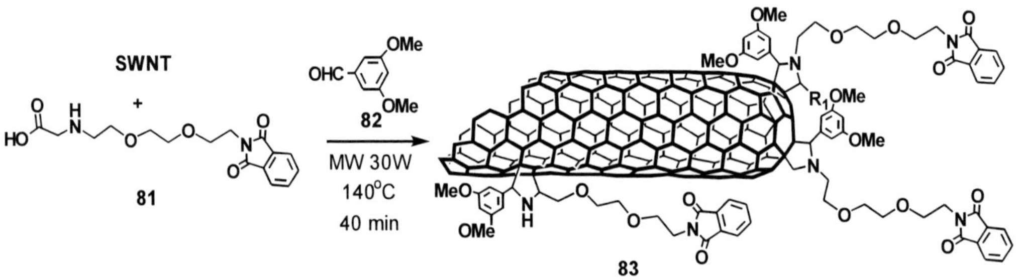

Scheme 23.

5.2. CARBON NANOTUBES

Recent experimental work has also explored the use of microwave irradiation to assist carbon nanotube (CNT) functionalizations. In the first account by Langa, cycloadditions to prefunctionalized CNTs were carried out. [26] The Diels-Alder cycloaddition of *o*-quinodimethane *75* generated *in situ* from 4,5-benzo-1,2-oxathiin-2-oxide *76* (sultine) onto the sidewalls of single-walled carbon nanotube (SWNT) *74* possessing ester end functionalization under microwave irradiation in *o*-dichlorobenzene takes place in 45 minutes (Scheme 21). This technique represents an expeditious access to modified SWNTs. Also rapid functionalization of SWNTs was achieved by 1,3-dipolar cycloaddition by Mitra. [27]

In Prato reaction, mixture of salicyladehyde *78*, methionine *79* and SWNTs in DMF was irradiated at 900 W, and at a pressure of 160 psi (Scheme 22). The final product *80* comprising of 1,3-dipolar cycloaddition functionalized SWNTs was highly soluble in $CHCl_3$, CH_2Cl_2 and dimethylformamide.

Multiple functionalizations of carbon nanotubes in microwave-induced reactions were carried out by Prato and Vázquez [28] (Scheme 23). Using α-aminoacid *81* and aromatic aldehyde *82* in solvent-free conditions of Prato reaction under microwave irradiation, carbon nanotubes have been readily functionalized affording functional derivative *83* characterized by orthogonally protected amino groups. When using classical heating, the reaction proceeded for 5 days under DMF reflux, while the aminoacid and the aldehyde were added portionwise every 24 h. Thermogravimetric analysis indicated one functional group every 128 carbon atoms of the nanotube, while functionalized CNTs *83* prepared under classical heating possess one functional group on every 270 carbon atoms of the nanotube framework.

Prato [29] achieved microwave-induced multiple functionalization of carbon nanotubes using 1,3-dipolar cycloaddition of azomethine ylides (Scheme 24). Reaction can be achieved in only 1 h using microwave irradiation instead of 5 days, necessary for the conventional thermal conditions, which are also severely limited by the great amount of DMF needed to disperse CNTs. Prato reaction using aldehydes *84a-d* and sarcosine *85* was carried out in solvent-free conditions under microwave irradiation, beacause in the absence of solvents the nanotubes display an extremely high microwave absorption. The short reaction times, the absence of harsh conditions, and the wide range of products *86a-d* that can be obtained open the way to a green chemical protocol. By using this procedure, it is possible to obtain multifunctional SWNTs with a variety of groups that could be customized to fit a desired application.

When functionalized SWNTs *87* were used in the the 1,3-dipolar cycloadditon reaction (with one functional group for every 56 SWNT carbons), only one functional group for every 216 carbons was introduced in double-functionalized SWNT *88* (Scheme 25). Reaction using the same aldehyde and the same amino acid but carried out with pristine nanotubes introduced one functional group for 87 SWNT carbons (Scheme 24). This indicates that the 1,3-dipolar cycloaddition proceeds to a much lesser extent on the functionalized SWNTs, because most of the reactive sites are saturated.

Microwave-assisted cycloaddition of aziridines to pristine CNT in solvent-free conditions was carried out by Prato. [30] Aziridines as well-known precursors to azomethine ylides readily react as 1,3-dipoles in cycloaddition reactions with SWNTs suspended in oily aziridine *89* (Scheme 26). When mixture was irradiated in microwave reactor functionalized SWNTs *90a* were obtained. Product *90a* was separated and to increase the degree of functionalization, the dipolar reaction/separation sequence was repeated another two times, giving rise to f-SWNTs *90b* and *90c*, respectively. By such a procedure, compounds *90a*, *90b*, and *90c* were obtained, which possess one functional group for about 76, 63, and 51 carbon atoms, respectively. For comparison, when SWNTs were functionalized with aziridine *89* by heating in DMF by standard procedure, the reaction proceeds in 5 days to obtain SWNTs possessing one functional group for about 138 carbon.

Grennberg [31] employed simple and powerful approach to fabricate azomethine ylides by the decarboxylation of iminium salts derived from the condensation of amino acids and aldehydes/ketones, *i.e.* Prato reaction (Scheme 27). One-pot reaction was carried out with aldehyde component 4-octadecyloxybenzaldehyde *93*, which, owing to its long flexible hydrocarbon chain, induces solubility to the addition product. The aldehyde *93* and cysteine *94a*, *N*-octyl glycine *94b* or glycine *94c* were reacted with insoluble MWNT *91* and with the soluble MWNT-ester *92*. The conventionally heated reactions yielded the soluble products both from *91* and *92*, whereas in the microwave heated reactions, only the soluble MWNT-ester *92* reacted to any considerable extent. The low reactivity observed for the MWNTs *91* is probably due to the insolubility of the material. When using conventional heating, the extended reaction times with constant stirring also lead to sufficient exposure of the inner parts of the MWNT bundles to the reagents, whereas only the outermost tubes of insoluble NT bundles can react when mechanical treatment is lacking during microwave heating. Microwave heating method was found to be highly efficient for MWNTs *92*, with the amount of sidewall functionalization by heating for 2-4 h being in the same range as that obtained from *91* or *92* after 4-5 days of conventional heating, *i.e.* 50-60 times faster.

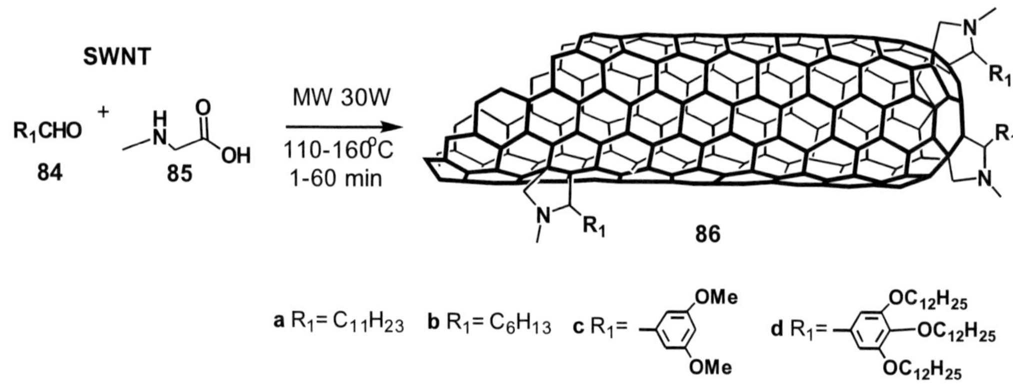

Scheme 24.

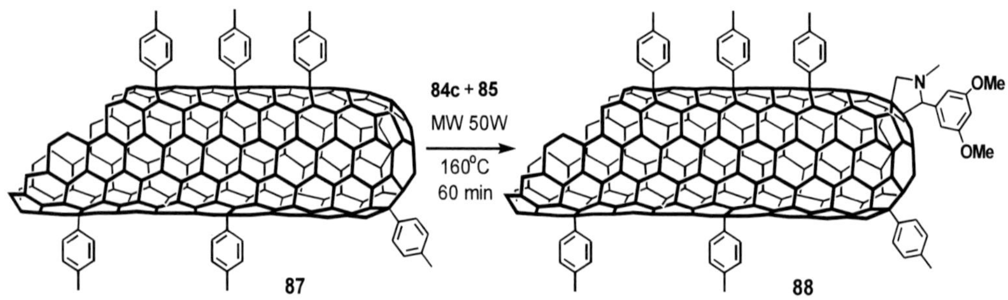

Scheme 25.

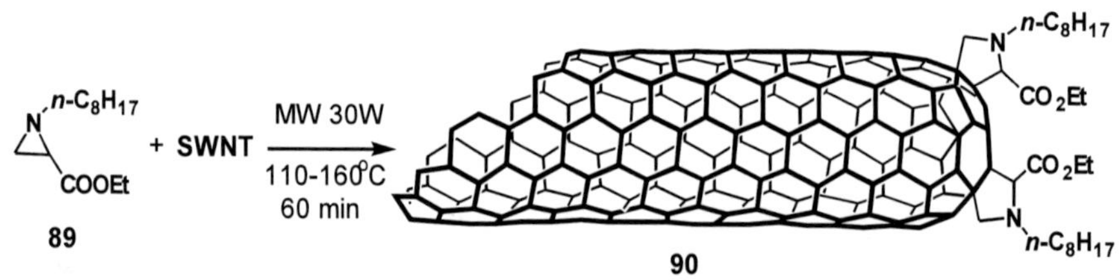

Scheme 26.

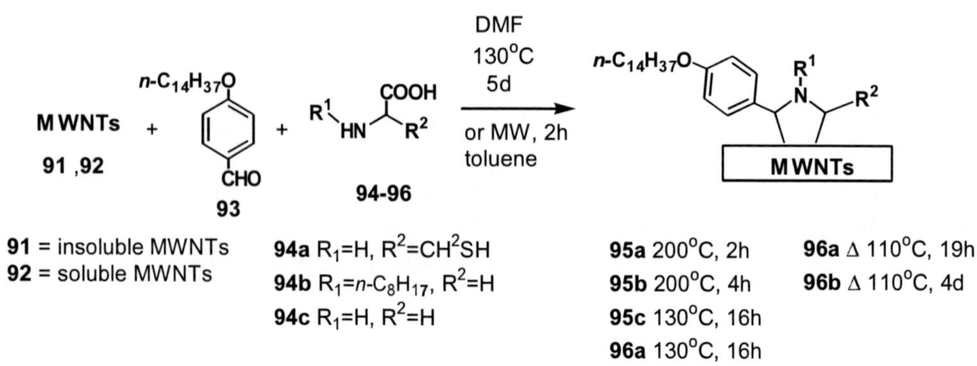

Scheme 27.

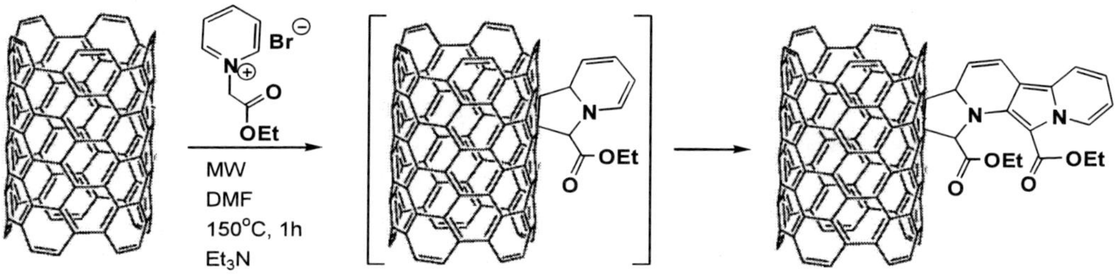

97 98 99

Scheme 28.

Coleman [32] applied a simple and convenient modification of SWNTs using 1,3-dipoar cycloaddition of pyridinium ylides (Scheme 28). When SWNTs acting as the dipolarophile were reacted with pyridinium ylide (generated *in situ* by the addition of base to the Kröhnke salt, *N*-(ethoxycarbonylmethyl)-pyridinium bromide *97*), a functionalized SWNTs *99* with indolizine groups covalently bound to the nanotube surface were obtained. Classical reaction required heating at reflux for 5 days in DMF (153 ^{o}C), while the application of microwave irradiation substantially shortens heating time to 1 h (150 ^{o}C at 2 bar pressure, 150 W for 5 min, followed by 20 W for 50 min). The 1,3-dipolar cycloaddition of the pyridinium ylide to the SWNT surface occurs with the pyridinium ylide first adding to the nanotube surface to form a pyrrolidine ring *98*, closely followed by the addition of a second ylide to the addendum on the nanotube surface to afford an indolizine *99*. The driving force for the double ylide addition is likely to be aromatization of the indolizine. The microwave conditions gives slightly higher levels of functionalization, the degree of functionalization corresponds to the presence of one functional group for every 133 and 96 carbon atoms (for microwave and classical heating, respectively).

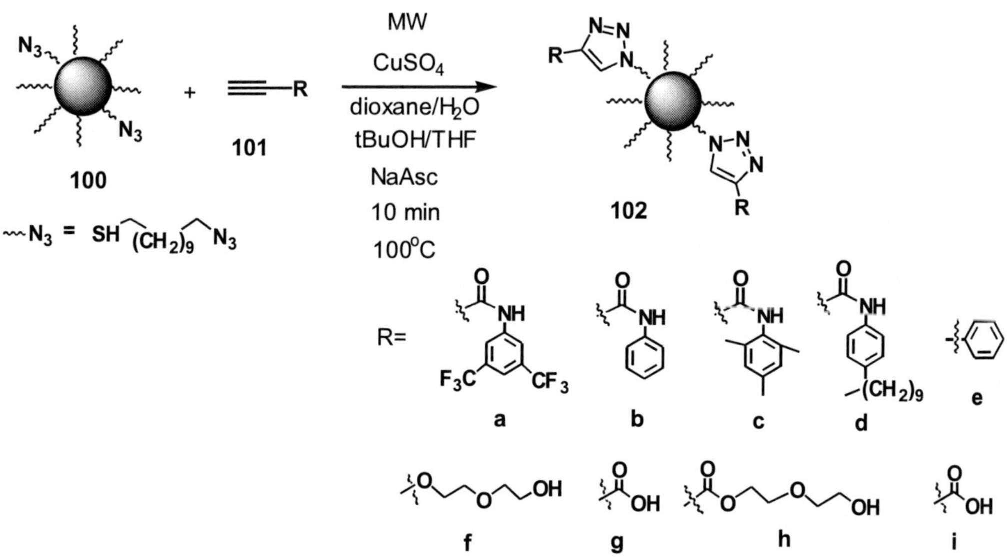

Scheme 29.

5.3. Gold Nanoparticles

Weck [33] developed procedure for facile functionalization of thiol-coated gold nanoparticles via microwave-assisted 1,3-dipolar cycloaddition. This procedure allows the attachment of terminal alkynes onto azide-containing gold nanoparticles in nearly quantitative conversions within minutes. The 1,3-dipolar cycloaddition of the alkyl thiol-protected, azide-functionalized nanoparticles *100* with acetylenes *101a-i* was carried out for 10 min at the temperature of 100°C to obtain triazole cycloadducts *102a-i* (Scheme 29). The optimal conversions were determined to be 10 mol % $CuSO_4$, 20 mol % sodium ascorbate, a mixture of dioxane, *t*-BuOH, and H_2O, or THF as solvent, and 10 equiv. of phenylacetylene.

References

[1] Yurovsakaya, M. A.; Trushkov, I. V. *Russ. Chem. Bull. Int. Ed. 2002*, *51*, 367-443.

[2] Dresselhaus, M. S.; Dresselhaus, G., Eklund, P. C. *Science of Fullerenes and Carbon Nanotubes*, Academic Press, San Diego, *1996*.

[3] Langa, F.; de la Cruz, P.; Espíldora, E.; de la Hoz, A. in Fullerenes, Vol. 9, *Functionalized Fullerenes: proceedings of an international symposium*, Martín, N.; Maggini, M.; Guldi, D. M. (Eds.), Electrochemical society, Pennington, New Jersey *2000*, 168-178.

[4] Illescas, B.; Martín, N.; Seoane, C.; de la Cruz, P.; Langa, F.; Wudl, F. *Tetrahedron Lett. 1995*, *36*, 8307-8310.

[5] Fernández-Paniagua, U. M.; Illescas, B.; Martín, N.; Seoane, C. *J. Org. Chem. 1997, 62,* 3705-3710.

[6] Chi, C.C.; Pai, F.; Chung, W.-S. *Tetrahedron, 2004*, *60*, 10869-10876.

[7] de la Cruz, P.; de la Hoz, A.; Langa, F.; Illescas, B.; Martín, N. *Tetrahedron 1997*, *53*, 2599-2608.

[8] de la Cruz, P.; de la Hoz, A.; Langa, F.; Illescas, B.; Martín, N.; Seoane, C. *Synth. Metals 1997*, *86*, 2283-2284

[9] Tovar, A.; Peña, U.; Hernández, G.; Portillo, R.; René Gutiérrez, R. *Synthesis 2007*, 22-24.

[10] de la Cruz, P.; Díaz-Ortiz, A.; García, J. J.; Gómez-Escalonilla, M. J.; de la Hoz, A.; Langa, F. *Tetrahedron Lett. 1999*, *49*, 1587-1590.

[11] Langa, F.; de la Cruz, P.; Espíldora, E.; de la Hoz, A.; Bourdelande, J. L.; Sánchez, L.; Martín, N. *J. Org. Chem. 2001, 66,* 5033-5041.

[12] de la Cruz, P.; Espíldora, E.; García, J. J.; de la Hoz, A.; Langa, F.; Martín, N.; Sánchez, L. *Tetrahedron Lett. 1999*, *40*, 4889-4892.

[13] Langa, F.; de la Cruz, P.; Espíldora, E.; González-Cortés, A.; López-Arza, V. *J. Org. Chem. 2000*, *65*, 8675-8684.

[14] Langa, F.; Gómez-Escalonilla, M.; Díez-Barra, E.; García-Martinez, J. C.; de la Hoz, A.; de la Cruz, P.; Rodríguez-López, J.; González-Cortés, A.; López-Arza, V. *Tetrahedron Lett. 2001*, *42*, 3435-3438.

[15] Armaroli, N.; Accorsi, G.; Gisselbrecht, J.-P.; Gross, M.; Krasnikov, V.; Tsamouras, D.; Hadziioannou, G.; Gómez-Escalonilla, M. J.; Langa, F.; Eckert, J.-F.; Nierengarten, J.-F. *J. Mater. Chem. 2002*, *12*, 2077-2087.

[16] Delgado, J. L.; de la Cruz, P.; López-Arza, V.; Langa, F.; Kimball, D. B.; Haley, M. M.; Araki, Y.; Ito, O. *J. Org. Chem. 2004*, *64*, 2661-2666.

[17] Oviedo, J. J.; de la Cruz, P.; Garín, J.; Orduna, J.; Langa, F. *Tetrahedron Lett., 2005*, *46*, 4781-4784.

[18] Wang, S.; Zhang, J.-M.; Song, L.-P.; Jiang, H.; Zhu, S.-Z. *J. Fluorine Chem. 2005*, *126*, 349-353.

[19] Zhang, J.; Yang, W.; Wang, S.; Ping, H.; Zhu, S. *J. Synth. Commun. 2005*, *35*, 89-96.

[20] Zeng, H.-P.; Wang, T.; Sandanayaka, A. S. D.; Araki, Y.; Ito, O. *J. Phys. Chem. A 2005*, *109*, 4713-4720.

[21] Troshin, P. A.; Peregudov, A. S.; Troyanov, S. I.; Lyubovskaya, R. N. *Russ. Chem. Bull., Int. Ed. 2008*, *57*, 887-912.

[22] Isobe, H.; Cho, K.; Solin, N.; Werz, D. B.; Seeberger, P. H.; Nakamura, E. *Org. Lett. 2007*, *9*, 4611-4614.

[23] Fazio, M. A.; Lee, O. P.; Schuster, D. I. *Org. Lett.* 2008, *10*, 4979-4982; Fazio, M. A.; Lee, O. P.; Schuster, D. I. *Org. Lett.* 2009, *10*, 1041-1043.

[24] Guryanov, I.; Montellano López, A.; Carraro, M.; Da Ros, T.; Scorrano, G.; Maggini, M.; Prato, M.; Bonchio, M. *Chem. Commun., 2009*, 3940-3942.

[25] Langa, F.; de la Cruz, P.; de la Hoz, A.; Espíldora, E.; Cossío, F. P.; Lecea, B. *J. Org. Chem. 2000, 65,* 2499-2507.

[26] Delgado, J. L.; de la Cruz, P.; Langa, F.; Urbina, A.; Casado, J.; Navarrete, J. T. L. *Chem. Commun. 2004*, 1734-1735.

[27] Wang, R.; Iqbal, Z.; Mitra, S. *Carbon 2005*, *43*, 1015-1020.

[28] Rubio, N.; Herrero, M. A.; de la Hoz, A.; Meneghetti, M.; Prato, M; Vázquez, E. *Org. Biomol. Chem. 2010*, *8*, 1936-1942.

[29] Brunetti, F. G.; Antonia Herrero, M.; de M. Muñoz, J.; Díaz-Ortiz, A.; Alfonsi, J.; Meneghetti, M.; Prato, M.; Vázquez, E. *J. Am. Chem. Soc. 2008*, *130*, 8094-8100.

[30] Brunetti, F. G.; Antonia Herrero, M.; de M. Muñoz, J.; Giordani, S.; Díaz-Ortiz, A.; Filippone, S.; Ruaro, G.; Meneghetti, M.; Prato, M.; Vázquez, E. *J. Am. Chem. Soc. 2007*, *129*, 14580-14581.

[31] Li, J.; Grennberg, H. *Chem.-Eur. J. 2006*, *12*, 3869-3875.

[32] Bayazit, M. K.; Coleman, K. S. *J. Am. Chem. Soc. 2009*, *131*, 10670-10676.

[33] Sommer, W. J.; Weck, M. *Langmuir 2007*, *23*, 11991-11995.

Chapter 6

CHOICE OF REACTION CONDITIONS FOR MICROWAVE ASSISTED CYCLOADDITIONS

Several reaction parameters have to be taken into consideration for the proper selection of reaction conditions for microwave assisted cycloadditions. [1] This choice depends on the available experimental information for the particular reaction. Namely, in the case of conversion of conventional thermal reactions to microwave conditions, initial reaction parameters are known (solvent, temperature, time). The other problem poses the development of new reactions in microwave reactor (when no experimental details exist for classical heating methods).

6.1. DEVELOPMENT OF NEW CYCLOADDITION REACTIONS

For the development of entirely new reactions, it is recommended that the first experiments are performed on a small scale, applying moderate temperatures, in order to avoid exceeding the operational limits of the instrument (temperature and pressure). [2] As a baseline when an unknown reaction is attempted, a collection of optimized conditions for related microwave reactions (such as data on cycloadditions presented in this book), can help to identify reaction parameters that are common in all microwave experiments. For instance, the microwave optimizations of the Diels-Alder reaction were found to range between 5 and 10 minutes and 125°C to 150°C. [3]

Solvent

Microwave assisted transformations could be carried out using the same solvent as for related classical reactions. For the proper solvent choice, it is important to keep in mind that solvents interact differently with microwaves, according to their dielectric properties, which are listed in Table 1 for the most common solvents used in organic reactions. The dielectric constant ε' is the permittivity of the material relative to that of vacuum, while the dielectric loss (ε'')' is the loss factor which reflects the conductance of the material. The loss tangent (tan δ=ε''/ε' is described as the energy dissipation factor, which is a convenient parameter for comparing the efficiency of conversion of microwave energy into thermal energy (a helpful parameter for comparing the heating rates). The higher is tan δ, the better is the conversion of microwave energy into heat and more effective the microwave heating.

When experiments in sealed vessel are carried out, the use of high boiling point solvents could be avoided. These are necessary to obtain a high reaction temperature for conventional reflux system. It is possible to increase the temperature of reaction in common organic solvents up to 100°C above the boiling point of the solvent. For example, ethanol has a conventional boiling point of 79°C, and microwave heating in a closed vessel can rapidly lead to temperatures of 164°C and a pressure of 12 atmospheres. [4] However, it should be noted that the use of the low-boiling solvents often results in a significant pressure development at higher temperatures, and it is crucial not to exceed the operational limits of used microwave equipment.

Table 1. Dielectric properties for common organic solvents[a], [5]

Solvent	Bp (°C)	ε'	tan δ	ε''
Ethylene glycol	197	37.0	1.350	49.950
Ethanol	78	24.3	.941	22.866
DMSO	189	45.0	.825	37.125
2-Propanol	97	18.3	.799	14.622
1-Propanol	97	20.1	.757	15.216
Formic acid	100	58.5	.722	42.237
Methanol	65	32.6	.659	21.483
Nitrobenzene	202	34.8	.589	20.497
1-Butanol	118	17.1	.571	9.764
Isobutanol	108	15.8	.522	8.248
2-Butanol	100	15.8	.447	7.063
2-Methoxyethanol	124	16.9	.410	6.929
o-Dichlorobenzene	180	9.9	.280	2.772
NMP	215	32.2	.275	8.855
Acetic acid	113	6.2	.174	1.079
DMF	153	37.7	.161	6.070
1,2-Dichloroethane	83	10.4	.127	1.321
Water	100	80.4	.123	9.889
Chlorobenzene	132	2.6	.101	0.263
Chloroform	61	4.8	.091	0.437
MEK	80	18.5	.079	1.462
Nitromethane	101	36.0	.064	2.304
Acetonitrile	82	37.5	.062	2.325
Ethyl acetate	77	6.0	.059	0.354
Acetone	56	20.7	.054	1.118
THF	66	7.4	.047	0.348
Dichloromethane	40	9.1	.042	0.382
Toluene	111	2.4	.040	0.096
Hexane	69	1.9	.020	0.038
o-Xylene	144	2.6	.018	0.047

[a] RT, 2450 GHz.

As an alternative, low absorbing solvents may be doped with small amounts of strongly absorbing solvents such as ionic liquids. [6] The addition of small amounts of a polar solvent with a large tan δ usually leads to high heating rates for the whole mixture. The energy transfer between the polar molecules which are coupling with the microwaves and the majority non-polar solvent is rapid and therefore this provides an effective mechanism for using non-polar solvents. To a difference of the classical organic solvents, ionic liquids are known to have a negligible vapor pressure below their decomposition temperature, they are non-flammable, and could be recycled. [7] In addition, they are thermally stable below temperatures of 350-450°C and soluble in most organic solvents, therefore mixtures could be used and generally considered to be a polar solvents. The choice of ionic liquid should be based on their dielectric properties and thermal stability of a variety of ionic liquids/solvent combinations under high-temperature microwave conditions. [8]

Alternatively, microwave reactions could be carried out in solvent-free conditions. Dry media reactions are different to those for reactions in solution, being advantageous as no solvent is involved, the pressure built-up is rather low and the concentration of reagents is much higher.

Temperature and Time

For setting up initial experiments of completely new reactions for which no thermal protocol is available, a feasible starting point is approximately 30-40°C above the boiling point of the solvent used. As the irradiation time is concerned, reaction carried out initially for 10 min generally gives a good overview of its progress. After the analysis of reaction progress, further modifications to the procedure could be carried out, according to obtained results: 1) no reaction, temperature is too low or non-suitable reaction system - increase temperature or change solvent; 2) partial conversion - reaction time should be extended and/or temperature increased; 3) decomposition of substrates - temperature is too high, it should be decreased; 4) completed reaction - time could be reduced until conversion is still complete. Generally, if a microwave reaction under sealed-vessel conditions is not completed within 60 min then the review of the reaction conditions is needed (solvent, catalyst, and molar ratios).

Another problem poses the scale-up of microwave single-mode heated reactions. Sealed vessel microwave conditions on small scale (2 mL) could be successfully scaled-up in multimode parallel batch reactors [9] on a 1 L scale producing quantities of 0.5 kg of products within less than one hour of overall processing time. Conversion from conventionally heated continuous flow and batch microwave scale-up protocols to scalable microwave continuous flow processes is more complex issue, in which besides solvent, temperature and time additional reaction parameters in a high-temperature/high-pressure stainless steel micro-capillary micro-reactor should be determined (such as residence time, flow rate, concentration and reactor volume). [10]

6.2. Conversion of Conventionally Heated Reactions to Microwave

When one needs to apply microwave heating to already known thermal reactions, the task of the selection of initial reaction parameters is somehow easier, and the best would be to start with known reaction conditions (solvent, temperature, time). As a general recommendation, initial temperature is usually by 20°C higher for transfer of thermal to microwave conditions. However, in some instances optimal temperature of microwave conducted reactions were lower than thermal. [11] After initial experiment, reaction conditions could be further optimized, with the aim of increasing the reaction yields and to shorten the reaction time.

References

[1] Kappe, C. O., Dallinger, D.; Murphree, S. S. *Practical Microwave Synthesis for Organic Chemists*, Wiley-VCH, Weinheim, *2009*.

[2] Kappe, C. O.; Stadler, A. *Microwaves in Organic and Medicinal Chemistry*, Wiley-VCH, Weinheim, *2005*, pp 91-106.

[3] Hutson, S.; James, T.; Saito, C.; Shelton, J.; Love, E.; Hijji, Y.; Benjamin, E.; Lack, J.; Benjamin, E. *12th International Electronic Conference on Synthetic Organic Chemistry (ECSOC-12)*, 1-30 November *2008*, http://www.usc.es/congresos/ecsoc/12/ECSOC12.htm & http://www.mdpi.org.ecsoc-12

[4] Gabriel, C.; Gabriel, S.; Grant, E. H.; Halstead, B. S. J.; Mingos, D. M. P. *Chem. Soc. Rev. 1998*, *27*, 213-223.

[5] Hayes, B. L., *Microwave Synthesis: Chemistry at the Speed of Light*, CEM Publishing, Matthews, *2002*.

[6] Van der Eycken, E.; Appukkuttan, P.; De Borggraeve, W.; Dehaen, W.; Dallinger, D.; Kappe, C. O. *J. Org. Chem. 2002*, *67*, 7904-7907.

[7] Hoffmann, J.; Nüchter, M.; Ondruschka, B.; Wasserscheid, P. *Green Chem. 2003*, *5*, 296-299.

[8] Leadbeater, N. E.; Torenius, H. M. *J. Org. Chem. 2002*, *67*, 3145-3148.

[9] Stadler, A.; Yousefi, B. H.; Dallinger, D.; Walla, P.; Van der Eycken, E.; Kaval, N.; Kappe, C. O. *Org. Proc. Res. Develop. 2003*, *7,* 707-716.

[10] Damm, M.; Glasnov, T. N.; Oliver Kappe, C. O. *Org. Proc. Res. Develop. 2010, 14,* 215-224.

[11] Lahue, B. R.; Lo, S.-M.; Wan, Z.-K.; Woo, G. H. C.; Snyder, J. K. *J. Org. Chem. 2004*, *69*, 7171-7182.

INDEX